Vehicle Refinement

Vehicle Refinement

Controlling Noise and Vibration in Road Vehicles

Matthew Harrison
Cranfield University, UK

ELSEVIER
BUTTERWORTH
HEINEMANN

AMSTERDAM • BOSTON • HEIDELBERG • LONDON • NEW YORK • OXFORD
PARIS • SAN DIEGO • SAN FRANCISCO • SINGAPORE • SYDNEY • TOKYO

Elsevier Butterworth-Heinemann
Linacre House, Jordan Hill, Oxford OX2 8DP
30 Corporate Drive, Burlington, MA 01803

First published 2004

British Library Cataloguing in Publication Data
A catalogue record for this book is available from the British Library

Library of Congress Cataloguing in Publication Data
A catalogue record for this book is available from the Library of Congress

ISBN   0 7506 6129 1

Cover images:

MTS Systems Corporation
Group Lotus plc

Typeset by Integra Software Services Pvt. Ltd, Pondicherry, India.
Printed and bound in Great Britain

Dedicated with love to Sue

Contents

Acknowledgements

I want to thank many people for teaching me different aspects of vehicle refinement. Some were my tutors, others colleagues and the rest have been students of mine. Particular mention is due to:

Peter Davies, Peter Fairhurst, the late Mike Russell, Frank Fahy, Chris Morfey, Phil Nelson, Joe Hammond, Neil Ferguson, Malcolm McDonald, Phil Crowther, Robin Sharp, John Robertson, the late Bob Wilson, Steve Hutchins, Neil McDougall, Ian Stothers, Paul Harvey, Melody Stokes, Tim Saunders, Dave Balcombe, Matteo Bettella, Iciar de Soto, Pedro Rubio Unzueta, Ruben Perez Arenas, Plamen Stanev, Sotiris Gritsis, Hazel Collins, John Shore, Alex Dunkley, Sotiris Alexiou, Sylvain Verstraeten, Tim D'Herde, Christian Haigerer and Andreas Dolinar.

Matthew Harrison
Cranfield
December 2003

Preface

This book has been written with three objectives in mind. Firstly, to emphasise the issues that the customer faces in vehicle refinement: those issues that directly influence the customer's purchasing decision. Secondly to include all of the mathematics needed to fully understand the subject along with as many full derivations of the key equations as practicable. Thirdly to construct a full reference list in order to direct further study.

As a result it should prove useful for the practising automotive engineer to read and then keep as a reference text. It should also prove useful as a teaching resource since much of the material is drawn from a lecture-series originally written for the Automotive Product Engineering MSc course at Cranfield University (UK).

The scope of the book is wide, covering all the main sources of noise and vibration in road vehicles. It also provides an introduction to both acoustics and dynamics so that a wider readership can gain an appreciation of the physical processes that cause, propagate and control noise and vibration.

The book is indexed and organised into six cross-referenced chapters, each one with its own reference list. A number of classroom-type demonstrations and group exercises are included.

Chapter 1 introduces the subject of vehicle refinement, setting out objectives and targets, and includes a case study.

Chapter 2 describes the physical behaviour of sound and gives practical advice on how it can be measured, recorded and analysed.

Chapter 3 describes how vehicle exterior noise can be assessed and controlled. Tyre noise, intake noise and exhaust noise are considered in detail.

Chapter 4 discusses the process of assessment and control of vehicle interior noise including discussion of road noise, engine noise, wind noise, brake noise and squeaks and rattles. The physics of sound absorption and sound encapsulation are explained. Chapters 3 and 4 together give the details of key vehicle refinement tests: noise homologation, noise source ranking, sound power measurement, interior noise measurement, engine noise testing, subjective rating, noise path analysis.

Chapter 5 describes the physical behaviour of vibration and advises on how it can be measured and analysed. A discussion on modes of vibration and resonance leads to a description of the modal analysis technique.

Chapter 6 discusses sources of vehicle vibration and the use of damping, isolation and absorption for their control. A detailed account of the sources of engine and drivetrain vibration is included along with a case study. The book concludes with a discussion of ride quality assessment.

About the author

Matthew Harrison is a Chartered Mechanical Engineer and a specialist in engineering dynamics, noise and vibration. He studied for both his BEng and his PhD at the Institute of Sound and Vibration, Southampton (UK), under the guidance of P.O.A.L. Davies. He has worked in the automotive industry and the railway industry and has taught in several Universities. He is the author of technical papers on vehicle refinement and acts as a consultant engineer to the automotive and the railway industries. He is currently a senior lecturer at Cranfield University (UK) and has been the Course Director for the Cranfield MSc in Automotive Product Engineering. He lives in Highgate, London, with his wife and children.

1

Vehicle refinement: purpose and targets

1.1 Introduction and definitions

The book opens with one man's thesis:

> "Styling and value sell cars – Quality keeps them sold". Lee Iacocca, *Iacocca: An Auto-biography**

To explain, it takes years and millions of US dollars to produce a new car. If the styling is attractive and the marketing is effective, and if the value for money is good then that car will probably sell moderately well for the first few months after launch. However, if the quality is bad then word will get round and sales will quickly drop off. It is vital that good sales are maintained for a significant period if the development costs for the car are going to be recouped.

Lee Iacocca was writing from Ford's and Chrysler's perspective. Both of these already had their own strong brand image and so branding was not included in his thesis, but it should be when comparing cars from different manufacturers.

To test this idea the following group activity can be tried. It has been found to work well with larger teams of people and needs a group of at least twenty to be effective:

Group exercise – Four corners

What is most important when deciding which new car to buy?

- styling?
- performance?
- value?
- brand?

Stand in the corner of the room associated with your choice. If you favour two of the above, stand midway between the appropriate corners. Be prepared to explain your decision.

At least ten minutes should be allocated to this exercise. A confident facilitator is required to interview team members as to the reasons for their choice.

---

*Lee Iacocca, former President of the Ford Motor Company, former President of Chrysler in *Iacocca: An Autobiography*, Lee Iacocca with William Novak, Bantam 1984.

The four-corners exercise can be used to obtain different effects from different groups of people. With a group of managers from the same automotive organisation the exercise can quickly identify if there is consensus on a marketing strategy (also it is interesting to watch peer-pressure at work as individual choice is being exercised in a very public way during this exercise). With a group of engineers the exercise can be used to remind them that their own work affects the customer's buying decision. With a group of engineering students, the exercise teaches them that the customer's buying decision is a multi-criteria choice and probably a compromise at the end.

It is useful to introduce some definitions at this point. The term *Vehicle Refinement** covers:

- noise, vibration and harshness** (NVH – a well-known umbrella term in the automotive industry);
- ride quality;
- driveability.

A refined vehicle has certain attributes,*** they being:

- high ride quality;
- good driveability;
- low wind noise;
- low road noise;
- low engine noise;
- idle refinement (low noise and vibration);
- cruising refinement (low noise and vibration, good ride quality);
- low transmission noise;
- low levels of shake and vibration;
- low levels of squeaks, rattles and tizzes;
- low level exterior noise of good quality;
- noise which is welcome as a 'feature'.

The term 'NVH' is usually taken to cover:

- noise suppression;
- noise design (altering the character of noise but not necessarily its level);
- vibration suppression;
- suppression of squeaks, rattles and 'tizzes'.

1.2 Scope of this book

The scope of this book is illustrated in Figure 1.1.

It has been designed to cover the core science, engineering and technology required by the NVH engineer with added material on ride quality and driveability. Wherever possible,

*Refine (vb)   to make or become free from coarse characteristics; make or become elegant or polished (Collins English Dictionary). Refinement is both a process (the act of refining) and a description of the eventual state (fineness, polish, etc.).
**Harsh (adj)   rough or grating to the senses (Collins English Dictionary).
*** Attribute (n) a property, quality or feature belonging to or representative of a person or thing (Collins English Dictionary).

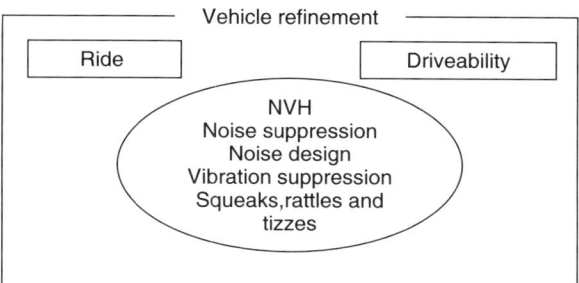

Figure 1.1 The scope of the term 'Vehicle Refinement'.

issues that affect the customer's buying decision have been emphasised. Refinement is a customer-facing subject as any refinement engineering undertaken affects the driving experience directly. Therefore, the reader is encouraged to think of refinement as being inextricably linked to the business of selling passenger cars (and increasingly this is being extended to vans, light trucks, trucks, buses, coaches and other road vehicles).

Some further definitions are offered for the sake of clarity. In this book the terms:

Noise   shall be used to describe audible sound, with particular attention paid to frequencies in the range of 30–4000 Hz.

Vibration   shall be used to describe tactile vibration, with particular attention paid to frequencies in the range of 30–200 Hz.

Other terms that will be encountered include:

Primary ride   taken to be the rigid body motion of the passenger compartment relative to the road. Typical frequency range – 0–6 Hz.

Secondary ride   taken to be the relatively large amplitude motion of sub-elements of the vehicle such as individual wheels, axles or elements of the powertrain. Typical frequency range – 6–30 Hz.

Structure-   being airborne noise radiated by a structural surface that is vibrating. *radiated noise*   Also known as 'structure borne noise'.

1.3 The purpose of vehicle refinement

Refinement helps manufacturers sell their vehicles. Brandl et al. (2000) published the results of a formal investigation of customer attitudes to vehicle refinement. Customers were asked to complete a questionnaire relating to their own vehicle. They were asked questions relating to their attitudes on vehicle prestige (brand by another name), performance, convenience, family friendliness, noise quality and cost. Although there was scatter in the results obtained, there was evidence of clustering with certain classes of customer showing predictable tastes with vehicle-refinement issues helping to define that taste.

Throughout the 1990s Rover Group in the UK (subsequently they traded under the MG badge) defined 'Refinement' as *the invisible feature* of their vehicles suggesting a strong commitment to it.

Refinement (or NVH) has always been a consideration for vehicle design and development. However, over the last 20–30 years it has assumed a greater importance (witness the advertisements by manufacturers that stress how quiet their cars are). Reasons for this include:

- *Legislation*   Since the adoption by Member States of Council Directive 70/157/EEC (1970) limiting the permissible levels of noise emitted by accelerating vehicles, the sale of new non-conforming vehicles is prohibited in the European Community. Similar legislation has been adopted in many non-EU countries, and certainly in all the main automotive territorial markets.
- *Marketing to new customers*   Refinement is a feature that may be used to distinguish a vehicle from its otherwise similar competitors, thus attracting customers not necessarily loyal to that particular brand.
- *Customer expectation*   Customers have come to expect continuous improvement in the new vehicles that they buy. They expect their new purchase to be better equipped, more comfortable, and perform better than the vehicle they just traded in (which may only be a few years old). The new vehicle may be better in all respects than the old one on paper, but if it lacks refinement then it *will not feel better* and the customer will not be fully satisfied. They may choose to take their loyalty to another vehicle manufacturer next time.
- *Marketing to existing customers*   'Trade your old model in for this year's model: it is loaded with features, more comfortable and more *refined*.' The modern car industry needs turnover to survive. People need to be encouraged to trade in regularly. The increase in vehicle leasing schemes for retail customers encourages this turnover.

The importance of vehicle refinement can be tested using the following group exercise where team members put a monetary price on refinement. This is an absolute test on the value that they place on refinement.

Group exercise – Line up

How much extra are you willing to pay for a B+ class vehicle (such as a small family sedan) that offers significant additional refinement compared to its rivals? Stand up and form a line – those willing to spend nothing extra at one end and those willing to pay a significant sum at the other.

You should interview those standing next to you in line to make sure that you are in the right place. You should consider the basis for your value judgement and be prepared to share it with the group.

At least ten minutes should be allocated to this exercise. A facilitator is required to interview a selection of people in front of their peers on the reasons for their choice.

1.4 How refinement can be achieved in the automotive industry

The traditional vehicle manufacturer is organised according to functional divisions being typically: Design, Engineering, Manufacturing, Marketing and Sales. Each division might be divided further into groups. For simplicity, an Engineering Division shall be divided into three groups: Powertrain, Vehicle and the Suppliers of components to the engineering effort as shown in Figure 1.2.

The refinement sub-group straddles the interface between the three main groups, having influence on each one in turn but not enjoying any decision-making authority. For example, consider the engine mounts: they are attached to both the powertrain and the vehicle and they are manufactured by a third party supplier. The refinement sub-group has an interest in their performance in order to improve NVH but no over-riding authority to broker compromises between the three main groups. Management of such an interface is never going to be easy, and inefficiencies, misunderstandings, mistakes or arguments that result may delay the development programme.

An alternative organisational structure is that of the 'Extended Enterprise' (Ashley, 1997) where suppliers assume greater responsibility for the design and development of their particular contribution to the whole vehicle. Such organisations are more fluid as illustrated in Figure 1.3.

Design authority is pushed out from the small client team to the various engineering functions (including a refinement function). With this fluid structure, design information is visible to all interested parties and compromises are brokered 'out in the open'. Responsibility for delivering refinement targets, defined by the client team in a PDS (Product Design Specification), is shared by all. However, in many cases the bulk of the responsibility falls to the Supplier, with the refinement function checking for compliance with the PDS.

The wider adoption of the Extended Enterprise structure within the global automotive industry has led to new opportunities for refinement engineers, particularly within component supply organisations used to manufacture but now required to design and engineer as well.

The Extended Enterprise also leads to consolidation in the industry, with the big organisations getting bigger by acquisition (Hibbert, 1999).

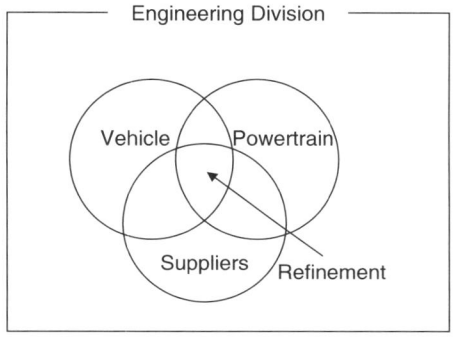

Figure 1.2 Refinement within the traditional vehicle manufacturer.

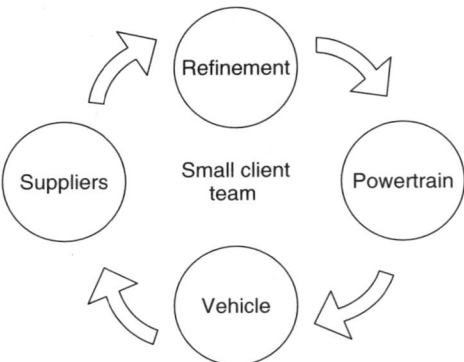

Figure 1.3 Refinement in the 'Extended Enterprise'.

1.5 The history of vehicle refinement: one representative 20-year example

In 1979, Vauxhall offered the Royale for sale in the UK – a 2.8-litre, six-cylinder, executive class (small) car for the on-the-road cost of £8354 (1979 prices). Motor magazine described it as being 'in general a refined car' (Vauxhall Royale – Star road test, Motor Magazine, 13 January 1979).

By 1989, that car had been replaced by the Vauxhall Senator 2.5i – a 2.5-litre, six-cylinder, executive class (small) car for the on-the-road cost of £16 529 (1989 prices). Autocar & Motor magazine attributed refinement as one of the car's strengths (Vauxhall Senator 2.5i – Test extra, Autocar & Motor, 21 September 1988).

By 1999, that car had been replaced by the Vauxhall Omega 2.5 V6 CD – a 2.5-litre, V6, executive class (small) car for the on-the-road cost of £21 145 (1999 prices). Autocar & Motor magazine described it as having 'one of the most refined, well mannered six-cylinder engines ...' (Vauxhall Omega 2.5I V6, Autocar & Motor, 1 June 1994).

By 1999, Vauxhall had also launched a smaller vehicle with what seems on paper to be the same degree of sophistication – the 2.5-litre V6 Vectra GSi. Autocar & Motor magazine (New cars – Vauxhall Vectra GSi, Autocar & Motor – 14 July 1999) were not so complimentary about its refinement as they had been with the Omega.

The specifications and relative costs of these four cars are compared in Table 1.1 (non-SI units are retained for historic authenticity).

This comparison of scaled historic costs with the costs in 1999 for new vehicles (99MY denotes the 1999 model year) is shown in more detail in Figure 1.5.

The above comparisons tell us something about vehicle refinement over the twenty-year period 1979–1999, although any conclusions drawn apply strictly only to the development of one series of vehicles produced by one manufacturer on sale in one territory. It can be seen that:

1. The term 'Refinement' has been used by motor journalists since the 1970s (and indeed probably quite sometime before that). Therefore, the term has been placed in the mind of the consumer as being a relevant factor in the decision-making process of buying a car for a generation.

Table 1.1 Some Vauxhall executive cars (small) 1979–1999

Vehicle	Peak power (bhp @ rpm)	Peak torque (lbft @ rpm)	Weight (kg)	Cost new	Estimate of real cost in 1999
1979 Royale (Ref.a)	140 @ 5200	161 @ 3400	1402	£8354	£28832*
1989 Senator 2.5i (Ref.b)	140 @ 5200	151 @ 4200	1465	£16528	£28333*
1999 Omega V6 2.5 CD (Ref.c)	168 @ 6000	167 @ 3200	1480	£21145	£21145
1999 Vectra V6 Gsi (Ref.d)	193 @ 6250	193 @ 3750	1431	£21700	£21700

Source: Autocar & Motor magazine

* An estimate of what the real cost of that particular vehicle would be in 1999 relative to the cost of other goods included in the Retail Price Index. It has been calculated by scaling the cost when new by data obtained for the percentage change in Retail Price Index in the period 1979–1999, shown in Figure 1.4.

References

(a) Vauxhall Royale – Star road test
   Motor Magazine, 13 January 1979
(b) Vauxhall Senator 2.5i – Test extra
   Autocar & Motor, 21 September 1988
(c) Vauxhall Omega 2.5I V6
   Autocar & Motor, 1 June 1994
(d) New cars – Vauxhall Vectra GSi
   Autocar & Motor – 14 July 1999

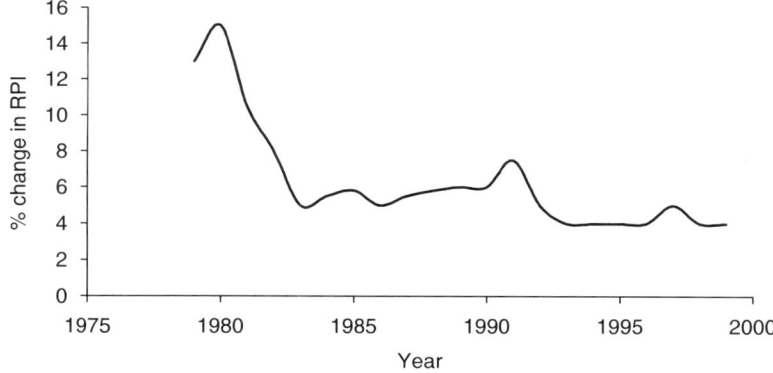

Figure 1.4 Change in Retail Price Index in the UK.
Source: www.biceznet.bris.ac.uk – University of Bristol.

2. The real cost of buying an executive class (small) Vauxhall car that has been commended for its refinement dropped significantly after 1979 – Table1.1 suggests that this drop might be as large as 25% by 1999. The drop seems to have occurred in the years between 1989 and 1999 (the scaled cost of the 1979 model year (79MY) Royale matches almost exactly the 1989 retail price for the Senator, but the scaled cost of the Senator is much greater than the 99MY Omega). During that ten-year period, the

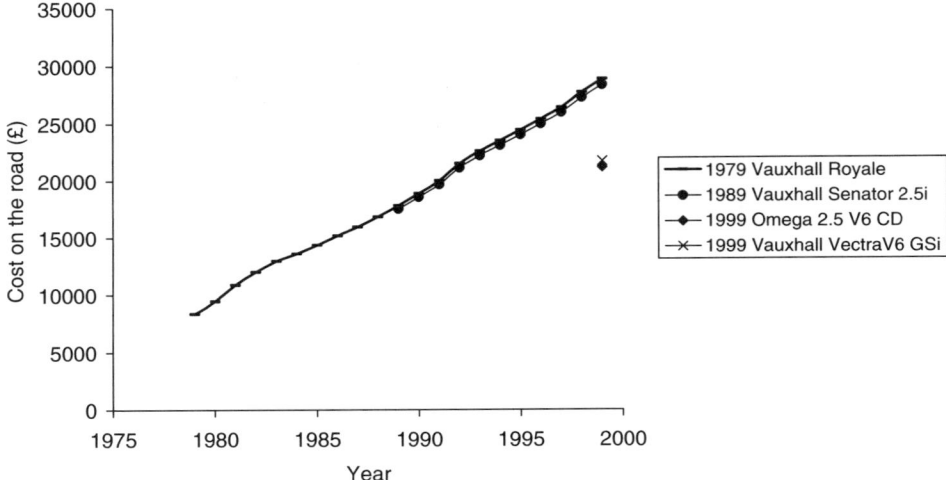

Figure 1.5 Comparing the cost of 99MY vehicles with the historic costs of equivalent 79MY and 89MY, scaled according to the UK Retail Price Index.
Data sources: References a – d in Table 1.1 and www.bizednet.bris.ac.uk – University of Bristol.

performance of the vehicle increased significantly in terms of specific engine torque and power (Table 1.1). In addition vehicle emissions dropped because of tougher legislation. The 99MY cars are more refined having to meet noise drive-by levels of 74 dBA compared with 82 dBA in 1979 (8 dBA is a significant drop in noise, equivalent to almost a halving in loudness). Also 99MY vehicles are generally better equipped than their predecessors (air conditioning, satellite navigation, CD-players, etc.).

3. The price of many new cars in the UK dropped by 10–15% in mid-2000 due to action by the UK Government and pressure from consumers. This makes the results discussed in (2) even more startling.

4. The consumer looking for a six-cylinder Vauxhall car in 1999 could choose between the Omega and the Vectra for virtually the same price. The Vectra seems to offer better performance, whilst the Omega is reported to offer more space and better refinement.

The situation described above is counter-intuitive – 99MY vehicles that have better performance, less emissions, better refinement and more equipment than 89MY vehicles but cost less in real terms. It is possible to speculate how the industry found itself in that position. It might have been due to:

- The effects of increased competition in a saturated market.
- Action by the UK Government (only applies to 2000MY prices).
- The effects of vehicle leasing schemes and manufacturer's vehicle finance schemes.
- The adoption of measures to improve organisational efficiency such as the 'Extended Enterprise' and technology employed in design and manufacturing activities.
- Macro political and economic factors affecting the automotive industry but not generated by that industry (globalisation, problems in Pacific Rim countries, etc.).

Table 1.2 Some six-cylinder Ford executive cars (small) 1979–1999

Vehicle	Peak power (bhp @ rpm)	Peak torque (lbft @ rpm)	Weight (kg)	Cost new	Estimate of real cost in 1999*
1979 (no comparable model)					
1989 Granada 2.4I V6 Ghia (Ref. e)	130 @ 5800	142 @ 3000	1340	£16 045	£25 996
1999 Mondeo 2.5 litre V6 Ghia X (Ref. f)	168 @ 6250	162 @ 4250	1408	£21 680	£21 680
1999 Mondeo 2.5 litre V6 ST24 (Ref. g)	168 @ 6250	162 @ 4250	1410	£19 680	£19 680

* Calculated in the same manner as Table 1.1.

References

(e) Test Update: Ford Granada 2.4i Ghia
     Autocar & Motor, 11 February 1987
(f) Autocar Twin Test: Ford Mondeo 24v Ghia X vs Peugeot 406 3.0 SVE
     Autocar & Motor, 6 November 1996
(g) Autocar Twin Test: Ford Mondeo ST24 vs Vauxhall
     Vectra Gsi
     Autocar & Motor, 22 April 1998

A similar trend can be shown with Ford, one of Vauxhall's traditional competitors. Data for six-cylinder executive (small) cars manufactured by Ford (broadly competing vehicles to the Senator/Omega/Vectra) are shown in Table 1.2 and these show similar trends to the data for the Vauxhall cars shown in Table 1.1.

1.6 Refinement targets

The setting of refinement targets is important for the successful operation of the so-called Extended Enterprise that is described in Section 1.4. Without these, individual system suppliers would determine their own interpretation of an appropriate level of refinement for their component and the final vehicle would most likely be truly refined only in some aspects and not in others. In addition, it should be noted that type approval testing (at present) is undertaken as 'whole vehicle' type approval (see Section 3.1.1) and therefore by definition it is only undertaken once the production intent vehicle is fully developed. If one component or sub-system causes the vehicle to fail its type approval test due to excessive noise then the cost implications are obviously serious.

The standard management tool for setting refinement (and other) targets is the PDS document. This is written by the brand holder and adherence to it becomes a conditions of contract for any supplier. A typical PDS will contain the following refinement targets:

• whole vehicle exterior noise targets;
• single component exterior noise targets;

- whole vehicle interior noise targets;
- ride quality targets (including tactile vibration targets).

The exterior noise targets will mostly be objective relating to the passing of the type approval noise test although occasionally subjective criteria might be included in addition (exhaust noise quality in particular for sports cars). The interior noise, ride quality and tactile vibration targets will be a mix of objective and subjective criteria.

1.6.1 Whole vehicle exterior noise targets

Whole vehicle noise targets are set in terms of drive pass noise levels for the type approval test required in the territory in which the vehicle will be offered for sale. Separate noise targets are commonly set for each sub-system on the vehicle. A typical set is given in Table 1.3 for vehicles offered for sale in the EU.

It is not possible to determine a single set of such targets for a given territory as every vehicle will have its own unique noise signature. This is made clear in Tables 3.2 and 3.3 (see Section 3.1.6) where it is clear that the EU passenger car fleet exhibits type approval noise levels that vary by as much as 10 dB. Notwithstanding this, Table 1.3 is offered as a starting point. If a particular vehicle is diesel powered, perhaps the engine noise target should be increased by 1–2 dB. If the vehicle is a sports-utility then perhaps the tyre noise and transmission noise targets should be raised by 1–2 dB apiece.

1.6.2 Single component exterior noise targets

In order to minimise the risk that the whole vehicle exterior noise targets are not met at the final type approval testing, the PDS will routinely include separate airborne noise targets for certain components or vehicle subsystems. The most common targets relate to engine-radiated noise, intake noise and exhaust noise.

1.6.2.1 Engine-radiated noise targets

These are normally set in terms of sound power level (see Section 2.5.1). European engine suppliers are now required by law to measure and declare the sound power emissions of their engines. This makes a comparison between competitor engines easy.

Table 1.3 Suggested target noise levels for achieving type approval under 9297/EEC. See Section 3.1.6 for further details

	Passenger car	Light truck	Heavy truck
	Target levels at 7.5 m, acceleration test (dBA)		
Engine	69	72	77
Exhaust	69	70	70
Intake	63	63	65
Tyres	68	69	75
Transmission	60	63	66
Other	60	72	65
Combined level	74.2	77.3	80.1

Engine-radiated noise targets are more important for diesel engines than for gasoline engines as the various type approval tests undertaken around the world demand only quite low engine speeds where gasoline engines are generally quiet and rarely cause failure of the test.

Again, it is not possible to adopt a single set of noise targets applicable to all potential engines, but an indicative set is offered in Table 1.4. This relates to a particular diesel engine used in light European trucks and sports-utility vehicles.

An alternative target for a 4.0 litre four-cylinder DI diesel is offered by Pettitt (1988) at $107\,dB$ re $10^{-12}\,W$. An alternative (and rather different) noise source ranking is offered for a six-cylinder diesel engine by Beidl et al. (1999).

1.6.2.2 Intake orifice–radiated noise targets

These are commonly set as maximum sound pressure levels to be recorded at a distance of 100 mm from the intake orifice at an angle of incidence of 90°. Common targets are:

- An overall A-weighted sound pressure level of 90 dBA at $1000\,rev\,min^{-1}$ wide-open throttle (full load) rising at a rate of 5 dBA per $1000\,rev\,min^{-1}$ to a maximum of 115 dBA at $6000\,rev\,min^{-1}$. The adoption of this target is likely to result in an intake level of 63 dBA during an EC type approval test (as required by the targets shown in Table 1.3) without relying on any attenuation offered by the vehicle bodyshell.
- A firing order sound pressure level (see Section 2.4.2.5) of 105 dB (lin) at low drive-away engine speeds (full load), a level of 100 dB (lin) at moderate engine speeds and a level of 105 dB (lin) at high engine speeds.
- Sound pressure levels for other higher orders of 95 dB (lin) at low drive-away engine speeds (full load), a level of 90 dB (lin) at moderate speeds and a level of 95 dB (lin) at high engine speeds.

1.6.2.3 Exhaust tailpipe–radiated noise targets

These are commonly set as maximum sound pressure levels to be recorded at a distance of 500 mm from the exhaust tailpipe at an angle of incidence of 45°. Common targets are:

Table 1.4 Typical sound power targets in dB re $10^{-12}W$ for a 4.0-litre I4 NA (naturally aspirated) IDI (in-direct injection) diesel used in light trucks and sports-utility vehicles

Component	Sound power level (dB re 10^{-12} W)
Sump	102
Block	104
Head	93
Exhaust	102
Intake	97
Fuel injection pump	96
Total	108

- An overall A-weighted sound pressure level of 82 dBA at 1000 rev min^{-1} wide-open throttle (full load) rising at a rate of 5 dBA per 1000 rev min^{-1} to a maximum of 107 dBA at 6000 rev min^{-1}. The adoption of this target is likely to result in an exhaust level of 69 dBA during a type approval test (as required by the targets shown in Table 1.3).
- A firing order sound pressure level (see Section 2.4.2.5) of 120 dB (lin) at low drive-away engine speeds (full load), a level of 100 dB (lin) at moderate speeds and a level of 115 dB (lin) at high engine speeds.
- Sound pressure levels for other higher orders of 105 dB (lin) at low drive-away engine speeds (full load), a level of 95 dB (lin) at moderate speeds and a level of 105 dB (lin) at high engine speeds.

1.6.3 Whole vehicle targets for interior noise

Interior noise levels are routinely measured at the driver's ear position (and elsewhere in the vehicle interior) in accordance with BS 6086 1981 (ISO 5128, 1980) (see Section 4.1.3 for details). Either of two basic schemes for objective interior noise targets can be adopted.

1.6.3.1 Interior noise targets: perceptible improvement in sound pressure level

In this scheme, sound pressure levels in the vehicle interior are measured in accordance with BS 6086 1981 (ISO 5128, 1980) in the competitor vehicle. Then a target is set for the vehicle under development in terms of relative improvement. Bies and Hansen (1996) summarise:

Change in apparent loudness	Change in sound pressure level (dB)
Just perceptible	−3
Clearly noticeable	−5
Half as loud	−10
Much quieter	−20

Recordings of interior noise can be analysed to obtain the following metrics:

- Overall sound pressure level. This would normally be A-weighted for sound pressure levels below 55 dB (re 20 micro-pascals) and C-weighted for sound pressure levels in the range of 55–85 dB (Bies and Hansen, 1996). Typical targets would be devised in terms of maximum sound pressure level at certain road speeds under cruise conditions and at certain engine speeds under full load acceleration.
- Sound pressure level at the engine firing frequency order (see Section 2.4.2.5).
- Sound pressure level at higher engine orders.

Perceptible improvement targets can be set for each of these metrics. Such targets are most likely to be 3 dB lower than the metrics measured in the competitor vehicle in order to achieve tangible improvement at minimum cost.

1.6.3.2 Interior noise targets: brand value

In this scheme, a definitive set of interior noise targets are adopted, irrespective of the relative performance of competitor vehicles, in order to make a particular brand statement. The targets might apply to either aural comfort, or more usually speech intelligibility. Such targets are obviously brand specific but some general rules do apply:

- American Standard ANSI S3.1–1977 shows that effective speech communication can take place at normal voice levels between two persons spaced 1 m apart (in a free acoustic field) providing the background noise level is less than 65 dBA (or the so-called speech interference level (SIL) being the arithmetic average sound pressure level in the 500, 1000 and 2000 Hz octave bands is less than 58 dB (lin)). These two targets are the upper limits for just acceptable speech communication. Bies and Hansen (1996) describe this as being 95% sentence intelligibility or 60% word-out-of-context recognition.

  In accordance with this, and assuming a highly sound absorbing vehicle interior, a rational target for interior noise level with a vehicle under steady-state cruise conditions would be an interior sound pressure level of 65 dBA at the driver's ear. When intelligibility of the female voice is of particular importance, this target might be lowered to 60 dBA. When front seat to back seat communication is of particular importance, the target might be lowered to 55 dBA. With the general adoption of hands-free mobile phones in vehicles, a target in the range of 60–65 dBA is prudent. A survey of a 2003 MY executive class sedan revealed interior cruise noise levels in the range of 60–65 dBA at 80 km hr^{-1} on B-class roads. Speech communication in the front seats in this vehicle should be adequate but front to back seat communication would be compromised. By way of historical comparison, Rust et al. (1989) suggested an interior noise level of 70 dBA at 80 km hr^{-1} for a 1980s direct injection (DI) diesel.

- Priede (Priede, 1974; Priede and Anderton, 1984) showed that typically, full-load engine noise increases by 5 dB per 1000 rev min^{-1} increase in engine speed. Challen and Crocker (1982) suggest that the maximum effect of engine load on engine noise levels is of the order of 1–2 dB. Priede (1974) suggests that intake and exhaust noise levels increase by 5 dB per 1000 rev min^{-1} and it is well known that intake noise is strongly affected by load (±15 dB) whereas exhaust noise levels are not so strongly affected (±10 dB). Underwood (1973) suggested that tyre noise increases by 9–13 dB per doubling of vehicle speed.

  In accordance with this guidance, any objective interior noise target for any metric should vary with engine speed, road speed and engine load as appropriate. A consistent set of interior noise targets for a mid-priced family car powered by a four-cylinder gasoline engine might be:

- 65 dBA at the driver's inner ear at 80 km hr^{-1} cruise and 70 dBA at 120 km hr^{-1}.
- Under full load acceleration in second or third gear, a sound pressure level at the driver's inner ear of 55 dBA at 1000 rev min^{-1} rising linearly to 80 dBA at 6000 rev min^{-1}. Under overrun (zero load) the targets should be 10 dB less.
- The level of any engine order should be at least 3 dB lower than the overall noise level.

For the case of an executive or luxury car, these targets should all be reduced by at least 5 dB.

Alternative brand-value interior noise targets can be set in terms of:

- (Zwicker) loudness in accordance with ISO 532 – 1975. This method is suitable for calculating the loudness level (in 'phons') of combinations of octave bands for random-like noise in which there are prominent tonal components. The loudness in phons is given by the set of equal loudness contours (BS 3383 – 1988, ISO 226 – 1987). Each contour is labelled with X phons, X being the sound level at 1000 Hz for that particular contour.
- Articulation index (AI) in the range of 0.5–0.6. The calculation method is given in ANSI S3.5 – 1969 and the long-term rms. speech level in 1/3 octave bands is compared against the long-term rms. masking noise level, and the 1/3 octave results so generated are aggregated using separate weighting functions for each 1/3 octave band.
- Speech transmission index (STI) and the more rapid RASTI (rapid speech transmission index) (BS 6840: part 16 – 1989) are commonly used alternatives to articulation index. Rust et al. (1989) suggest a RASTI of 0.8 at 50 km hr^{-1} reducing linearly to 0.4 at 125 km hr^{-1}.

1.6.3.3 Interior noise: subjective targets

There are several different strategies for the subjective assessment of interior noise. These are discussed, along with target levels in Section 4.1.4. An engineering method for subjective appraisal involves a panel of people driving and riding in the vehicle(s) along a pre-determined test route on public roads and rating the following noise (and vibration) attributes:

- wind noise;
- road noise;
- engine noise;
- idle refinement;
- cruising refinement;
- transmission noise;
- general shakes and vibrations;
- squeaks, rattles and tizzes;
- ride quality;
- driveability;
- noise that is a 'feature' (sporty exhaust notes, etc.).

Table 1.5 Common subjective rating scheme

1	2	3	4	5	6	7	8	9	10
Not acceptable		Objectionable		Requires improvement	Medium	Light	Very light	Trace	No trace

The ratings are made to a common scale from 1 to 10 as per Table 1.5.

- A rating of less than 4 is unacceptable for any attribute.
- A rating of 5 or 6 is borderline.
- A rating of 7 or more on any attribute is acceptable.

Most new passenger cars are launched with a subjective rating of 7 or 8 on most attributes.

1.6.4 Targets for ride quality (including tactile vibration)

Ride quality is taken here to be the subjective response to a low-frequency vibration phenomenon. There are several different strategies for the assessment of in-vehicle vibration levels and these are discussed in Section 6.5. A summary is measured vibration levels are rated according to objective criteria, and the most commonly used criteria are offered in:

- ISO 2631 Part 1 (1985);
- BS 6841 (1987);
- NASA discomfort level index (1984) (Leatherwood and Barker, 1984).

Experience of using all three has led to the conclusion that a family class vehicle for either the European or the US Federal markets will be ready for sale when the appropriately frequency-weighted seat rail vibration levels measured at $80\,\mathrm{km\,hr^{-1}}$ on a straight road with 5–10-year-old tarmac and a few spot repairs (in other words a typical B-class inter-urban road) are:

- Close to the four-hour reduced comfort boundary in the vertical direction as defined in ISO 2631 Part 1 (1985).
- Have an rms. level less than $0.63\,\mathrm{m\,s^{-2}}$ (classed as better than 'a little uncomfortable' according to BS 6841 (1987)).
- Have a NASA discomfort rating below 4.0 (Bosworth et al., 1995).

References

Ashley, S. Keys to Chrysler's comeback, Mechanical Engineering, November 1997

Beidl, C.V., Rust, A., Rasser, M. Key steps and methods in the design and development of low noise engines, SAE Paper No. 1999-01-1745 1999

Bies, D.A., Hansen, C.H. Engineering noise control – theory and practice, Second edition, E&FN Spon, London 1996

Bosworth, R., Trinick, J., Smith, T., Horswill, S. Rover's system approach to achieving first class ride comfort for the new, Rover 400, IMEchE Paper No. C498/25/111/95, 1995

Brandl, F.K., Biermayer, W., Thomann, S., Pfluger, M., von Hofe, R. Objective description of the required interior sound character for exclusive passenger cars, IMechE Paper No. C577/013/2000, 2000

Challen, B.J., Crocker, M.D. A review of recent progress in diesel engine noise reduction, SAE paper No. 820517, 1982

Hibbert, L., Last ones standing are the winners, Professional Engineering, 3 November 1999

Leatherwood, J.D., Barker, L.M. A user-orientated and computerized model for estimating vehicle ride quality, NASA Technical Paper 2299, 1984

Pettitt, R.A. Noise reduction of a four litre direct injection diesel engine, IMechE Paper No. C22/88, 1988

Priede, T. Effect of operating parameters on sources of vehicle noise, Symposium on noise in transportation, University of Southampton, 1974

Priede, T., Anderton, D. Likely advances in mechanics, cooling, vibration and noise of automotive engines, Proceedings of the Institution of Mechanical Engineers, Vol. 198D, No. 7, 95–106, 1984

Rust, A., Schiffbaenker, H., Brandl, F.K. Complete NVH optimisation of a passenger vehicle with DI diesel engine to meet subjective market demands and future legislative requirements, SAE Paper No. 890125, 1985

Underwood, M.C.P. A preliminary investigation into lorry tyre noise, Transport Research Laboratory Paper LR 601, Crowthorne, UK, 1973

2

The measurement and behaviour of sound

2.1 How sound is created and how it propagates

2.1.1 Sound

The Penguin Dictionary of Physics (Illingworth, 1990) defines sound as:

> Periodic mechanical vibrations of a medium by means of which sound energy is carried through that medium.

From this it is clear that:

- Sound is a periodic process (at least at the particle level within the medium).
- Sound involves energy transport.

To understand sound more fully, it must also be appreciated that:

- The propagation of sound may involve the transfer of energy but it does not (on its own) cause the transport of matter. Particles of the medium are excited into oscillation about their usual position of rest but they do not get swept along with the passing sound. By way of illustration, the molecules of air within the speaker's mouth do not need to find their way to the listener's ear for them to hear the speaker's voice.
- In order that particles may oscillate about their usual position of rest, the medium through which sound propagates must have both elasticity (so that a restoring force is imposed on a displaced particle) and inertia (so that a returning particle overshoots its usual position of rest and oscillates to and fro).
- In the case of a fluid, the inertial qualities are given by its mass density ρ ($\mathrm{kg\,m^{-3}}$). The fluid is under uniform pressure P ($\mathrm{N\,m^{-2}}$) and is at a uniform temperature T (K). The three quantities are connected by an equation of state given by (Morse and Ingard, 1968)

$$\kappa_{\mathrm{T}} = -\frac{1}{V}\left(\frac{\partial V}{\partial P}\right)_{\mathrm{T}} = \frac{1}{\rho}\left(\frac{\partial \rho}{\partial P}\right)_{\mathrm{T}} \tag{2.1}$$

$$\beta = \frac{1}{V}\left(\frac{\partial V}{\partial T}\right)_{\mathrm{P}} = -\frac{1}{\rho}\left(\frac{\partial \rho}{\partial T}\right)_{\mathrm{P}} \tag{2.2}$$

Equation (2.1) gives the fractional rate change of density or volume with pressure at constant temperature and is called the isothermal compressibility of the fluid. Equation (2.2) gives the coefficient of thermal expansion of the fluid.

As one requirement for the passage of sound through a fluid is that the medium exhibit some elasticity, equation (2.1) shows that the fluctuations in P commonly understood to occur with the sound wave are in fact accompanied by corresponding fluctuations in ρ. Equation (2.2) shows that these fluctuations will have an effect on T.

The acoustic disturbance (denoted by p', ρ' and T') being fluctuations of P, ρ and T about their normal values at rest (denoted by P_0, ρ_0, T_0)

$$P = P_0 + p' \tag{2.3}$$

$$\rho = \rho_0 + \rho' \tag{2.4}$$

$$T = T_0 + T' \tag{2.5}$$

propagates with a certain velocity $c\,(\mathrm{m\,s^{-1}})$ which is known as the speed of sound.

For the case of a perfect gas (having no viscosity and hence no resistance to a change in shape), the speed of sound is given by (Morse and Ingard, 1968)

$$c = \sqrt{\frac{1}{\kappa\rho}} \tag{2.6}$$

$$c = \sqrt{\gamma R T} \tag{2.7}$$

where

κ = compressibility of the gas ($1/P$ for the perfect gas)
γ = ratio of specific heats $(c_\mathrm{p}/c_\mathrm{v})\gamma = 1.4$ for air at standard temperature and pressure
c_p = specific heat capacity at constant pressure ($\mathrm{J\,kg^{-1}\,K^{-1}}$)
c_v = specific heat capacity at constant volume ($\mathrm{J\,kg^{-1}\,K^{-1}}$)
R = specific gas constant ($\mathrm{kJ\,kg^{-1}\,K^{-1}}$)
   $R = 287\,\mathrm{kJ\,kg^{-1}\,K^{-1}}$ for air

$$R = c_\mathrm{p} - c_\mathrm{v} \tag{2.8}$$

Therefore, the speed of sound is around $343\,\mathrm{m\,s^{-1}}$ at standard temperature and pressure.

The values of both c_p and c_v for air rise with temperature by similar amounts leading to very little change in R. The ratio of specific heats γ drops with increasing temperature; however, the net effect is that the speed of sound in air increases with temperature.

At 600 K $c = 487\,\mathrm{m\,s^{-1}}$ for air

At 900 K $c = 590\,\mathrm{m\,s^{-1}}$ for air

The speed of sound also changes with altitude (Thompson, 1988):
Standard Atmosphere – Metres above sea level:

0 m	$T = 288$ K	$P = 1.000$ atm	$\rho = 1.2250$ kg m^{-3}	$c = 340.3$ m s^{-1}
1000 m	$T = 281$ K	$P = 0.887$ atm	$\rho = 1.1117$ kg m^{-3}	$c = 336.4$ m s^{-1}
10 000 m	$T = 223$ K	$P = 0.262$ atm	$\rho = 0.4135$ kg m^{-3}	$c = 299.5$ m s^{-1}

The preceding information indicates that the speed of sound in a gas is dependent on its composition (varying the specific heat capacities) and on its temperature. It is also dependent on the compressibility of the gas that is in turn dependent on the strength of the inter-molecular bonds in the material.

One can expect therefore that materials with high strength molecular bonds and therefore low compressibility will have high speeds of sound. This is borne out in practice with, for example Kinsler et al. (1982):

Material	Speed of sound
Water (fresh)	1481 m s^{-1}
Water (salt)	1500 m s^{-1}
Concrete	3100 m s^{-1}
Steel	6100 m s^{-1}

The notion that the speed of sound is a function of the compressibility of the medium helps in the appreciation of the physical process through which sound propagates.

The common model adopted for this process is that of a piston being used to compress air in a tube (for example, Morse and Ingard (1968), Dowling and Ffowcs Williams (1983), Fahy and Walker (1998). Pierce (1994) takes a different approach).

Picture the piston at rest within the tube. Allow the piston to suddenly start moving with a constant velocity u. The motion of the piston gradually compresses more and more of the air in front of it. The velocity of this air rises up towards u. The leading edge of this compression propagates down the tube at the speed of sound c.

So, now there are two velocities to consider – the particle velocity u (m s^{-1}) and the speed of sound c (m s^{-1}). Commonly,

$$u \ll c$$

If the piston is then brought to rest, the fluid in direct contact with it is decelerated and the local pressure drops. This forms a trailing edge to the pressure pulse set up earlier. Both the leading and trailing edges propagate down the tube at the same speed c.

If the piston is made to oscillate sinusoidally about its normal position of rest, then both compressions and rarefractions will occur directly in front of the piston. Both propagate down the tube with the same speed and a sound wave is produced.

The sound wave is a function of both time and distance down the tube. Consider a 100-Hz sinusoid with a peak pressure of 1 Pa. At a given position down the tube, its time history looks like that shown in Figure 2.1.

The period of oscillation T is given by the reciprocal of the frequency which is equal to 0.01 s.

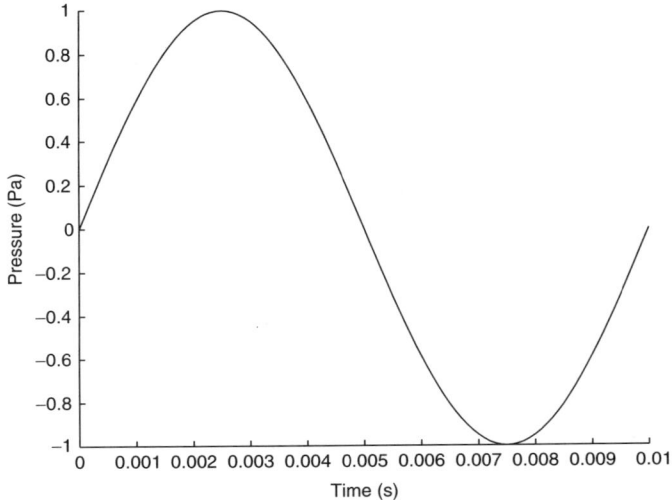

Figure 2.1 Time history for a 100-Hz sinusoid of peak amplitude 1 Pa.

At a given moment in time, the pressure is distributed down the tube as shown in Figure 2.2.

The wavelength λ is $3.43\,\mathrm{m\,s^{-1}}$. The frequency and the wavelength are related by:

$$c = f\lambda \tag{2.9}$$

A time history for the particle velocity as well as one for the pressure can be produced. The net force applied to the piston considered earlier is given by pS (assuming a uniform

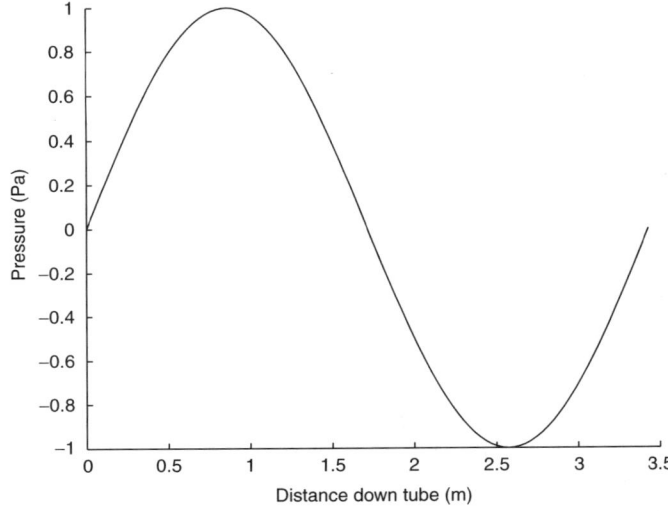

Figure 2.2 Pressure distribution down a tube of a 100-Hz sinusoid of peak amplitude 1 Pa.

pressure distribution across the area of the piston face S – this is a plane wave). This equates to a rate of change of momentum of the fluid (Morse and Ingard, 1968):

$$pS = \frac{\mathrm{d}}{\mathrm{d}t}(\rho_0 c_0 t S u)$$

$$p = \rho_0 c_0 u \tag{2.10}$$

The term $\rho_0 c_0$ is known as the characteristic specific impedance of the fluid. Air at 1 atm and 20°C has characteristic impedance of $415\,\mathrm{kg\,m^{-2}\,s^{-1}}$. This information can be used to produce a particle velocity time-history for the 100-Hz sinusoid considered earlier as shown in Figure 2.3.

Note that the particle velocity is a function of pressure but not of frequency (equation 2.10). This means that the displacement of an oscillating particle must reduce as the frequency rises (if not, u would rise with frequency for constant p, contrary to equation 2.10).

It is useful at this point to look at some numbers:

Lowest frequency audible to the typical human subject	20 Hz	Wavelength in air	17.15 m
Highest frequency audible to the typical human subject	20 000 Hz	Wavelength in air	17.15 mm
Lowest amplitude audible to the typical human subject	$20e^{-6}$ Pa	Particle velocity in air	$0.00005\,\mathrm{mm\,s^{-1}}$
Highest amplitude audible to the typical human subject in air without pain	200 Pa	Particle velocity in air	$481\,\mathrm{mm\,s^{-1}}$

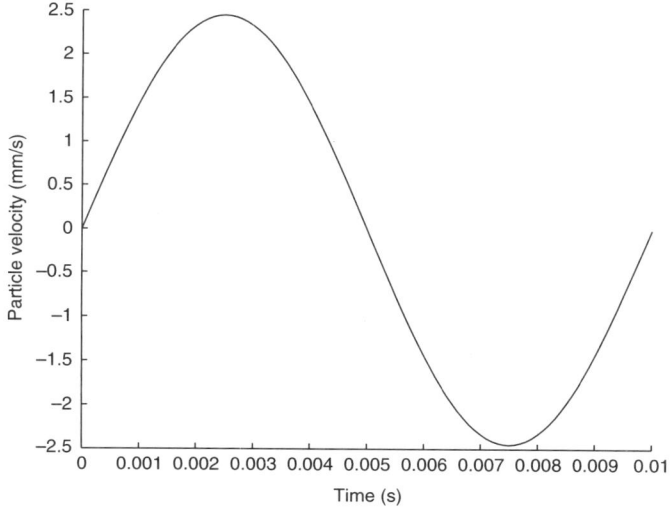

Figure 2.3 Particle velocity time-history for a 100-Hz sinusoid of peak pressure 1 Pa in air.

Demonstrations: some single tones

100 Hz tone, generated by Matlab™ and played on Microsoft™ MediaPlayer.
1000 Hz tone, generated by Matlab™ and played on Microsoft™ MediaPlayer.
10 000 Hz tone, generated by Matlab™ and played on Microsoft™ MediaPlayer.
   The tones can easily be created using the short Matlab™ script shown below and the results played using a PC and Windows MediaPlayer™

```
clear
clf

fs=30000;
t=[0:1/fs:3.0-(1/fs)];

a1=0.9;
f0=100; % frequency (Hz): change from 100 to 1000 to 10,000

a=a1*sin((2*pi*f0*t));

wavwrite(a',fs,'c:\hz100')
```

In order to explore sound further, attention should now be given to the subject of *pitch*. Lord Rayleigh gives a well-known discussion of pitch (Rayleigh, 1894). The highlights are:

> When passing from any note to its octave, the frequency of vibration is doubled . . .

> At the Stuttgard Conference in 1834, $c' = 264$ complete vibrations per second was recommended [c' is one octave above the lower c of the tenor voice].

With $c' = 256\,\text{Hz}$, all the c's have frequencies represented by powers of 2 ($256 = 2^8$). This choice for c' corresponds to a' being 440 Hz. Tuning forks are often found with a tuned frequency of 440 Hz. The French tuning at the time of the Stuttgard conference was $a' = 435\,\text{Hz}$. Handel tuned even lower.

Demonstrations: two middle A's

The tones at 440 Hz and 435 Hz are generated by Matlab™ and played on Microsoft™ MediaPlayer.
   The tones can easily be created using the short Matlab™ script shown below and the results played using a PC and Windows MediaPlayer™

```
clear
clf

fs=30000;
t=[0:1/fs:1.0-(1/fs)];
```

```
a1=0.9;
a2=0.9;

f0=440; % Hz
f2=435; % Hz

a=a1*sin((2*pi*f0*t));
b=a2*sin((2*pi*f2*t));

x=[a';b';a';b';a'];

wavwrite(x',fs,':\h440_435')
```

The twelve notes of the chromatic scale according to the system of equal temperament was given by Zamminer (1855) and also by Rayleigh (1894):

c		1.000
c#	$2^{1/12}$	1.05946
d	$2^{2/12}$	1.12246
d#	$2^{3/12}$	1.18921
e	$2^{4/12}$	1.25992
f	$2^{5/12}$	1.33484
f#	$2^{6/12}$	1.41421
g	$2^{7/12}$	1.49831
g#	$2^{8/12}$	1.58740
a	$2^{9/12}$	1.68179
a#	$2^{10/12}$	1.78180
b	$2^{11/12}$	1.88775
c'	$2^{12/12}$	2.000

The mathematical description of sound is attributed (at least in part) to Earnshaw (1860). However, more modern derivations of the linearised acoustic equations are probably easier to follow (for instance Morse and Ingard (1968) or Kinsler et al. (1982)).

The derivation of the linear plane wave equation shall now be demonstrated as a means of gaining further insight into the propagation of sound.

2.1.2 The linear plane wave equation

Consider a one-dimensional (1-D) section of fluid-filled pipe as shown in Figure 2.4. There is a slug of fluid of length dx between points (x) and $(x+dx)$. Suppose there is a finite displacement of length ξ at the left-hand end of the fluid slug, this will results in:

- the slug of fluid contracting in size to length $(dx - \xi)$;
- the pressure of the fluid in the slug increasing by the acoustic (sound) pressure p from P_0 to P;
- the density of the fluid in slug increasing by $\Delta\rho$ from ρ_0 to ρ.

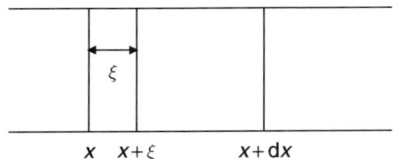

Figure 2.4 A one-dimensional section of fluid-filled pipe.

The fractional change in density is very small (0.000 175 at the threshold of pain), and is described as the condensation (s) of the fluid.

$$s = \frac{\Delta\rho}{\rho_0} = \frac{\rho - \rho_0}{\rho_0} \approx \frac{-\partial\xi}{\partial x} \tag{2.11}$$

so that

$$\rho = \rho_0 (1 + s) \tag{2.12}$$

The rise in pressure of the fluid in the slug must be related to the change in the volume of the slug. If the pipe walls were very thick, being of a material which is a very efficient heat conductor and with a high heat capacity, then any heat generated within the fluid slug under compression would quickly disperse into the pipe wall, and the fluid would essentially retain its original temperature. Such a process is called isothermal, and the following relation holds:

$$P_1 V_1 = P_2 V_2 = \text{constant} \tag{2.13}$$

However, it is generally known that acoustic disturbances are generally near adiabatic (they do not lose or gain heat) and isentropic (they retain a constant level of entropy). Such processes can be defined using the relations:

$$PV^\gamma = \text{constant} \tag{2.14}$$

$$\frac{P}{P_0} = \left(\frac{\rho}{\rho_0}\right)^\gamma \tag{2.15}$$

where γ is the ratio of the specific heats c_p/c_v of the fluid.
    Now from equation (2.3) under acoustic compression

$$p = P - P_0 \tag{2.16}$$

and re-arranging equation (2.16) and then substituting equation (2.15) gives

$$p = [P - P_0] = P_0 \left[\frac{P}{P_0} - 1\right] = P_0 \left[\left(\frac{\rho}{\rho_0}\right)^\gamma - 1\right] \tag{2.17}$$

Now substituting equation (2.12) into equation (2.17) gives

$$p = P_0 [(1 + s)^\gamma - 1] \tag{2.18}$$

The term $(1+s)^\gamma$ may be expanded using the binomial theorem

$$(a+b)^n = a^n + \left(\frac{n}{1}a^{n-1}b\right) + \left(\frac{n \times (n-1)}{1 \times 2}a^{n-1}b^2\right) + \left(\frac{n \times (n-1) \times (n-2)}{1 \times 2 \times 3}a^{n-3}b^3\right)$$
$$+ \left(\frac{n \times (n-1) \times (n-2) \times \cdots \times 2 \times 1}{1 \times 2 \times \cdots \times (n-2) \times (n-1) \times n}b^n\right)$$

(where the last term is included for completeness, and has value 1) to give

$$(1+s)^\gamma - 1 = 1 + \gamma s + \left(\frac{\gamma(\gamma-1)}{2}s^2\right) + \left(\frac{\gamma(\gamma-1)(\gamma-2)}{6}s^3\right) + \cdots - 1 \qquad (2.19)$$

Now as the value of s for air at normal acoustic levels is very small, the higher-order terms may be neglected (linearisation) giving

$$(1+s)^\gamma - 1 \approx \gamma s$$

therefore from equations (2.18) and (2.11)

$$p \approx P_0 \gamma s \approx -P_0 \gamma \frac{\delta\xi}{\delta x} \qquad (2.20)$$

Now consider the slug of fluid which has a cross-sectional area S. The force on the element at position x required to maintain equilibrium is

$$F_x = +S \cdot p$$

and at position $x + \delta x$, the equilibrium force is

$$F_x + \delta_x = -S\left[p + \frac{\delta p}{\delta x}\delta x\right]$$

and thus the net force across the slug of fluid is

$$F = -S\frac{\delta p}{\delta x}\delta x \qquad (2.21)$$

Now consider Newton's second law of motion

$$F = m \cdot a \qquad (2.22)$$

where m is mass (kg) and a is acceleration (m s^{-2}). If the motion of the slug of fluid under a small displacement is assumed to be planar, and the effects of viscosity are assumed to be negligible, then the acceleration of the fluid may be defined as

$$a = \frac{\delta^2\xi}{\delta t^2}$$

and Newton's second law for the slug of fluid as

$$-S\frac{\delta p}{\delta x}\delta x = \rho_0 S \delta x \frac{\delta^2 \xi}{\delta t^2}$$

$$-\frac{\delta p}{\delta x} = \rho_0 \frac{\delta^2 \xi}{\delta t^2} \tag{2.23}$$

Now differentiating equation (2.20) with respect to x

$$\frac{\delta p}{\delta x} = -P_0\gamma\frac{\delta^2 \xi}{\delta x^2} \tag{2.24}$$

and substituting equation (2.24) into equation (2.23) the linearised, 1-D plane wave equation in terms of displacement ξ is obtained

$$\frac{\delta^2 \xi}{\delta t^2} = \frac{\gamma P_0}{\rho_0}\frac{\delta^2 \xi}{\delta x^2} \tag{2.25}$$

It is more customary to perform the above analysis in terms of pressure p, particle velocity u and condensation s.

Such analysis is started by the double differentiation of equation (2.20) with respect to t

$$\frac{\delta p}{\delta t} = -P_0\gamma\frac{\delta^2 \xi}{\delta x\, \delta t}$$

$$\frac{\delta^2 p}{\delta t^2} = -P_0\gamma\frac{\delta^3 \xi}{\delta x\, \delta t^2} \tag{2.26}$$

Then equation (2.23) is differentiated with respect to x

$$\frac{\delta^2 p}{\delta x^2} = -\rho_0\frac{\delta^3 \xi}{\delta t^2\, \delta x} \tag{2.27}$$

and $\dfrac{\delta^3 \xi}{\delta t^2\, \delta x}$ from equation (2.27) is substituted into equation (2.26) to give

$$\frac{\delta^2 p}{\delta t^2} = \frac{P_0\gamma}{\rho_0}\frac{\delta^2 p}{\delta x^2} \tag{2.28}$$

Now from (2.7) and from the ideal gas equation $P_0 = \rho_0 RT$

$$c^2 = \frac{\gamma P_0}{\rho_0} \tag{2.29}$$

where c is the speed of sound in the fluid (m s^{-1}), and thus the linear plane wave equation becomes

$$\frac{\delta^2 p}{\delta t^2} = c^2\frac{\delta^2 p}{\delta x^2} \tag{2.30}$$

2.1.3 The linear wave equation in spherical coordinates

Cartesian coordinates (Figure 2.5) may be transformed into spherical coordinates thus:

$$x = r \sin\theta \cos\psi$$

$$y = r \sin\theta \sin\psi$$

$$z = r \cos\theta$$

$$r^2 = x_L^2 + y_L^2 + z_L^2$$

$$r^2 = x_i^2$$

In the case of a simple acoustic source in free space, the wave propagation depends only on the radial term (as no angular dependence is implied) and on the time term.

Consider the spatial derivative (in three axes) of the pressure, $\dfrac{\delta}{\delta x_i} p(r, t)$.

Now $r^2 = x_i^2$ so the derivative of a 'function of a function' is sought which calls for the use of the 'chain rule'.

$$\text{Chain rule} \quad y = f(g(x)); \qquad y' = \frac{df}{dg} g'(x)$$

So using the chain rule

$$\frac{\delta}{\delta x_i} p(r, t) = \frac{\delta p}{\delta r} \frac{\delta r}{\delta x_i} = \frac{x_i}{r} \frac{\delta p}{\delta r} \tag{2.31}$$

(see Pierce (1994)).

Consider the second spatial derivative of p, for which he product rule of differentiation is required:
where

$$y = u(x) \cdot v(x),$$

$$y' = uv' + vu'$$

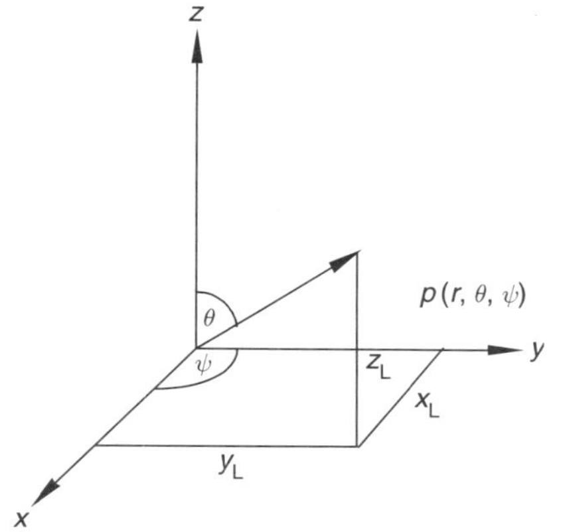

Figure 2.5 The transformation of Cartesian coordinates to spherical coordinates.

so

$$\frac{\delta}{\delta x_i}\left(\frac{x_i}{r}\frac{\delta p}{\delta r}\right) = \frac{\delta^2}{\delta x_i^2}p(r,\ t) = \frac{1}{r}\frac{\delta p}{\delta r} + \frac{x_i}{r}\frac{\delta}{\delta x_i}\left(\frac{\delta p}{\delta r}\right) \tag{2.32}$$

Now as $x_i^2 = r^2$

$$\frac{\delta^2}{\delta x_i^2}p(r,\ t) = \frac{1}{r}\frac{\delta p}{\delta r} + \frac{x_i^2}{r^2}\frac{\delta}{\delta r}\left(\frac{\delta p}{\delta r}\right) \tag{2.33}$$

and equally

$$\frac{\delta^2}{\delta x_i^2}p(r,\ t) = \frac{1}{r}\frac{\delta p}{\delta r} + \frac{x_i^2}{r}\frac{\delta}{\delta r}\left(\frac{1}{r}\frac{\delta p}{\delta r}\right) \tag{2.34}$$

Now

$$\nabla^2 F = \frac{\delta^2 F}{\delta x^2} + \frac{\delta^2 F}{\delta y^2} + \frac{\delta^2 F}{\delta z^2}$$

By inspection, equation (2.34) can be used for each axis and the results added together (note that $\sum x_i^2 = r^2$)

$$\nabla^2 p(r,\ t) = \frac{3}{r}\frac{\delta p}{\delta r} + r\frac{\delta}{\delta r}\left(\frac{1}{r}\frac{\delta p}{\delta r}\right) \tag{2.35}$$

If the product rule of differentiation is applied to the last derivative

$$\nabla^2 p(r,\ t) = \frac{3}{r}\frac{\delta p}{\delta r} + r\left[-\frac{1}{r^2}\frac{\delta p}{\delta r} + \frac{1}{r}\frac{\delta^2 p}{\delta r^2}\right] \tag{2.36}$$

so

$$\nabla^2 p(r,\ t) = \frac{2}{r}\frac{\delta p}{\delta r} + \frac{\delta^2 p}{\delta r^2} \tag{2.37}$$

Now, looking ahead, if the product rule were applied to rp

$$\frac{\delta rp}{\delta r} = r\frac{\delta p}{\delta r} + p$$

The second differential of rp is therefore:

$$\frac{\delta^2 rp}{\delta r^2} = \frac{\delta p}{\delta r} + \frac{\delta p}{\delta r} + r\frac{\delta^2 p}{\delta r^2} \tag{2.38}$$

and by inspection the division of equation (2.38) by r yields the right-hand side of equation (2.37). So

$$\nabla^2 p(r,\ t) = \frac{1}{r}\frac{\delta^2 rp}{\delta r^2} \tag{2.39}$$

Consequently the wave equation (2.30)

$$\frac{\delta^2 p}{\delta t^2} = c^2 \frac{\delta^2 p}{\delta x^2} \tag{2.30}$$

after substitution of equation (2.39) becomes:

$$\frac{\delta^2 p}{\delta t^2} = \frac{c^2}{r} \frac{\delta^2 rp}{\delta r^2}$$

$$\frac{1}{r} \frac{\delta^2 rp}{\delta r^2} = \frac{1}{c^2} \frac{\delta^2 p}{\delta t^2} \tag{2.40}$$

Fortunately, most acoustic phenomena involve small perturbations of pressure around a much greater static pressure. For instance, the pressure fluctuation associated with a sound pressure level of 94 dB is 1 Pa, which is 0.001% of 1 atm. In cases such as these, assumptions can be made, resulting in linear simplifications leading to the linearised acoustic wave equation (2.30 and 2.40) to which analytical solutions may be found.

There are of course conditions where the linearisation is not appropriate such as cases involving:

- high amplitude waves with large particle velocities;
- acoustic waves superimposed on high mean speed flows;
- transmission of sound through small holes where significant losses arise due to visco thermal effects.

2.1.4 Solutions to the 1-D linear plane wave equation

It has been shown that

$$\frac{\delta^2 p}{\delta t^2} = c^2 \frac{\delta^2 p}{\delta x^2}$$

This is a 'second order hyperbolic partial differential equation'(Duchateau and Zachmann, 1986). Such equations link the temporal variation in a quantity (pressure, density and temperature in this case) with a spatial variation by a finite speed of propagation. They exhibit wave-like properties.

The plane wave equation (2.30) can be solved for $p(x, t)$ using a numerical integration technique such as the finite difference method (Duchateau and Zachmann, 1986). However, an analytical solution is available for hyperbolic PDE which is put to use in Section 3.3 of this book for the design of intake and exhaust systems.

$$P(x, t) = p^+ e^{i(\omega t - kx)} + p^- e^{i(\omega t + kx)} \tag{2.41}$$

where

$\omega =$ radial frequency $(2\pi f - \mathrm{rad\ s}^{-1})$
$f =$ frequency (Hz)
$k =$ wave number (ω/c)
$c =$ sound speed $(\mathrm{m\ s}^{-1})$

This form of solution is an extension of the D'Alembert's solution (Weltner et al., 1986) which can be thus derived (Temkin, 1981). Start with a general form of the 1D plane wave equation

$$\frac{\partial^2 \phi}{\partial t^2} = c^2 \frac{\partial^2 \phi}{\partial x^2} \tag{2.42}$$

where, ϕ is a function of x, t because there is a value for ϕ for every pair of (x, t).

Now introduce two new independent variables ξ and η that have the same dimensions as x, where

$$\xi = x - ct \tag{2.43}$$

$$\eta = x + ct \tag{2.44}$$

Here ϕ becomes a function of ξ and η and the total differential in space of the function ϕ is (Weltner et al., 1986)

$$d\phi = \frac{\partial \phi}{\partial \xi} d\xi + \frac{\partial \phi}{\partial \eta} d\eta \tag{2.45}$$

The total derivative of function ϕ with respect to x is (Weltner et al., 1986)

$$\frac{\partial \phi}{\partial x} = \frac{\partial \phi}{\partial \xi} \frac{\partial \xi}{\partial x} + \frac{\partial \phi}{\partial \eta} \frac{\partial \eta}{\partial x} \tag{2.46}$$

Now from equations (2.43) and (2.44)

$$\frac{\partial \xi}{\partial x} = 1 \tag{2.47}$$

$$\frac{\partial \eta}{\partial x} = 1 \tag{2.48}$$

so equation (2.46) becomes

$$\frac{\partial \phi}{\partial x} = \frac{\partial \phi}{\partial \xi} + \frac{\partial \phi}{\partial \eta} \tag{2.49}$$

The second partial derivative can be found by repeating the above, using equation (2.46) as an operator on equation (2.49), thus

$$\frac{\partial^2 \phi}{\partial x^2} = \frac{\partial}{\partial \xi}\left(\frac{\partial \phi}{\partial \xi} + \frac{\partial \phi}{\partial \eta}\right) \times \frac{\partial \xi}{\partial x} + \frac{\partial}{\partial \eta}\left(\frac{\partial \phi}{\partial \xi} + \frac{\partial \phi}{\partial \eta}\right) \times \frac{\partial \eta}{\partial x} \tag{2.50}$$

Substituting equations (2.47) and (2.48) into equation (2.50)

$$\frac{\partial^2 \phi}{\partial x^2} = \frac{\partial^2 \phi}{\partial \xi^2} + 2\frac{\partial^2 \phi}{\partial \xi \partial \eta} + \frac{\partial^2 \phi}{\partial \eta^2} \tag{2.51}$$

Stages (2.46) to (2.51) can be repeated but with respect to time thus

$$\frac{\partial \phi}{\partial t} = \frac{\partial \phi}{\partial \xi} \frac{\partial \xi}{\partial t} + \frac{\partial \phi}{\partial \eta} \frac{\partial \eta}{\partial t} \tag{2.52}$$

$$\frac{\partial \xi}{\partial t} = -c \tag{2.53}$$

$$\frac{\partial \eta}{\partial t} = c \tag{2.54}$$

so substituting equations (2.53) and (2.54) into equation (2.52)

$$\frac{\partial \phi}{\partial t} = -c\frac{\partial \phi}{\partial \xi} + c\frac{\partial \phi}{\partial \eta} \tag{2.55}$$

The second partial derivative with respect to time is thus found

$$\frac{\partial^2 \phi}{\partial t^2} = \frac{\partial}{\partial \xi}\left(c\frac{\partial \phi}{\partial \eta} - c\frac{\partial \phi}{\partial \xi}\right) \times \frac{\partial \xi}{\partial t} + \frac{\partial}{\partial \eta}\left(c\frac{\partial \phi}{\partial \eta} - c\frac{\partial \phi}{\partial \xi}\right) \times \frac{\partial \eta}{\partial t} \tag{2.56}$$

which on substituting equation (2.53) becomes

$$\frac{\partial^2 \phi}{\partial t^2} = -c^2\frac{\partial^2 \phi}{\partial \xi \partial \eta} + c^2\frac{\partial^2 \phi}{\partial \xi^2} + c^2\frac{\partial^2 \phi}{\partial \eta^2} - c^2\frac{\partial^2 \phi}{\partial \xi \partial \eta}$$

$$\frac{\partial^2 \phi}{\partial t^2} = c^2\left[\frac{\partial^2 \phi}{\partial \xi^2} + \frac{\partial^2 \phi}{\partial \eta^2}\right] - 2c^2\frac{\partial^2 \phi}{\partial \xi \partial \eta} \tag{2.57}$$

Substituting equation (2.51) and (2.57) into the wave equation (2.42) gives

$$c^2\left[\frac{\partial^2 \phi}{\partial \xi^2} + \frac{\partial^2 \phi}{\partial \eta^2}\right] - 2c^2\frac{\partial^2 \phi}{\partial \xi \partial \eta} = c^2\left[\frac{\partial^2 \phi}{\partial \xi^2} + \frac{\partial^2 \phi}{\partial \eta^2}\right]$$

thus the wave equation reduces to

$$-2c^2\frac{\partial^2 \phi}{\partial \xi \partial \eta} = 0$$

or

$$\frac{\partial^2 \phi}{\partial \xi \partial \eta} = 0 \tag{2.58}$$

Integration of equation (2.58) with respect to η gives

$$\frac{\partial \phi}{\partial \xi} = F(\xi)$$

and integrating again with respect to ξ gives the general form of the D'Alembert solution

$$\phi = f(\xi) + g(\eta)$$

or

$$\phi = f(x - ct) + g(x + ct) \tag{2.59}$$

The functions f and g could take one of many forms depending on circumstances, but for the case of travelling harmonic waves, a complex exponential offers a compact way of describing the temporal and spatial distribution of ϕ thus: $e^{i(x \pm ct)}$. If the exponent is multiplied by the wave number k, $e^{i(kx \pm \omega t)}$ is obtained which is more commonly written as $e^{i(\omega t \pm kx)}$.

Notice that there would be two exponential terms in the D'Alembert solution. The first represents the effects of a wave travelling in the direction in which x is positive. The second represents the effects of a wave travelling in the direction in which x is negative. Both of these travelling wave components will exist simultaneously in a given acoustic field. This is the principal outcome of the D'Alambert solution.

The reason that a complex exponential form is useful for harmonic waves is explained by the relation given by Euler (Weltner et al., 1986)

$$e^{i\alpha} = \cos \alpha + i \sin \alpha \tag{2.60}$$

so that

$$e^{i(\omega t - kx)} = \cos(\omega t - kx) + i \sin(\omega t - kx) \tag{2.61}$$

which are the real and imaginary parts of the complex number that describes the variation of ϕ with space and time. The magnitude of that complex number varies in the form of a harmonic function (a sine or cosine) as does the magnitude of the acoustic wave, so the complex exponential is a perfect descriptor of the physical process.

If ϕ is being used to describe the variation in acoustic pressure, then the D'Alembert solution can be applied along with equation (2.61) to give

$$p(x, t) = p^+ e^{i(\omega t - kx)} + p^- e^{i(\omega t + kx)} \tag{2.62}$$

where

p^+ is the complex magnitude of the positive travelling wave component (travelling in the positive direction)
p^- is the complex magnitude of the negative travelling wave component (travelling in the negative direction)
$i(\omega t \pm kx)$ yields the relative phase of the two travelling wave components.

The development of the D'Alembert solution to the useful harmonic form used in equation (2.62) and is given in Bies and Hansen (1996) and Fahy and Walker (1998).

It is easy to demonstrate that the D'Alembert solution is indeed a solution to the wave equation. Take one part of the solution and differentiate twice with respect to t

$$\frac{\partial \phi}{\partial t} = \frac{\partial \phi}{\partial \xi} \frac{\partial \xi}{\partial t} = -c \frac{\partial \phi}{\partial \xi}$$

$$\frac{\partial^2 \phi}{\partial t^2} = \frac{\partial}{\partial \xi}\left(-c\frac{\partial \phi}{\partial \xi}\right) \times \frac{\partial \xi}{\partial t} = c^2 \frac{\partial^2 \phi}{\partial \xi^2}$$

and then with respect to x

$$\frac{\partial \phi}{\partial x} = \frac{\partial \phi}{\partial \xi}\frac{\partial \xi}{\partial x} = \frac{\partial \phi}{\partial \xi}$$

$$\frac{\partial^2 \phi}{\partial x^2} = \frac{\partial}{\partial \xi}\left(\frac{\partial \phi}{\partial \xi}\right) \times \frac{\partial \xi}{\partial x} = \frac{\partial^2 \phi}{\partial \xi^2}$$

Combining these two equations gives the wave equation

$$\frac{\partial^2 \phi}{\partial t^2} = c^2 \frac{\partial^2 \phi}{\partial x^2} \tag{2.63}$$

The same analysis can be applied to the variable η. It is possible to take account of the effects of temperature gradients, attenuation down pipes and mean flow by using a complex sound speed to produce a complex wave number β.

As the wave equation is linear, the principles of additivity and homogeneity apply. The first property is used to sum the contributions from different sound sources. The second dictates that if the input to a system is multiplied by a factor, then the output will multiply by the same factor (Sinha, 1991). If the input is cyclic, then it is easy to transfer from a temporal solution to the wave equation to a frequency-specific solution using the Fourier transform (Sinha, 1991).

2.2 Making basic noise measurements: the sound level meter, recording sound

Summary: Sound is an oscillatory disturbance of a medium from its usual position of rest. This acoustic disturbance is in fact fluctuations of three variables – pressure, density and temperature (denoted by p', ρ' and T') about their normal values at rest (denoted by p_0, ρ_0, T_0) as described by equations (2.3), (2.4) and (2.5).

The three disturbances propagate with a certain velocity c ($\mathrm{m\,s^{-1}}$) which is known as the speed of sound. The speed of sound in air is commonly around $340\text{–}345\,\mathrm{m\,s^{-1}}$ at the earth's surface. Individual particles undergoing oscillatory motion have a cyclic velocity of their own about their usual position of rest – this is the particle velocity and is generally much lower than the speed of sound.

It is not easy to measure fluctuating density or temperature accurately. Therefore, sound is commonly detected by the fluctuating pressure using a microphone (or other pressure transducer). With two microphones positioned at a known spacing, both the acoustic pressure and the particle velocity can be measured. An acoustic field is fully characterised by these two variables if obtained simultaneously.

Occasionally, some literature states that the vibration of a solid surface is needed to produce sound. That is not true. Sound is indeed produced in many instances by the vibration of a solid body (violin string, violin casing, engine sump (!)) but there are other mechanisms for sound production.

Sound may be generated in many different ways. However, these various mechanisms may be classed into two groups, which for convenience will be labelled:

1. Direct sound generation mechanisms whereby sound is a direct by-product of a physical process such as combustion or mass transfer.
2. Indirect sound generation mechanisms where sound is a result of a force producing the vibration response of a structure which in turn radiates sound.

2.2.1 Direct sound generation mechanisms

This is an arbitrary classification of noise sources and is illustrated in Figure 2.6.
   The physical mechanism could be any of the following:

- net displacement of fluid mass or volume (loudspeaker, exhaust tailpipe);
- unsteady combustion (engine, furnace, boiler);
- unsteady fluid transport (flow noise);
- fluctuating force on a fluid (fans, wires in a flow);
- fluctuating shear force (shear layer noise in a jet).

When controlling the sound-power output from a direct sound generation mechanism at source, only a limited number of strategies may be followed:

(i) Reduce the source strength by altering the parameters within the physical process itself such as the rate of combustion, flow rate, the force imported to the fluid, etc.
(ii) Reduce the physical size of the source as a means of reducing the source strength (i.e. use a smaller machine if possible).

2.2.2 Indirect sound generating mechanism

This is the second arbitrary classification of noise sources and is illustrated in Figure 2.7.
   This classification covers many sound sources commonly found on vehicles, such as:

- casing vibration
- bearing noise
- electric motor or generator noise

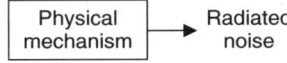

Figure 2.6 The direct sound generation mechanism.

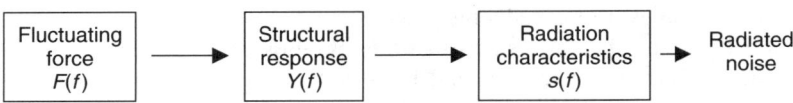

Figure 2.7 The indirect sound generation mechanism.

- belt, chain or gear drives
- internal combustion (IC) engine structural noise.

There is a wider scope for controlling noise, at source, due to indirect sound generation mechanisms than due to direct sound generation mechanisms. The noise control strategies fall into three groups:

1. Reducing the amplitude of the force, or where there are a number of independent cyclic forces, arranging their phasing so as to obtain cancellation.
2. Reducing the vibration response of the structure to a given input force.
3. Reducing the acoustic radiation efficiency of the structure at a given frequency by altering the critical frequency of the radiating component.

A note is offered on the terms 'sound' and 'noise'. The terms 'sound' and 'noise' are probably interchangeable although generally the term 'noise' is taken to describe unwanted sound. This is supported by the fact that the job description 'noise engineer' usually describes someone dedicated to reducing noise levels whereas 'sound engineer' is usually someone involved in recording music or speech for broadcast or entertainment.

Some demonstrations

Equipment:

- loudspeaker and amplifier
- signal generator
- oscilloscope
- microphone, amplifier and cable
- vibration exciter or small shaker
- cardboard box lid
- balloons
- sharp pencil.

Demonstration 1
The periodic nature of sound can be seen from the voltage output of a microphone displayed on an oscilloscope, with a loudspeaker producing single tones of varying frequency.

Demonstration 2
Automotive noise sources: Second-order engine-related noise, 30–200 Hz (see Section 2.4.2.5). The division between airborne and structure-radiated road noise is 500 Hz (see Section 4.1.2).

Demonstration 3
Direct sound sources: A loudspeaker, and bursting of a balloon with the sharp pencil.

Demonstration 4
Indirect sound sources: The vibration exciter placed on a cardboard box lid.

2.2.3 Measuring sound (or noise) levels

Sound levels are measured using a microphone attached to some form of electrical signal conditioning and analysis equipment. There are many types of microphone with different operational means of converting fluctuating pressure (or pressure difference) into an electrical signal. The main types are:

Carbon   (cheapest and lowest quality);
Dynamic   (moving coil, operating like a loudspeaker in reverse);
Ribbon   (pressure difference across a moving ribbon);
Electret   (very small, usually cheap and durable yet reasonable quality);
Condenser   (the precision instrument).

For engineering and scientific sound level measurements, the precision of the condenser microphone is usually required. Therefore, it is the only type to be discussed further here.

2.2.3.1 Construction of the condenser microphone

As the name suggests, the operating principle of a condenser microphone is to use a diaphragm as the moving electrode of a parallel-plate air capacitor ('condenser' now being the obsolete term for a capacitor).

Bruel & Kjaer (B&K) condenser microphones (B&K data handbook on condenser microphones, 1982) have a nickel diaphragm, $1.6–6.5\,\mu m$ thick, that acts as a moving electrode, and a rigid monel backplate as the fixed electrode. The backplate is mounted on an electrical insulator of quartz, synthetic sapphire or synthetic ruby. The air gap between the electrodes is around $20\,\mu m$.

The diaphragm will exhibit resonance (at rather high frequencies) that is controlled by air movement between the electrodes. A small pressure-equalisation vent allows the air pressure between the two electrodes to equalise with that on the exterior face of the diaphragm (thus preventing any movement due solely to atmospheric conditions). The pressure equalisation vent is small, thus preventing the passage of sound at frequencies above around 2 Hz.

When manufacturers such as B&K talk of 'condenser microphones' they are referring only to the small housing that holds the two electrodes. This produces a small electrical charge that needs immediate signal conditioning before that electrical signal can be communicated over any distance along wires. Therefore, the microphone (or 'microphone capsule' or 'microphone element' or 'microphone cartridge') must be screwed onto a preamplifier. It is the preamplifier that gives the stick-like shape to what most people think of as a 'microphone'. The preamplifier allows the electrical signal to be driven only a few metres along a 'microphone cable' before it must be amplified further.

In order for the two electrodes to act together as a capacitor, they must have a polarisation voltage put across them. Sometimes this is an externally applied DC voltage of 200 V. In this case it takes about a minute for the correct charge to build up on the electrodes – during this time the microphone will not operate as normally expected. Other condenser microphones are internally charged by a pre-polarised electret layer deposited on the backplate. This is sometimes known as 0 V polarisation.

Beware!   Applying an external polarisation voltage to an internally charged microphone will immediately damage it beyond repair.

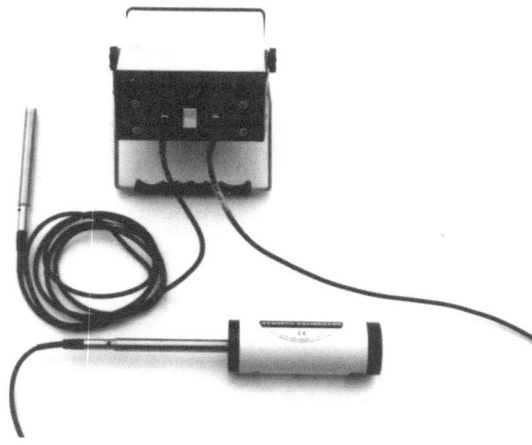

Figure 2.8 Laboratory grade condenser microphone equipment.

Figure 2.8 shows a photograph of a microphone capsule, a preamplifier and a power supply/signal amplifier along with an acoustic calibrator.

2.2.3.2 Some characteristics of the condenser microphone: sensitivity

As might be expected the larger the electrodes within the microphone, the greater is the charge produced by a given deflection of the diaphragm (caused by fluctuating pressure), the greater is the sensitivity of the microphone.

Precision condenser microphones usually come in four sizes, denoted by their (approximate) external diameters:

1. one inch;
2. half inch;
3. quarter inch;
4. one-eighth inch.

The open-circuit sensitivity is usually quoted for direct comparison between microphones. This is the voltage (mV) produced per Pascal of pressure at 250 Hz with 200 V polarisation (except for a few microphones requiring 28 V or 0 V polarisation).

This open-circuit sensitivity is the sensitivity of the capsule on its own before it is electrically loaded by being attached to a preamplifier (in that case the loaded sensitivity is required). The sensitivity of the microphone varies with temperature and with atmospheric pressure.

2.2.3.3 Frequency response

At low frequencies (above say 20 Hz) the open-circuit sensitivity of a precision condenser microphone remains constant with changing frequency. However, the sensitivity does become frequency-dependent at higher frequencies. The change in sensitivity with frequency is known as the frequency response of the microphone.

The frequency response of a condenser microphone is determined by four main factors (B&K data handbook on condenser microphones, 1982):

1. diaphragm stiffness;
2. mechanical damping of the diaphragm due to viscous resistance to air movement between the two electrodes;
3. diaphragm mass;
4. interference and diffraction effects at frequencies where the microphone diameter becomes of the same order as the wavelength of the impinging sound.

At frequencies below the mechanical resonance of the diaphragm, the motion of the diaphragm is most strongly controlled by its stiffness. Around resonance, the damping governs the motion. Above the resonant frequency it is the mass of the diaphragm that imposes the greatest limit on the excursion of the diaphragm for a given pressure (and hence the sensitivity of the microphone) and this effect increases with increasing frequency. The resonant frequency of the 1" microphone is expected to be lower than that of the 1/2" microphone and this is expected to limit the sensitivity of the 1" microphone at higher frequencies.

The interference and diffraction effects mentioned earlier result in free-field correction curves being applied to the microphone response. When the physical size of the microphone is small compared with the wavelength of impinging sound, its presence does not affect the local sound field and it measures the true fluctuating pressure known as the free-field pressure (P_0) as if the microphone was not there. The response of the microphone in these conditions is known as the free-field response (V_0/P_0). At higher frequencies, near the resonant frequency of the microphone, the impinging sound is reflected and diffracted by the presence of the microphone, and as a result the pressure at the diaphragm is increased from its free-field value (P_0) to P_1. The response of the microphone to this (artificially) higher pressure is the pressure response (V_0/P_1).

By altering the damping characteristics, it is possible to engineer a microphone that has a free-field response that remains independent of frequency up to relatively high frequencies (known as a linear, free-field response microphone). The response is only linear at zero degrees incidence to the impinging sound, so such microphones should be pointed at the sound source. It is also possible to construct microphones that respond linearly with frequency (up to a limiting frequency) to the actual pressure at the diaphragm. These are known as pressure response microphones.

The difference between the free field and the pressure response (in dB) with frequency is known as the free-field correction. It is a function of the type of microphone and the angle of incidence with the impinging sound. A family of frequency curves is usually published for each type of microphone with different curves for selected angles of incidence (0−180°). Individual precision microphone capsules are delivered with their own calibration curves – typically open-circuit free-field response at 0° incidence and open-circuit pressure response.

For environmental noise, one is usually interested in the free-field response (note that a pressure microphone with its diaphragm at 90° incidence will behave almost as a free-field microphone). When making measurements in a diffuse field, it is the random incidence response that is of interest and a random incidence microphone, pressure microphone or a smaller free-field microphone should be used (more omni-directional response) or else a random incidence corrector should be attached to the free-field capsule. When making measurements within small cavities or couplers or when the microphone diaphragm is mounted flush with a hard surface, it is the pressure response that is of interest.

2.2.3.4 Directional characteristics

As discussed earlier, the response of a condenser microphone is dependent (in part) on the angle of incidence of the impinging sound. This gives rise to its directional characteristics. The directional characteristics of condenser microphones are dependent on the physical size of the unit once coupled to its preamplifier and so many variants of a particular type of microphone will have the same directional characteristics.

The 1" microphone is the most directional of all (particularly above 10 kHz). The effect is much less marked (even up to 20 kHz) in the more commonly used 1/2" microphone. The omni-directionality of the 1" microphone can be improved by fitting a nose cone or the special random incidence corrector.

2.2.3.5 Dynamic range

The lower limit of the dynamic range of the condenser microphone and preamplifier combination is determined by the levels of internal (electrical) noise. The upper limit is determined by distortion.

Typical dynamic ranges are (approximate – for guidance only):

1" microphone	12–150 dBA
1/2" microphone	25–155 dBA
1/4" microphone	40–170 dBA
1/8" microphone	55–175 dBA

Significant levels of distortion can be expected with any size of microphone at very high sound levels – 140 dB and above (200 Pa).

2.2.3.6 Preamplifiers

Preamplifiers are designed to have high input impedance of around 10–50 GΩ. The input capacitance is usually around 0.2 pF which is rather small compared with the capacitance of the polarised microphone capsule (around 3–65 pF at 250 Hz, with the smallest values for 1/8" microphones and the largest for 1" microphones). The high input impedance produces a reasonable voltage level from the charge output of the microphone capsule.

The noise-floor of the preamplifier is dependent on the capacitance load imposed by the microphone capsule. In general, larger capsules with the highest capacitance yield the lowest noise.

Preamplifiers are designed to have low output impedance (around 25–100 Ω) in order to preserve high frequency response (usually flat in the range 1–200 kHz). The capacitive output load given by the microphone cable and the input impedance of the next device in the signal chain also determines the frequency response. For this reason, microphone cables are restricted in length to a few metres. Typical preamplifiers have a gain of 0 dB reflecting their role in impedance conversion rather than voltage amplification in the usual sense.

2.2.3.7 Power supplies

The stabilised polarisation voltage (200 V or 28 V) is provided by the microphone power supply: a large battery-operated box (or mains via an adapter) used within a few metres of the preamplifier. A low output impedance permits the use of long(er) cables (usually BNC coaxial type) between the power supply and the next device in the signal chain.

2.2.3.8 Measuring amplifiers

Measuring amplifiers are available that provide both the power supply and also convert the fluctuating electrical output of the preamplifier into either:

- amplified voltage for tape recording or digital storage;
- rms level (with optional time and frequency weightings) for quantifying noise levels (after appropriate calibration).

2.2.3.9 Sound level meters

If a microphone capsule, a preamplifier, a power supply, a measuring circuit (such as found in a measuring amplifier), are put together into one portable box, then a sound level meter is created. If that box includes filter networks (beyond the usual frequency and time weightings) a sound level analyser is created. A typical example is shown in Figure 2.9.

IEC 804 (1985), IEC 651 (1979) and ANSI S1.4 (1983) describe four grades of sound level meter (summarised in Bies and Hansen (1996)):

Type 0   Laboratory reference standard, intended entirely for calibration of other sound level meters.

Type 1   Precision sound level meter, intended for laboratory use or field use where accurate measurements are required.

Figure 2.9 A typical sound level analyser.

Type 2   General purpose sound level meter intended for general use and for recording noise level data for later frequency analysis.

Type 3   Survey sound level meter, intended for preliminary investigations such as the determination of whether noise environments are unduly bad.

The more complex sound level meters are of the integrating type. These allow the classification of time varying noise levels using:

L_{EQ} The equivalent continuous sound pressure level for a given period $T(s)$ expressed in dB. The period T should be quoted whenever L_{EQ} levels are quoted.

$$L_{EQ} = 10 \; \log_{10} \left[\frac{1}{T} \int_0^T 10^{\frac{L(t)}{10}} \; dt \right] dB \qquad (2.64)$$

L_x The sound pressure level (dB) that is exceeded for $x\%$ of the time.

Both of these indices are commonly used in the assessment of environmental noise.

2.2.3.10   Calibration

Each microphone cartridge is supplied with an individual calibration chart that includes a complete frequency response curve along with sensitivity data. The reference sensitivity for B&K equipment is -26 dB re 1 V per Pa (i.e. 50 mV per Pa (B&K data handbook on condenser microphones, 1982)). A correction factor K may be calculated from:

- the open-circuit sensitivity S_0(dB re 1 V per Pa);
- the preamplifier gain g(dB);
- the microphone capacitance C_t(pF);
- the preamplifier capacitance C_i(pF).

$$K(dB) = -26 - S_0 - \left[g + 20 \; \log_{10} \left(\frac{C_t}{C_t + C_i} \right) \right] \qquad (2.65)$$

where K may be added to the reference sensitivity to yield the calibration factor for a particular microphone and preamplifier pair (dB re 1 V per Pa).

In the field, the entire signal chain from microphone capsule to analyser display can be calibrated using either:

- An acoustic calibrator – a small cylindrical device that fits over the microphone capsule and produces a reference sound pressure level at a given frequency (usually 94 dB at 1000 Hz, see Figure 2.10). The level of accuracy is usually ±0.2 dB for a Type 1 calibrator.
- A pistonphone – a larger device operating at 124 dB and 250 Hz. Full pistonphone kits are supplied with a barometer to allow compensation for atmospheric pressure. The level of accuracy is ±0.09 dB under reference conditions.

It is routine to calibrate the signal chain at the start of every measurement session. Many practitioners also confirm the calibration at the end of the measurement session.

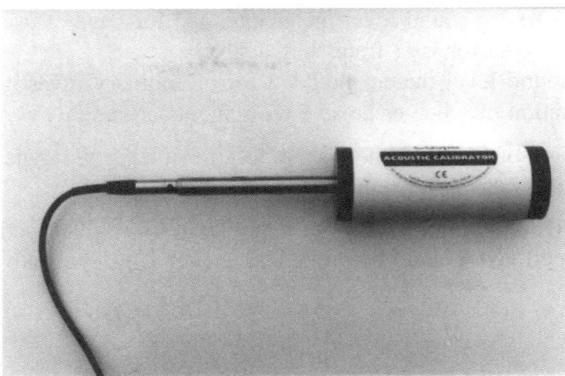

Figure 2.10 A Type 1 acoustic calibrator.

2.2.3.11 *Recording sound*

Sound recording allows the refinement engineer to keep a record of noise events for subsequent analysis in the laboratory. This is particularly important for narrow band frequency analysis as the equipment involved is often too bulky to be truly portable.

Making high-quality sound recordings is a skill that must be acquired with practice. However, there are some important points worthy of note here:

- Choose an appropriate recording medium. Common choices include Digital acquisition (straight onto the hard disk of a PC), DAT (digital audio tape), Minidisk (Digital) and regular audio tape. The first three are digital techniques with high signal-to-noise ratio (around 90 dB) low wow and flutter (often below measurable limits) and wide frequency response (10 Hz–20 kHz is common). A note of caution is given regarding the Minidisk format. This employs a particular form of data compression that may alter spectral content and level and therefore this format should not be used for laboratory standard work. The performance of regular audio tape is poor by comparison (50–80 dB signal-to-noise ratio for a professional grade unit, 35–40 dB for a consumer unit, 0.05% wow and flutter, 25 Hz–20 kHz frequency response for a professional unit).
- Ensure that record levels are set carefully to avoid overloads – even with occasional transient noises. This means choosing a recording medium and a recording level that suits the dynamic range of the sound. The use of 'A' weighting can limit the influence of 'bump' type noises but obviously this means that all subsequent analysis will be 'A' weighted. Select the highest possible record level that does not result in overloads to preserve signal-to-noise ratio.
- Record ten seconds of calibration tone before gathering data. This will allow the data to be calibrated when analysed. The calibration of sound recordings basically limits the choice of microphones to condenser microphones for which there are acoustic calibrators and pistonphones available. Many sound level meters have AC outputs which allow for connection with a recorder and this makes a versatile and relatively inexpensive setup for many automotive applications. If a sound level meter or micro-

phone power supply is being used that has more than one range setting, ensure that the calibration tone is recorded each time a change is made in range setting.
- 'Lock' all record levels to avoid them being disturbed. If they are disturbed, the calibration will no longer be true.

2.3 Making basic noise measurements: the decibel scale, frequency and time weightings

2.3.1 Decibel scales

Sound levels are commonly described in terms of the sound power (W) output of noise sources and the sound pressure (Pa) amplitude at a given location. However, decibel scales (dB) are useful due to the wide range of sound powers and sound pressure amplitudes that can be encountered.

Table 2.1 shows that the sound power outputs of everyday machines lie in the 0.001–1 000 W range (a factor of one million times) and the human voice has a power output in the range of 0.000 000 001–0.001 (also a factor of one million times).

Table 2.2 shows that everyday noise environments relate to pressure amplitudes in the 0.00002–20 Pa range (a factor of one million times). The human ear can detect sound pressure amplitudes over this one million factor range (beyond in fact $20 \times 10^{-6} - 60$ Pa).

To simplify the situation sound levels are usually described using the decibel scale.

$$L = 10 \ \log_{10} \left[\frac{x}{x_{\text{ref}}} \right] \text{dB}$$

where

$$x = \frac{L}{10} \times 10^{\frac{L}{10}}$$

Table 2.1 Typical sound power levels for different sources (Hassall and Zaveri, 1979)

Source	Sound Power Output (Watts)
Saturn rocket	25–40 million
Military jet engine with afterburner	100 000
Turbojet engine	1000–10 000
4 propeller airliner	100–1000
75 piece orchestra	10–1000
Small aircraft engine	1–10
Piano	0.1–1
Car alarm	0.1
Hi-Fi	0.01
Moving car	0.001–0.01
Fan	0.001
Shouting voice	0.0001–0.001
Conversational voice	0.000 01
Whispering voice	0.000 000 001

Table 2.2 Typical sound pressure levels for different sources (Bies and Hansen, 1996)

Environment	Sound Pressure Amplitude (Pa)	Description
100 m from Saturn rocket	200	Beyond pain threshold
At the front of a rock concert	20	Potentially damaging
Noisiest factory	2	Harmful
Next to a road or railway	0.2	Noisy
Busy restaurant	0.02	
Quiet suburban street	0.002	Quiet
Recording studio	0.0002	Very quiet
	0.00002	Threshold of hearing

The sound pressure level (L_p) is expressed as a ratio of the squared sound pressure amplitude to the threshold of hearing.

$$L_p = 10 \ \log_{10} \left[\frac{p^2}{p_{\text{ref}}^2} \right] = 20 \ \log_{10} \left[\frac{p}{p_{\text{ref}}} \right] \text{dB} \qquad (2.66)$$

where

$$p_{\text{ref}} = 20 \times 10^{-6} \, \text{Pa}.$$

According to this scale, everyday noise environments fall in the sound pressure level range of 0–120 dB.

By way of comparison, the typical operating pressures associated with internal combustion engines are as follows:

Peak cylinder pressure	$\approx$60 bar	$\approx$230 dB
Exhaust pressure wave	$\approx$0.8 bar	$\approx$192 dB
Intake pressure wave	$\approx$0.2 bar	$\approx$180 dB

Some further demonstrations

Equipment:

- loudspeaker and amplifier
- signal generator
- oscilloscope
- microphone, amplifier and cable;
- sound level meter
- BNC leads
- Type 1 field calibrator.

Demonstration 1
The calibrator can produce 94 dB re $20e^{-6}$ Pa (rms). This is 1 Pa rms. The waveform from a microphone inserted in the calibrator can be viewed on an oscilloscope. Set the input sensitivity of the oscilloscope to produce 1 cm peak-to-peak on the screen.

The calibrator can also produce 104 dB re 20e^{-6} Pa. This is 3.1698 Pa rms. This can also be viewed on the oscilloscope screen and the peak-to-peak is 3.1 cm.
Demonstration 2
With the signal generator producing a 100 Hz tone, compare A and linear outputs of the microphone amplifier. Also compare A- and C-weighted sound pressure levels.

The sound power level (L_w) is expressed as a ratio of the sound power to a reference power of 10^{-12} W.

$$L_W = 10 \; \log_{10} \left[\frac{W}{W_{ref}} \right] \; dB$$

where

$$W_{ref} = 10^{-12} \, W \tag{2.67}$$

As a result, everyday machine power outputs fall in a sound power range of 70–160 dB and the human voice produces sound power levels in the 30–70 dB range.

Group exercise: voice levels

Two members of the group will stand 1 m apart. One will talk to the other at a normal conversational voice level. A third group member will measure the sound pressure level at the ear of the listening person with a sound level meter.

Next the speaker will talk in a raised voice and the sound pressure level will be measured again.

Finally, the speaker will shout and the sound pressure level will be measured.

At least ten minutes should be allocated to this exercise.

2.3.2 Decibel arithmetic

Decibel addition is undertaken as follows:

$$L_1 + L_2 = 10 \; \log_{10} \left[10^{(L_1/10)} + 10^{(L_2/10)} \right] \; dB \tag{2.68}$$

Decibel subtraction follows the same procedure.

$$L_1 - L_2 = 10 \; \log_{10} \left[10^{(L_1/10)} - 10^{(L_2/10)} \right] \; dB \tag{2.69}$$

The combination of two identical sound levels produces a sum which is 3 dB greater than the individual levels. Combining a sound level with another, which is 10 dB lesser in magnitude, produces a sum that is negligibly greater than the highest sound level.

2.3.3 The significance of decibel differences

Subjectively, to a young person with normal hearing:

1 dB change in the level of a tone is barely perceptible;
3 dB change in the level of a tone is clearly perceptible;
10 dB change in the level of a tone appears as a doubling or halving of loudness.

In engineering terms,

1 dB change in the level of noise represents a 21% reduction in sound power;
3 dB change in the level of noise represents a 50% reduction in sound power;
10 dB change in the level of noise represents a 90% reduction in sound power;
20 dB change in the level of noise represents a 99% reduction in sound power.

Seeking reductions in sound level from industrial machines and vehicles of more than 1 or 2 dB requires very significant engineering effort.

Group exercise: multiple sources of sound power

Eight members of a group will stand together. One will start humming at a low pitch. A ninth member will measure the sound pressure level at some distance from the hummer using a sound level meter.

Next, the hummer will continue and one person next to them will join in. The sound pressure level will be measured again. Then four people in total will start humming and the sound pressure level will be measured again.

Finally, all eight people will hum together and the sound pressure level will be measured.

What happens to the sound pressure level when the number of roughly equivalent sound sources doubles?

Ten minutes should be allocated to this exercise.

2.3.4 Time and frequency weightings for sound levels

2.3.4.1 Frequency weightings

Almost all sound level meters are equipped with an 'A-weighting' setting in addition to the un-weighted 'lin' or 'linear' setting. The A-weighting is the standard weighting for outdoor community noise measurements and is commonly used for noise measurements within architectural spaces and within vehicles.

The A-weighting reduces the sensitivity of the measuring instrument to both low and very high frequency sounds. It approximately follows the inverted shape of the equal loudness contour passing through 40 dB at 1 kHz. Therefore, sound of a particular A-weighted level should appear equally loud regardless of frequency (providing that the corresponding sound level at 1000 Hz is 40 dB).

The equal loudness contours become flatter and wider with increasing sound levels. Therefore, two further frequency weightings are provided:

- The 'B'-weighting which approximately follows the inverted shape of the equal loudness contour passing through 70 dB at 1 kHz.
- The 'C'-weighting which approximately follows the inverted shape of the equal loudness contour passing through 100 dB at 1 kHz.

The three frequency weighting curves are shown in Figure 2.11 and are tabulated in Table 2.3.

Sound levels within road vehicles are typically well above 40 dB but below 100 dB. Therefore, the 'B'-weighted sound pressure level L_B (dBB) is adopted by some vehicle manufacturers as their preferred unit for measuring interior noise. However, others use the A-weighted sound pressure level L_A (dBA) as it is prescribed for use in the exterior noise type approval test and it serves the useful function of suppressing the unwanted influence of both wind noise around the microphone and low frequency 'thumps' (caused by handling the microphone or bumps in the road).

2.3.4.2 Time weightings

Sound level meters are commonly fitted with three time weightings:

- *Fast* having an exponential time constant of 125 ms, corresponding approximately to the integration time of the ear (sounds of duration less than around 125 ms do not register their full loudness with the average human subject).
- *Slow* having an exponential time constant of 1 s to allow for the average level to be estimated by eye with greater precision.
- *Peak* having an exponential time constant of below 100 μs to respond as quickly as possible to the true peak level of transient sounds.

They may also be fitted with one further time weighting:
Impulse having a 35 ms exponential rise but a much longer decay time, thought to mirror the ear's response to impulsive sound.

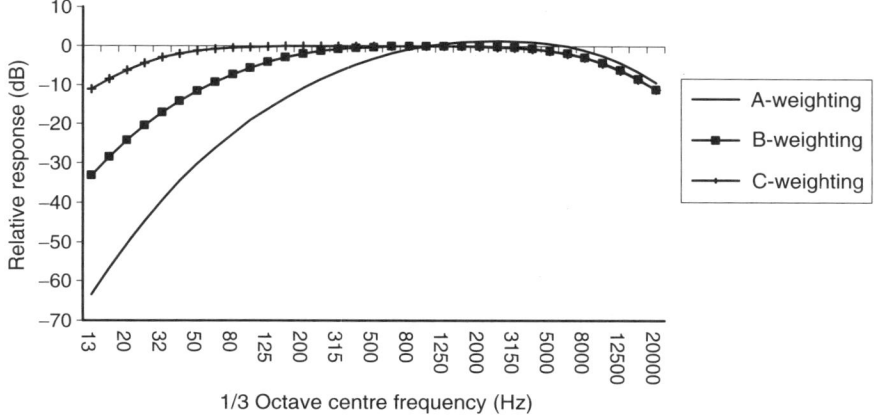

Figure 2.11 The A, B and C frequency weighting curves (Beranek, 1971).

Table 2.3 Frequency weightings (Beranek, 1971)

1/3 octave band (Hz)	Weighting relative response (dB)		
	A	B	C
12.5	−63.4	−33.2	−11.2
16.0	−56.7	−28.5	−8.5
20.0	−50.5	−24.2	−6.2
25.0	−44.7	−20.4	−4.4
32.0	−39.4	−17.1	−3.0
40.0	−34.6	−14.2	−2.0
50.0	−30.2	−11.6	−1.3
63.0	−26.2	−9.3	−0.8
80.0	−22.5	−7.4	−0.5
100.0	−19.1	−5.6	−0.3
125.0	−16.1	−4.2	−0.2
160.0	−13.4	−3.0	−0.1
200.0	−10.9	−2.0	0.0
250.0	−8.6	−1.3	0.0
315.0	−6.6	−0.8	0.0
400.0	−4.8	−0.5	0.0
500.0	−3.2	−0.3	0.0
630.0	−1.9	−0.1	0.0
800.0	−0.8	0.0	0.0
1000.0	0.0	0.0	0.0
1250.0	0.6	0.0	0.0
1600.0	1.0	0.0	−0.1
2000.0	1.2	−0.1	−0.2
2500.0	1.3	−0.2	−0.3
3150.0	1.2	−0.4	−0.5
4000.0	1.0	−0.7	−0.8
5000.0	0.5	−1.2	−1.3
6300.0	−0.1	−1.9	−2.0
8000.0	−1.1	−2.9	−3.0
10000.0	−2.5	−4.3	−4.4
12500.0	−4.3	−6.1	−6.2
16000.0	−6.6	−8.4	−8.5
20000.0	−9.3	−11.1	−11.2

2.4 Analysis and presentation of noise data

There are many different ways of analysing sound data. The methods broadly fall into two categories:

1. single-value indices
2. frequency-dependent indices.

2.4.1   Single-value index methods

2.4.1.1 Pressure time history

This is a two-dimensional plot of calibrated pressure (Pa) on the vertical axis against time (s) on the horizontal axis. Such plots are useful as a preliminary check on the quality of the data (checking for 'clipping' of the peaks, etc.). Such plots are most commonly used to publish data from cyclic processes involving relatively low frequency sound (IC engine intake and exhaust noise, for example).

2.4.1.2 Root mean square pressure (P_{rms})

Take a sine wave (Beranek, 1988)

$$y(t) = A \sin \omega t \tag{2.70}$$

where

$\omega = 2\pi f (\text{rad s}^{-1})$
$f = \text{frequency (Hz)}$

The time average $\bar{y}(T)$ over period $T(s)$ of the signal $y(t)$ is given by

$$\bar{y}(T) = \frac{1}{T} \int_0^T y(t) dt \tag{2.71}$$

So the mean square value y_{rms}^2 is

$$y_{rms}^2 = \frac{1}{T} \int_0^T y^2(t) dt$$

$$= \frac{1}{T} \int_0^T A^2 \sin^2 \omega t \, dt \tag{2.72}$$

Now remembering that

$$\cos^2 \theta + \sin^2 \theta = 1$$

$$\sin^2 \theta = 1 - \cos^2 \theta \tag{2.73}$$

and substituting equation (2.73) into equation (2.72)

$$y_{rms}^2 = \frac{1}{T} \int_0^T A^2 \left(1 - \cos^2 \omega t \right) dt$$

$$y_{rms}^2 = \frac{A^2}{T} \int_0^T 1 - \cos^2 \omega t \, dt \tag{2.74}$$

Now, remembering that

$$\cos 2\theta = 2\cos^2 \theta - 1$$

$$\cos^2 \theta = \frac{1}{2}[\cos 2\theta + 1] \tag{2.75}$$

is obtained.

So on substitution of equation (2.75), equation (2.74) becomes

$$y^2_{rms} = \frac{A^2}{T} \int_0^T 1 - \frac{1}{2}[\cos 2\omega t + 1]\, dt \qquad (2.76)$$

$$y^2_{rms} = \frac{A^2}{2T} \int_0^T (2 - \cos 2\omega t - 1)\, dt$$

$$y^2_{rms} = \frac{A^2}{2T} \int_0^T (1 - \cos 2\omega t)\, dt$$

$$y^2_{rms} = \frac{A^2}{2T} \left[t - \frac{\sin 2\omega t}{2\omega} \right]_0^T$$

$$y^2_{rms} = \frac{A^2}{2T} \left[T - \frac{\sin 2\omega T}{2\omega} - 1 \right]$$

$$y^2_{rms} = \frac{A^2}{2} - \frac{A^2 \sin(2\omega T)}{4\omega T} \qquad (2.77)$$

As $T \to \infty$ the second (oscillatory) term in equation (2.77) tends to zero and so as $T \to \infty$, the time average of y^2_{rms} becomes

$$\overline{y^2_{rms}} = \frac{A^2}{2} \qquad \text{or} \qquad \overline{y_{rms}} = \frac{A}{\sqrt{2}} \qquad (2.78)$$

The second term in equation (2.77) is an oscillation of maximum amplitude $A^2/4\omega T$ that decreases with increasing averaging period T.

For a signal measured over a time period $\ll \infty$ there is obvious uncertainty over how much the true rms deviates from $A/\sqrt{2}$.

The maximum uncertainty is the maximum error over the assumed rms, given by:

$$\text{Maximum uncertainty} = \frac{A^2/4\omega T}{A^2/2} = \frac{1}{2\omega T} \qquad (2.79)$$

So, at lower frequencies, longer averaging times are required to reduce the uncertainty of the measurements to an acceptable level. Equivalent uncertainties may be derived for other non-sinusoidal waveforms (Beranek, 1988).

General Note: The precise location relative to the sound source at which a pressure measurement is made must always be quoted for that measurement to be meaningful. This is relaxed when quoting sound power data.

2.4.1.3 Sound pressure level time history

The sound pressure level is given by

$$L_p = 20 \, \log_{10} \left[\frac{P_{rms}}{P_{ref}} \right] \text{dB} \qquad (2.80)$$

where

$$P_{ref} = 20\, \mu \text{Pa rms} \, (20 \times 10^{-6} \, \text{Pa rms}).$$

The time varying sound pressure level offers a compact means of displaying a measure of a fluctuating sound field on a two-dimensional plot. It is commonly used for both interior and exterior vehicle noise levels as well as for IC engines and other machines with a wide operating speed range (cooling fans, alternators, pumps, injectors, etc.).

Frequency weightings may be applied to the data (A, B and C weightings). It is usual to use the fast time weighting (rather than impulse or slow). In any case, it is important to state the use of any weightings in the vertical axis label – the usual form being for instance $L_{Pf}dB(A)$ @ 1 m from source (fast and A-weightings applied to the raw data).

2.4.1.4 Continuous equivalent level – L_{EQ}

The L_{EQ} over period T (s) is given by

$$L_{EQ,T} = 10 \ \log_{10} \left[\frac{1}{T} \int_0^T 10^{\frac{L_p(t)}{10}} \, dt \right] dB \qquad (2.81)$$

This is the continuous steady noise level which would have the same energy over the period T as the actual fluctuating noise level measured during that period.

Again, A, B and C weightings may be applied but the use of the fast weighting is implied.

The period T should be stated along with any weightings used when quoting L_{EQ} – the usual form being.

$L_{EQ,18 \ hours} \ dB(A)$@1 m from source   (A-weighted and measured over an 18-hour period)

The one-second L_{EQ} is given the special name of the single event level (SEL or L_{Ax} if A-weighted). Again any frequency weighting applied should be stated – usual form being SEL dB(A)@1 m from source   (A–weighted).

The relationship between SEL and longer period L_{EQ} is as follows:

$$L_{EQ} = 10 \ \log_{10} \left[\frac{1}{T} \sum_{i=1}^{n} 10^{\frac{SEL_i}{10}} \right] dB \qquad (2.82)$$

or for an event that repeats itself often (n times in a period of t seconds) and a reasonable arithmetic average of the SEL may be achieved (denoted with an overscore)

$$L_{EQ} \approx \overline{SEL} + 10 \ \log_{10} n - 10 \ \log_{10} t \qquad (2.83)$$

The procedure for combining LEQ_i from n contiguous samples, each with sample period T_i, is as follows:

$$L_{EQ} = 10 \ \log_{10} \left[\frac{1}{T_T} \sum_{i=1}^{n} T_i 10^{\frac{LEQ_i}{10}} \right] dB \qquad (2.84)$$

where

$$T_T = \sum_{i=1}^{n} T_i \ (s)$$

2.4.1.5 Statistical levels L_x

The sound pressure level exceeding $x\%$ of the time is denoted by L_x (dB).

L_{90} dB   is usually taken as a measure of the background noise level (a 10 dB variation in background level is commonly experienced over the day-to-night period).

L_{10} dB   is usually taken as a reliable measure of the noise from a stream of road traffic (up to 40–dB variation in L_p is commonly experienced over a short period as individual vehicles pass by).

It can be seen that L_{90} (dB) and L_{10} (dB) are used to separate the slowly varying background noise levels from short-duration noise events.

Confusion often arises between ambient noise levels and background noise levels. In BS 4142 (1990), background noise levels are defined as the L_{90} (dB) and ambient noise levels as the level of *totally encompassing sound in a given situation at a given time – usually composed of sound from many sources near and far.*

Any frequency weighting applied to the data should be declared – usual form

$$L_{90} \text{ dB}(A) \quad \text{(A-weighted)}$$

2.4.2 Frequency-dependent index methods

Before discussing frequency-dependent indices, the noise bandwidth B_n of a narrow band filter must be defined (Beranek, 1988).

The noise bandwidth B_n is defined as the bandwidth of the ideal filter that would pass the same signal power as the real filter when each is driven by stationary random noise.

$$B_n = f_1 - f_2 = \int_0^\infty |H(f)|^2 \mathrm{d}f \tag{2.85}$$

In the ideal filter, the modulus of the amplitude transfer function $H(f)$ is zero outside the pass band (given by $f_1 - f_2$) and unity within the pass band.

$$H(f) = \frac{\overline{x_{\text{out}}}}{\overline{x_{\text{in}}}} \tag{2.86}$$

where the overscore denotes a complex quantity.

A real filter will have an amplitude transfer function that is not unity right across the pass band and does not go immediately to zero outside the pass band. A stationary random signal is one that has statistical properties that do not change with time.

There are two common classes of narrow band filter used for analysis of sound data (Beranek, 1988):

1. The constant bandwidth type where B_n is the same for all filter centre frequencies f_c.
2. The constant percentage bandwidth type where B_n is a constant percentage of f_c throughout the frequency range.

The constant percentage bandwidth class will be considered first. Commonly available filters of this type have widths of one octave, 1/3 octave, 1/12 octave and 1/24 octave. The preferred centre frequencies for the common octave and 1/3 octave filters are given in ANSI S1.16 (reproduced in Beranek, 1988).

2.4.2.1 Octave filter

The centre, lower and upper frequencies of the octave filter set may be obtained from:

$$f_c = 10^{\frac{n}{10}} \qquad (2.87)$$

$$f_2 = 10^{\frac{n-1.5}{10}} \qquad (2.88)$$

$$f_1 = 10^{\frac{n+1.5}{10}} \qquad (2.89)$$

where n = band numbers. For the audio range, the octave filter is given by values of n between 12 and 43 that are integer multiples of the number 3.

The bandwidth is a consistent 69% (68.75%) of the centre frequency. The preferred centre frequencies given by ANSI S 1.16 (reproduced in Beranek, 1988) vary slightly from those obtained from the numerical relations above and these are shown in Table 2.4.

Table 2.4 Preferred frequencies for octave filters (ANSI S 1.16, reproduced in Beranek, 1988)

Band number	f_2	F_c	f_1
12	11	16	22
15	22	31.5	44
18	44	63	88
21	88	125	177
24	177	250	355
27	355	500	710
30	710	1 000	1 420
33	1420	2 000	2 840
36	2 840	4 000	5 680
39	5 680	8 000	11 360
42	11 360	16 000	22 720

2.4.2.2 1/3 octave filter

The centre, lower and upper frequencies of the 1/3 octave filter set may be obtained from:

$$f_c = 10^{\frac{n}{10}} \qquad (2.90)$$

$$f_2 = 10^{\frac{n-0.5}{10}} \qquad (2.91)$$

$$f_1 = 10^{\frac{n+0.5}{10}} \qquad (2.92)$$

where n = band numbers. For the audio range, the 1/3 octave filter is given by all integer values of n between 12 and 43.

Table 2.5 Preferred frequencies for 1/3 octave filters (ANSI S 1.16, reproduced in Beranek, 1988)

Band number	f_2	f_c	f_1
12	14.1	16	17.8
13	17.8	20	22.4
14	22.4	25	28.2
15	28.2	31.5	35.5
16	35.5	40	44.7
17	44.7	50	56.2
18	56.2	63	70.8
19	70.8	80	89.1
20	89.1	100	112
21	112	125	141
22	141	160	178
23	178	200	224
24	224	250	282
25	282	315	355
26	355	400	447
27	447	500	562
28	562	630	708
29	708	800	891
30	891	1 000	1 122
31	1 122	1 250	1 413
32	1 413	1 600	1 778
33	1 778	2 000	2 239
34	2 239	2 500	2 818
35	2 818	3 150	3 548
36	3 548	4 000	4 467
37	4 467	5 000	5 623
38	5 623	6 300	7 079
39	7 079	8 000	8 913
40	8 913	10 000	11 220
41	11 220	12 500	14 130
42	14 130	16 000	17 780
43	17 780	20 000	22 390

The bandwidth is a consistent 23% (23.075%) of the centre frequency. The preferred centre frequencies given by ANSI S 1.16, (reproduced in Beranek, 1988) vary slightly from those obtained from the numerical relations above, and these are shown in Table 2.5.

Figure 2.12 shows an example of a 1/3 octave spectrum.

2.4.2.3 1/12 octave filter

The centre, lower and upper frequencies of the 1/12 octave filter set may be obtained from:

$$f_c = 10^{\frac{n+0.5}{40}} \tag{2.93}$$

$$f_2 = 10^{\frac{n}{40}} \tag{2.94}$$

$$f_1 = 10^{\frac{n+1}{40}} \tag{2.95}$$

where n = band numbers. For the audio range, the 1/12 octave filter is given by all integer values of n between 48 and 172. The bandwidth is a consistent 6% (5.756%) of the centre frequency.

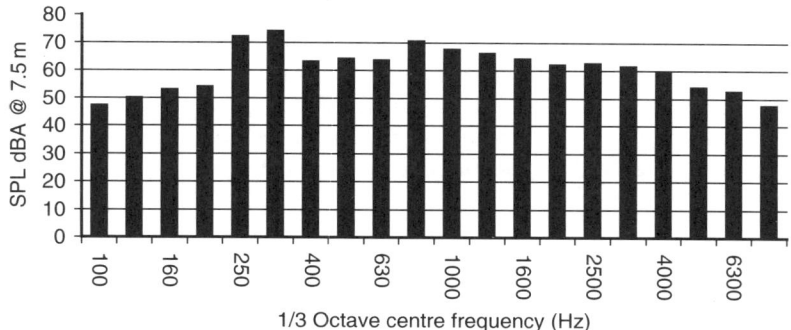

Figure 2.12 An example of a constant percentage bandwidth spectrum (1/3 octave) - noise levels 7.5 m from an accelerating sedan, second gear full load.

2.4.2.4 1/24 octave filter

The centre, lower and upper frequencies of the 1/24 octave filter set may be obtained from:

$$f_c = 10^{\frac{n+0.5}{80}} \tag{2.96}$$

$$f_2 = 10^{\frac{n}{80}} \tag{2.97}$$

$$f_1 = 10^{\frac{n+1}{80}} \tag{2.98}$$

where n = band numbers. For the audio range, the 1/24 octave filter is given by all integer values of n between 96 and 344. The bandwidth is a consistent 3% (2.878%) of the centre frequency.

The levels of narrow bands that fit within the bandwidth of coarser filters (for instance the three 1/3 octave bands that fit within the bandwidth of the one octave filter) may be combined to give the band level of the coarser filter. The combination of band levels must be thus done by logarithmic addition:

$$L_{\text{Total}} = 10 \, \log_{10} \left[\sum_{i=1}^{n} 10^{\frac{L_i}{10}} \right] \text{dB} \tag{2.99}$$

Following this logic, 1/24 octave bands may be combined to give a 1/12 octave spectra. The 1/12 bands may be combined to give a 1/3 octave spectra and so on until the overall level is obtained (single index).

2.4.2.5 Order tracking

When analysing the sound from rotating machines (such as IC engines) it is common to use the order-tracking technique. The now obsolete but instructive analogue method was as follows:

1. Obtain an electrical signal that is proportional in some way to the speed of rotation of the machine (for instance a tachometer signal). Calculate the rotational frequency of the machine f_0 (Hz).

2. Set a constant percentage bandwidth filter (6% is common for 1/12 octave) to have f_c equal to the rotational frequency of the machine and organise by electrical means for it to follow (track) changes in the rotational frequency f_0.
3. Set other constant percentage bandwidth filters (6% is common for 1/12 octave) to follow (track) changes in the rotational frequency, each one with f_c set to a different order (or harmonic) of the rotational frequency of the machine f_0:

$$f_c = nf_0 \quad n > 0, \text{ not necessarily an integer value} \tag{2.100}$$

4. Plot the output from each filter against the rotational speed of the machine as shown in the example in Figure 2.13.

Often, in order to reduce the number of tracking filters required, the raw data was recorded onto tape or magnetic disk and passed several times through the same filter set up to track a different order on each occasion. Modern digital techniques have of course replaced this process but an understanding of the analogue method helps the understanding of the digital method.

2.4.2.6 *The constant bandwidth class of frequency analysis methods*

One of the most commonly used outputs from a constant bandwidth spectrum analyser is the power spectral density (psd).

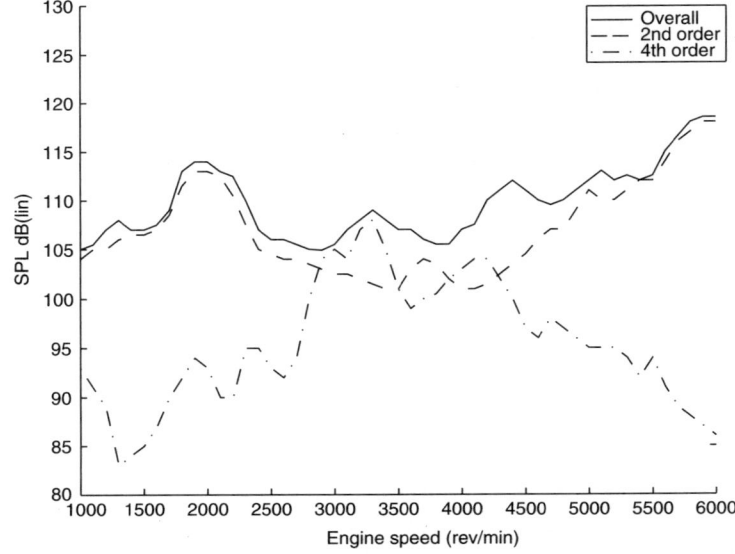

Figure 2.13 An example of order tracking analysis — noise levels 100 mm from an intake snorkel.

In this case, the word power is perhaps a misnomer as in general the psd has units of Volts^2/Hz rather than W/Hz (Beranek, 1988). It may have true units of power if calibrated accordingly.

For instance in the far, free acoustic field

$$W = \frac{P_{rms}^2 S}{\rho_0 c} \tag{2.101}$$

with S being the surface area of the wavefront (m^2), ρ_0 the undisturbed density (kg m^{-3}) and c the speed of sound (m s^{-1}).

The power in each spectral band i is given by:

$$w_i = y_{rms(i)}^2 = \frac{1}{T} \int_0^\infty y_i^2(t) dt \tag{2.102}$$

$$W = \sum_{i=1}^n w_i \tag{2.103}$$

$$psd_i = \lim B_i \to 0, \frac{w_i}{B_i} \tag{2.104}$$

with B_i being the bandwidth of the band i.

The psd of the signal from a stationary random process is a smooth continuous function of frequency. For a cyclic process the psd is not smooth, consisting of a series of harmonically related peaks. An example of this is given in Figure 2.14.

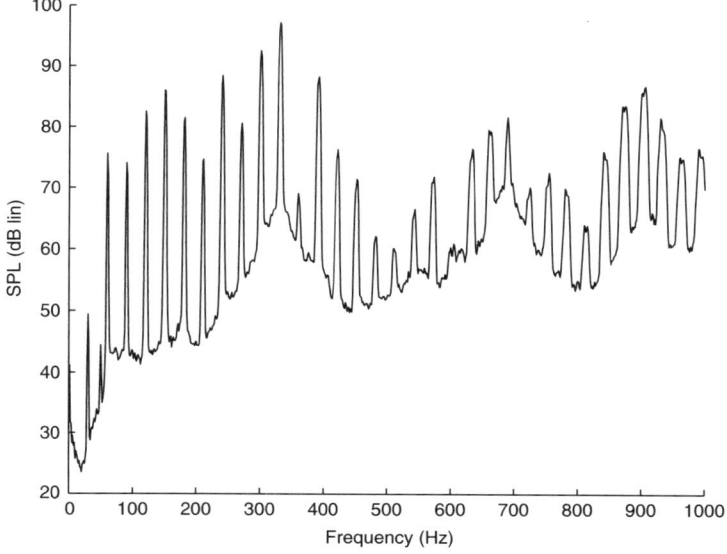

Figure 2.14 An example of a narrow band (constant bandwidth) spectrum — noise levels 100 mm from an intake snorkel with the engine running at full load, 3600 rev min^{-1}.

Spectral analysis may be performed using either:

- contiguous filters (analogue or digital)
- fourier analysis.

Many cyclic processes operate over a wide range of speeds (cycle rates). One can deduce things about the nature of the process from its psd (Fahy and Walker, 1998):

- The cycle rate (events per second) determines the fundamental frequency of the harmonic series that makes up the psd.
- The time history of each event determines the overall shape (envelope) of the spectrum and the frequency range of significant energy.
- The dependence on cycle rate of the total power held in the radiated sound psd is strongly influenced by the steepest slope in the spectral envelope of the forcing. For instance, a 40 dB-per-decade slope in the envelope of the forcing spectrum suggests a 40 dB-per-decade dependence on cycle rate of the sound power radiated.

2.4.2.7 Waterfall plot

Some spectrum analysers allow the use of a tachometer signal to produce a three-dimensional plot of frequency spectra against time (or machine speed). These are known as waterfall plots (or Campbell diagrams) and an example is shown in Figure 2.15.

Each horizontal 'slice' is an individual spectrum, gathered over a user-defined averaging period. Beware, with rapid changes in machine speed, the averaging time for each spectrum will have to be short, and therefore individual band levels will only be estimates of the true band level.

Vertical lines of peaks signify resonances – high amplitudes at particular frequencies that are independent of machine speed (see around 750 Hz on Figure 2.15). Diverging lines of peaks signify orders (see left-hand side of Figure 2.15).

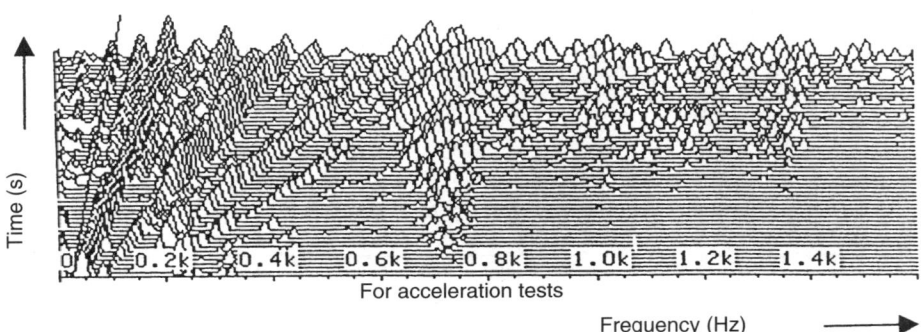

Figure 2.15 An example of a waterfall plot – noise levels 100 mm from an intake snorkel.

2.5 Sound power level, sound intensity level, sound pressure level

2.5.1 Radiation of sound power sources outdoors and in enclosed spaces

In open, non-reverberant spaces, the sound pressure level at a particular point in space is a function of:

- the sound power level of the source;
- the directivity of the source;
- the distance between the source and the receiver;
- excess attenuation due to air ground absorption and barrier attenuation.

World wide, many machines, products and processes are supplied along with details of the typical sound power level produced by their operation. This is in part due to EC Directive 89/392/EEC.

In an environment in which there are no reflecting surfaces, the sound pressure at any point in any type of freely travelling (plane, cylindrical, spherical, etc.) wave is related to the maximum intensity I_{max} by (Bies and Hansen, 1996):

$$p_{rms}^2 = I_{max}\rho c \tag{2.105}$$

which comes from (spherical coordinates) $I(r, t) = \langle p(r, t)\, u(r, t)\rangle$ (Bies and Hansen, 1996) where $\langle\ \rangle$ denotes a time average, and assuming great enough distance from the source for plane waves to form we also have $p = \rho c u$.

Now the intensity level L_I is

$$L_I = 10\ \log_{10} \frac{I}{I_{ref}}\ \text{(dB)} \tag{2.106}$$

where $I_{ref} = 10^{-12}\ \text{W m}^{-2}$
Substituting equation (2.105) into equation (2.106) yields

$$L_I = 10\ \log_{10} \frac{p_{rms}^2}{\rho c I_{ref}} \tag{2.107}$$

Now the sound pressure level L_p is

$$L_P = 10\ \log_{10} \left[\frac{p_{rms}}{p_{ref}}\right]^2\ \text{(dB)} \tag{2.108}$$

where $p_{ref} = 20 \times 10^{-6}\ \text{Pa}$

Dividing both the numerator and the denominator of equation (2.107) by p_{ref}^2 and substituting equation (2.108) gives

$$L_I = 10\ \log_{10} \left[\frac{p_{rms}}{p_{ref}}\right]^2 - 10\ \log_{10} \left[\frac{\rho c I_{ref}}{p_{ref}^2}\right] \tag{2.109}$$

or

$$L_I = L_p - 10 \ \log_{10} K \tag{2.110}$$

where

$$K = \frac{\rho c I_{ref}}{p_{ref}^2} \tag{2.111}$$

At room temperature,

$$\rho \simeq 1.2 \, \text{kg m}^{-3}$$
$$c \simeq 343 \, \text{m/s}$$
$$\rho c \simeq 412 \, \text{kg m}^{-2} \, \text{s}^{-1}$$

And thus

$$10 \ \log_{10} K = 10 \ \log_{10} \left[\frac{412 \times 10^{-12}}{20 \times 10^{-6} \times 20 \times 10^{-6}} \right] = 0.128 \ \text{dB}$$

Therefore, in open space with a free travelling wave at room temperatures and pressures and for approximate calculations $L_p \simeq L_I$

Now the relationship between sound power (W) and sound intensity is

$$W = I \cdot S \ (\text{W}) \tag{2.112}$$

where S is the area through which intensity I is passing.

Therefore, in order to relate sound power to sound intensity and hence sound intensity to sound pressure the area S must be known, which is related to the radiation pattern of the source.

For a point source with spherical propagation

$$S = 4\pi a^2 \tag{2.113}$$

where a is the distance between source and receiver.

For a line source with cylindrical propagation

$$S = 2\pi a l \tag{2.114}$$

where l is source length.

Now the sound power level L_W is

$$L_W = 10 \ \log_{10} \frac{W}{W_{ref}} \tag{2.115}$$

where commonly $W_{ref} = 10^{-12} \, \text{W}$

Now from equations (2.106), (2.112) and (2.115)

$$L_W = L_I + 10 \ \log_{10}[S] \tag{2.116}$$

and from equation (2.110)

$$L_p = L_I + 10 \ \log_{10} K \qquad (2.117)$$

and from

$$L_I = L_W - 10 \ \log_{10}[S] \qquad (2.118)$$

and thus

$$L_p = L_W - 10 \ \log_{10} S + 10 \ \log_{10} K \qquad (2.119)$$

In some cases, the radiation from the source may not be adequately described by a simple spherical or cylindrical model. For instance, consider a sound source placed on the ground (hemispherical propagation). When viewed from above the ground the layout might look like that illustrated in Figure 2.16 and say the radiation pattern (plan view) is known to be that shown in Figure 2.17.

Therefore, the sound power radiated towards the receiver positioned at angle $\theta = 45$ degrees is going to be 20% less (0.8 rather than unity) than that radiated towards the axial position. In this case, the sound pressure level at the receiver $L_{p\theta}$ would be

$$L_{p\theta} = L_W + 10 \ \log_{10}[0.8] - 10 \ \log_{10}[2\pi r^2] + 10 \ \log_{10} K \qquad (2.120)$$

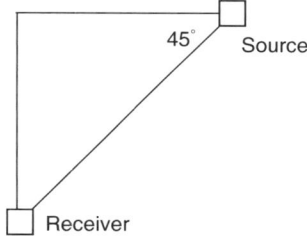

Figure 2.16 Plan view of a hemispherical source.

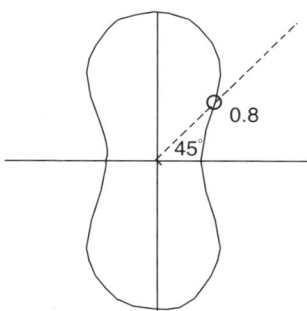

Figure 2.17 Polar radiation diagram for the notional hemispherical source of figure 2.16.

The treatment so far has implicitly assumed that the attenuation of sound pressure level with distance is only due to the geometric spreading of the wavefront. Although this is often the most significant attenuation mechanism, it is not the only one as in practical outdoor situations; additional attenuation is provided by the air and the ground. The last equation would therefore be a more realistic propagation model in this form (Bies and Hansen, 1996):

$$L_{p\theta} = L_W - 10 \, \log_{10}[4\pi r^2] + 10 \, \log_{10} D + 10 \, \log_{10} K - A_E \qquad (2.121)$$

where

r = propagation distance
D = directivity, $D = I_\theta / <I>$, $\langle I \rangle = W/4\pi r^2$, the mean intensity
$K = \dfrac{\rho c I_{ref}}{p_{ref}^2} = 0.128 \, \text{dB}$
A_E = excess attenuation

The excess attenuation comprises attenuation due to:

- air absorption;
- ground reflection and absorption;
- meteorological effects;
- natural and man-made barriers to sound propagation.

Air absorption is a function of temperature, humidity and frequency. In general, air absorption increases with frequency, decreases with humidity and generally decreases with temperature (except in certain frequency bands). Therefore, sound propagates most effectively through the air on hot, humid days.

It is usually reasonable to assume that at the propagation distances usually considered by the refinement engineer (<7.5 m), air absorption is significant only at frequencies above 1000 Hz.

Sutherland et al. (1974) published a method for the prediction of air absorption which is reproduced here in the form given by Bies and Hansen (1996).

$$\text{Air Absorption} = mx \, (\text{dB}) \qquad (2.122)$$

where x is propagation distance and m is taken from Table 2.6.

Taking proper account of the effect of ground attenuation requires a more complex analysis. Ground attenuation is a function of:

- the impedance of the ground surface;
- the area proportions of ground present with different surface impedance;
- propagation distance;
- height of direct propagation above the ground;
- frequency.

Table 2.6 Air absorption coefficients after Sutherland et al. (1974).

Relative Humidity in %	Temperature in °C	m (dB per 1000 m)							
		63 Hz	125 Hz	250 Hz	500 Hz	1 kHz	2 kHz	4 kHz	8 kHz
25	15	0.2	0.6	1.3	2.4	5.9	19.3	66.9	198.0
	20	0.2	0.6	1.5	2.6	5.4	15.5	53.7	180.0
	25	0.2	0.6	1.6	3.1	5.6	13.5	43.6	153.4
	30	0.1	0.5	1.7	3.7	6.5	13.0	37.0	128.2
50	15	0.1	0.4	1.2	2.4	4.3	10.3	33.2	118.4
	20	0.1	0.4	1.2	2.8	5.0	10.0	28.1	97.4
	25	0.1	0.3	1.2	3.2	6.2	10.8	25.6	82.2
	30	0.1	0.3	1.1	3.4	7.4	12.8	25.4	72.4
75	15	0.1	0.3	1.0	2.4	4.5	8.7	23.7	81.6
	20	0.1	0.3	0.9	2.7	5.5	9.6	22.0	69.1
	25	0.1	0.2	0.9	2.8	6.5	11.5	22.4	61.5
	30	0.1	0.2	0.8	2.7	7.4	14.2	24.0	58.4

There are three basic models for considering ground absorption, each one more complex than the last (Bies and Hansen, 1996). These models are, in order of simplicity:

1. the hard/soft model;
2. the incoherent addition of a direct wave and a ground-reflected wave model;
3. the coherent addition of a direct wave and a ground-reflected wave model.

The hard/soft model assumes that a sound source with spherical propagation will produce sound pressure levels in the far field which are 3 dB greater when the source is placed above hard ground than when it is placed over soft ground. This model assumes incoherent addition of direct and reflected waves when the source is more than a quarter wavelength above hard ground and the complete absence of a reflected wave over soft ground. Therefore, the ground attenuation is −3 dB for hard ground and 0 dB for soft ground and is independent of propagation distance. This is obviously an oversimplified model, but can produce fairly realistic results when only one ground type is present and the propagation distances are small.

The second ground attenuation model involves the incoherent addition of a ground-reflected wave with the direct wave. As the addition is incoherent, the squares of the pressure amplitudes (i.e. the intensities) add. It should be noted that incoherent addition is only likely at higher frequencies above 500 Hz. At low frequencies coherent addition is likely with the addition of sound pressures resulting in a 6 dB increase in sound pressure levels. At the frequencies near the junction of coherent and incoherent addition, where the propagation distances for the direct and reflected waves are multiples of a half wavelength, different, destructive interference could in theory produce zero acoustic pressure and hence a sound pressure level of minus infinity as illustrated in Figure 2.18.

A minus infinity discontinuity is shown in Figure 2.18. However, in practice, destructive interference is seldom complete and the trough will not be so deep. Regions of interference occur at higher frequencies causing the dips in the curve, but these tend to smooth out once wider band signal analysis such as third octave are used.

The destructive interference effect is often seen in wayside noise measurements of vehicles as a dip in the noise around 500 Hz. This effect is often attributed to frequency-selective ground absorption which is not the case.

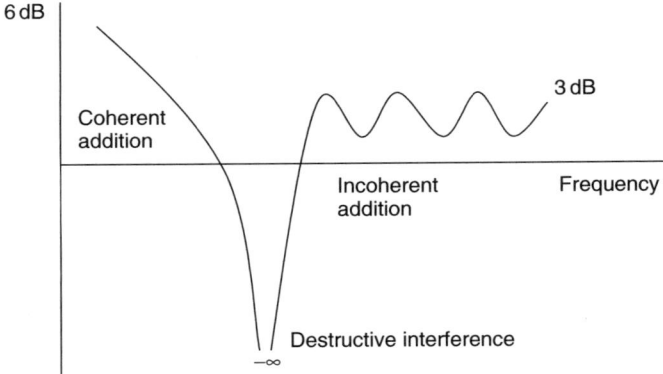

Figure 2.18 The effect of combining direct sound waves and those that are reflected.

In the incoherent addition model, the amplitude of the reflected wave is obtained from the complex pressure reflection coefficient (Bies and Hansen, 1996).

$$R = \frac{Z \sin \beta - \rho c}{Z \sin \beta + \rho c} \tag{2.123}$$

The reflected wave makes an angle β with the ground surface (see Figure 2.19), and Z is the surface impedance ($Pa \cdot s m^{-1}$) which in turn may be obtained from the flow resistivity of the surface ($Pa \cdot s m^{-2}$).

The ground attenuation A_g is calculated from (Bies and Hansen, 1996)

$$A_g = -10 \, \log_{10}[1 + 10^{-A_R/10}] \tag{2.124}$$

where $A_R = -20 \, \log_{10} |R|$.

The third method for the assessment of ground attenuation involves coherent addition. It will not be discussed here further.

For practical noise control situations, the incoherent model is appropriate at high frequencies and long propagation distances whereas the coherent model is appropriate at low frequencies propagating over shorter distances.

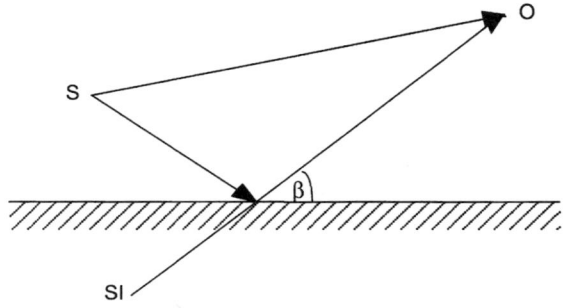

Figure 2.19 Direct and reflected sound paths.

Meteorological effects on sound propagation are hard to predict. In general (Bies and Hansen, 1996):

1. A positive temperature gradient (cold near the earth, warmer higher up) will diffract sound downwards resulting in increased noise levels.
2. A negative temperature gradient will diffract sound upwards.
3. Wind travelling towards a receiver from a sound source will diffract sound downwards and increase noise levels (Figure 2.20).
4. Wind travelling away from a receiver towards a sound source will diffract sound upwards and decrease noise levels.

In enclosed spaces, the relationship between the sound power emitted by the source and the sound pressure field achieved in the space is different from the outdoor, free-field case previously discussed. Whenever a continuous source of sound is present in a space, two sound fields are said to be produced. One is the 'direct sound field' or the 'direct arrival' from the source. The other is the 'reverberant sound field' which is produced by the reflections from surfaces within the space.

The (rms) pressure amplitude p_d due to the direct field at any point in the space is calculated assuming direct spherical radiation from a point source thus,

$$p_d^2 = \frac{\rho c W}{4\pi r^2} \tag{2.125}$$

where $r(m)$ is the radial distance from the effective centre of the source and W is the source power in Watts.

The relationship between the sound power of a source and the pressure distribution in the reverberant field is more complex than for the case of a direct field. If a source of sound is operated continuously (i.e. continuously emitting energy) in an enclosure, absorption in the air and at the surfaces of the enclosure prevents the sound pressure from becoming infinitely large. In small and medium-sized enclosures the absorption of the air is negligible compared to the absorption provided by the surfaces, and so the rate at which the sound pressure amplitude grows in the enclosure after switching on a sound source and the rate at which it decays after switching the sound sources off are controlled by surface absorption.

If the total sound absorption is large then the pressure amplitude quickly reaches a maximum value only slightly in excess of that produced by the direct field alone. Such spaces are known to be 'acoustically dead' or non-reverberant.

By contrast, if the absorption is small, then considerable time will elapse while the pressure amplitude grows to a maximum value significantly greater than that produced by the direct field alone. Spaces of this type are known as 'acoustically live' or reverberant.

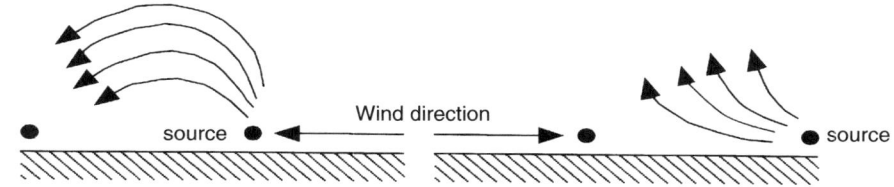

Figure 2.20 Meteorological effects on sound propagation.

When a source of sound is started in a live space, reflections from the walls produce a sound energy distribution that becomes more and more uniform with increasing time.

Eventually, except close to the source or to the absorbing surfaces, this energy may be assumed to be completely uniform with essentially random local directions of propagation.

A simple model for such a diffuse field can be created from first principles (not shown here, see Kinsler et al. (1982) for further details) which yields the following relationship between the sound power of the source in watts and the pressure amplitude in the reverberant field:

$$p_r^2 = \frac{4\rho c W}{A} \qquad (2.126)$$

where A is the total absorption in the enclosure. Now the total sound pressure distribution in the space is equal to

$$p^2 = p_d^2 + p_r^2$$
$$p^2 = \rho c W \left(\frac{1}{4\pi r^2} + \frac{4}{A} \right) \qquad (2.127)$$

It can be shown that the ratio of the reverberant intensity I_r to the direct field intensity I_d is (Kutruff, 1979; Kinsler et al., 1982):

$$\frac{I_r}{I_d} = \frac{4\pi r^2}{A/4} = \frac{16\pi r^2}{A} \qquad (2.128)$$

This equation shows that for locations very close to the source where $4\pi r^2 \ll (A/4)$

$$I_r = \frac{4\pi r^2}{A/4} I_d \qquad (2.129)$$

At such locations, I_r is a small fraction of I_d and hence the shape or acoustic treatment of the space will have little influence on the measured sound pressure levels. This local sound field is dominated by the direct field, and the sound pressure level will drop by up to 6 dB per doubling of distance from the source.

At greater distances from the source when

$$4\pi r^2 \gg A/4$$

then

$$I_d = \frac{(A/4)}{4\pi r^2} I_r \qquad (2.130)$$

This time I_d is a small fraction of I_r and hence the sound pressure field will become increasingly uniform (the initial 6 dB drop in sound pressure level per doubling of distance from the source will reduce to zero drop per doubling of distance) and this uniform sound pressure level will be reduced by 3 dB per doubling of the total sound absorption A.

As a consequence, a person near a noisy machine will receive little benefit from increasing the total absorption of the space. However, people further away would receive some benefit from such treatment.

In order to calculate the sound pressure level due to the reverberant field, the total absorption A of the space must be known.

The total absorption of the enclosed space may be found experimentally by measuring the reverberation time of the space and applying Sabines equation (for instance Kinsler et al. (1982)):

$$T = \frac{0.161V}{A} \tag{2.131}$$

where V is the volume of the space (m³) and T is the reverberation time (s) defined as the time required for the level of the sound to drop by 60 dB. The initial work by Sabine on reverberation was limited to a single frequency of 512 Hz. Custom has attached so much importance to the 512 Hz (and later 500 Hz) frequency that when the expression 'reverberation time' is used without specifying a frequency it usually refers to 500 Hz (Kutruff, 1979).

If the surface area of the enclosed space is S (m⁻²) then the average Sabine absorptivity $\bar{a}$ is defined by (Kinsler et al., 1982):

$$\bar{a} = \frac{A}{S} \tag{2.132}$$

and therefore

$$T = \frac{0.161\ V}{S\bar{a}} \tag{2.133}$$

Sabine adopted the simplifying assumption that the total sound absorption of the enclosed space is the sum of the absorptions A_i of the individual surfaces.

$$A = \sum_i A_i = \sum_i S_i a_i$$

and thus

$$\bar{a} = \frac{1}{S} \sum_i S_i a_i \tag{2.134}$$

Sabine absorptivities are measured using standard methods with a large sample of the material in a reverberation chamber (Bies and Hansen, 1982).

When the chamber is empty:

$$T = \frac{0.161\ V}{S\bar{a}} \tag{2.135}$$

Now if a sample of material with surface area S_p is mounted on the walls of the chamber the reverberation time becomes:

$$T_p = \frac{0.161\ V}{S\bar{a} - S_p a_o + S_p a_p} \tag{2.136}$$

where a_o is the absorptivity of the now covered portion of wall and a_p is the unknown absorptivity of the material.

Thus combining this last equation with equation (2.135) yields:

$$a_p = a_o + \frac{0.161V}{S_p}\left(\frac{1}{T_p} - \frac{1}{T}\right) \tag{2.137}$$

The frequency varying absorptivities of different materials are published widely.

In summary, it has been shown in this section that the total absorption of a space may be derived from measurements of reverberation time or estimated through the summation of the absorption of all the surfaces in the space (equations 2.133 and 2.134).

There is a third method which is more approximate (Kinsler et al., 1982). A combination of

$$T = \frac{0.161\ V}{A}$$

and

$$\frac{I_r}{I_d} = \frac{4\pi r^2}{A/4} = \frac{16\pi r^2}{A}$$

yields

$$\frac{I_r}{I_d} = 312\frac{r^2 T}{V} \tag{2.138}$$

The energy of the direct sound falls off as the square of the distance from the source (energy is proportional to r^2). Therefore, it is impossible to have a constant ratio of reverberant to direct sound throughout an enclosed space.

The interior sound propagation case is further complicated by the presence of acoustic modes which occur at the natural frequencies of the space. A knowledge of the natural frequencies of a space is essential for a complete understanding of its acoustic properties. The space will respond strongly to sound having frequencies in the immediate vicinity of any of these natural frequencies. Each natural frequency has a corresponding mode shape – a spatial distribution of nodes and anti-nodes (pressure minima and maxima respectively).

Every space superimposes its own characteristics on those of any sound source present so that if sound pressure measurements are taken throughout the space the true sound radiation characteristics of the source may be concealed (unless the total absorptivity of the space is greater than 0.99 and then the reverberant field is negligible compared to the direct field).

There are different labels for certain types of acoustic mode, and often take the form of subscripts l, m and n to the natural frequency f_{lmn}.

If only one of the integer subscripts is non-zero then the mode is termed axial because the propagation vector is parallel to one of the axes (this could be a standing wave between two parallel sides to a vehicle interior or between the floor and the roof). Axial modes run parallel to all three pairs of surfaces of the space, and touch two parallel surfaces.

If two of the three integer subscripts are non-zero then the mode is termed tangential as it runs parallel to one pair of surfaces and touches the other two pairs of parallel surfaces. If all three integer subscripts are non-zero then the mode is termed oblique and the standing waves are not parallel to any surfaces and all six surfaces of a rectangular space are touched.

Each of the individual modes in an enclosure can be excited to its fullest extent only by a sound source located in regions where the particular mode shape has a pressure antinode. The pressure amplitudes of all mode shapes in a rectangular space are maximised in the corners of the space. Therefore, if a source is at the corner of a space it will be able to excite all allowable modes to their fullest extent. Correspondingly, a microphone placed in the corner of a space will measure the peak pressure of every mode excited. By contrast, a source located at a pressure node will only weakly (if at all) excite a mode. For instance, a loudspeaker located at the centre of a rectangular space will only excite modes having even numbers simultaneously for l, m, n and hence only about one in ten modes will be excited.

Acoustic modes are set up at natural frequencies of the space. These natural frequencies can be predicted using the equation (Bies and Hansen, 1996; Kinsler et al., 1982):

$$f_{lmn} = \frac{c}{2\pi}\sqrt{(l\pi/L_x)^2 + (m\pi/L_y)^2 + (n\pi/L_z)^2} \tag{2.139}$$

where l, m and n are integer mode labels and L_x, L_y and L_z are the length, width and height of the space in metres.

Table 2.7 shows the idealised acoustic modes for a space $3.12 \times 4.69 \times 6.24\,\mathrm{m}^3$. As the frequency increases, the mode shapes overlap more and more so that eventually the response of the space becomes quite smooth. Each of these mode shapes has different sets of direction cosines and hence at high frequencies direction vectors in the sound field become increasingly random. This explains why reverberation equations based on diffuse fields agree best with experimental results at higher frequencies.

The response of a space becomes less uniform as its symmetry increases. This is a result of the increase in degenerate modes having different l, m, n but the same natural frequency.

The distribution of acoustic pressure in a rectangular space with rigid walls is given by (Bies and Hansen, 1996)

$$p = p_0 \cos(l\pi x/L_x)\cos(m\pi y/L_y)\cos(n\pi z/L_z)e^{i\omega t} \tag{2.140}$$

Table 2.7 Table of acoustic modes for a space $3.12 \times 4.69 \times 6.24$ m

l	m	n	$f(Hz)$	l	m	n	$f(Hz)$
0	0	1	27.5	1	0	2	77.5
0	1	0	36.6	0	2	1	78.5
0	1	1	45.9	0	0	3	82.5
1	0	0	55.0	1	1	2	86.0
0	0	2	55.0	0	1	3	90.2
1	0	1	61.5	0	2	2	91.5
0	1	2	66.0	1	2	0	91.5
1	1	0	66.0	1	2	1	95.5
1	1	1	71.5	1	0	3	99.0
0	2	0	73.2				

where

$$\omega = 2\pi f$$

The frequency distribution of normal modes and their corresponding mode shapes have important bearing on noise control in enclosed spaces such as vehicle interiors:

1. The sound pressure level at a particular frequency and at a particular point in a space due to a sound source may be up to 20 dB higher than expected from simple calculations if the sound source is well coupled to a mode and the monitoring position is at a pressure antinode.
2. Sound sources near walls and in the corners of spaces will couple to modes more effectively than sound sources at the centre of spaces.
3. Spaces will be particularly sensitive to excitation at frequencies where degenerate modes occur.

The effects of acoustic modes can be damped by applying absorptive material to the surfaces of the space. An absorbing surface is most effective in damping a normal mode if it is located in a region of maximum mean square pressure. Since all normal modes have pressure maxima at the corners of a space, absorbing material placed there will be more effective than anywhere else.

In an enclosed space whose every surface is covered in absorptive material, oblique modes are absorbed the fastest, followed by tangential modes and then axial modes. This means that axial modes have the longest reverberation times and oblique modes have the shortest reverberation times.

At low resonant frequencies therefore, the reverberation time is related to the damping of particular modes. At higher resonant frequencies, the rate of decay of sound may vary with time: decaying quickly at first as the oblique modes are damped faster, and then decaying more slowly as the tangential and axial modes begin to damp further. The actual length of time required for the sound pressure level to drop by 60 dB will therefore depend on the relative amounts of energy stored in the various types of mode.

At high frequencies, most of the energy is held in oblique modes and so more of the early decay is dominated by these, and for longer. The slope of this early decay can be used to determine the reverberation time associated with the major portion of energy in the space. Consequently, it is only in this frequency range that the concept of reverberation time is in itself a sufficient criterion for judging the acoustic characteristics of an enclosure.

References

ANSI S 1.16, Preferred frequencies, frequency levels and band numbers for acoustical measurements, Acoustical Society of America

BS 4142: 1990, Rating industrial noise affecting mixed residential and industrial areas, BSI, 1990 (withdrawn, current revision dated 1997)

Beranek, L. (ed.), Noise and vibration control, McGraw Hill Inc, 1971

Beranek, L.L. (ed.), Noise and Vibration Control, Revised edition, Institute of Noise Control Engineering, 1988

Bies, D.A., Hansen, C.H., Engineering noise control – theory and practice, Second edition, E&FN Spon, 1996

Bruel & Kjaer Data Handbook, Condenser microphones and microphone preamplifiers for acoustic *measurements*, Bruel & Kjaer, Revision September 1982

Dowling, A.P., Ffowcs Williams, J.E., Sound and sources of sound, Ellis Horwood Limited, 1983

Duchateau, P., Zachmann, D.W., Schaum's outline series – theory and problems of partial differential equations, McGraw-Hill Book Company, 1986

Rev. Earnshaw, S., On the mathematical theory of sound, Trans. R. Soc. London, Vol. 150, 133–148, 1860

Fahy, F., Walker, J. (eds), Fundamentals of noise and vibration, E&FN Spon, 1998

Hassall, J.R., Zaveri, K., Acoustic noise measurements – Fourth edition, Bruel & Kjaer, 1979

Illingworth, V. (ed.), The Penguin Dictionary of Physics – Second edition, Penguin Books Ltd, London, 1990

Kinsler, L.E., Frey, A.R., Coppens, A.B., Sanders, J.V., Fundamentals of acoustics, Third edition, John Wiley & Sons Inc., 1982

Kuttruff, H., Room Acoustics, Second edition, E&FN Spon, 1979

Morse, P.M., Ingard, K.U., Theoretical Acoustics, McGraw-Hill Book Co, New York, 1968

Pierce, A.D., Acoustics – an introduction to its physical principles and applications, Acoustical Society of America, 1994

Rayleigh, J.W.S., The theory of sound, Vol. 1, First published in 1877, Second revised edition, 1894, also Dover Publications, 1945

Sinha, N.K., Linear systems, John Wiley & Sons, 1991

Sutherland, L.C., Piercy, J.F., Bass, H.E., Evans, L.B., Method for calculating the absorption of sound by the atmosphere, Journal of the Acoustical Society of America, 66, 885–894, 1974

Temkin, S., Elements of acoustics, John Wiley & Sons, 1981

Thompson, P.A., Compressible Fluid Dynamics, 1988

Weltner, K., Grosjean, J., Shuster, P., Weber, W.J., Mathematics for engineers and scientists, Stanley Thornes, 1986

Zamminer (Reported by Rayleigh, vol. 1, 1894), Die musik und die musikalischen instrumente, Giessen, 1855

Appendix 2A: Introduction to logarithms

A logarithm is an exponent. To resolve an equation of the type

$$a^x = c \tag{A2.1}$$

two steps are required.

1. Write both sides of the equation as powers to the same base i.e.

$$a^x = a^{(\log_a c)} \tag{A2.2}$$

2. Compare exponents

$$x = \log_a c \tag{A2.3}$$

The definition of a logarithm may be written as the following equation:

$$a^{(\log_a c)} = c \tag{A2.4}$$

Notation for logarithms

To avoid writing the base below the log as a subscript, the following notation has been adopted:

Base 10   Logarithms to the base 10 are known as 'common logs'. Base 10 is mainly used in numerical calculations involving a wide spread of number magnitudes. Logarithms to the base 10 are written as:

$$\log_{10} = \lg \text{ or } \log$$

Base 2   Base 2 is mainly used in data processing. Logarithms to the base 2 are written as:

$$\log_2 = ld$$

Base e   Logarithms to the base 'e' are called 'natural logarithms' or 'Napierian logarithms'. They are frequently used in calculations relating to physical problems. Logarithms to the base 'e' are written as:

$$\log_e = ln$$

Operations with logarithms

Operations with logarithms follow the power rules since logarithms are exponents

$$\log_a AB = \log_a A + \log_a B \tag{A2.5}$$

$$\log_a \frac{A}{B} = \log_a A - \log_a B \tag{A2.6}$$

$$\log(A+B) \neq \log A + \log B \tag{A2.7}$$

$$\log_a A^m = m \log_a A \tag{A2.8}$$

$$\log_a \sqrt[m]{A} = \log_a A^{1/m} = \frac{1}{m} \log_a A \tag{A2.9}$$

$$y = \ln x \tag{A2.10}$$

$$\frac{dy}{dx} = \frac{1}{x} \tag{A2.11}$$

3

Exterior noise: assessment and control

3.1 Pass-by noise homologation

3.1.1 Background to homologation

Most countries restrict the types of vehicle that can operate legally on their roads by some form of national legislation. Taking the United Kingdom (UK) as an example, the relevant legislation is the Road Vehicles (Construction and Use) Regulations 1986, sections of which have been amended many times since first coming into operation on 11 August 1986. The 1986 Regulations specify minimum requirements for the construction (and hence design), maintenance and use of road vehicles (including heavy goods vehicles, passenger service vehicles and track-laying vehicles) for the following:

- dimensions and maneuverability;
- brakes;
- wheels, springs, tyres and tracks;
- steering;
- vision;
- instruments and equipments;
- fuel;
- control of emissions (including noise);
- laden weight;
- the use of trailers;
- the avoidance of danger when in use.

Compliance with the requirements of the 1986 Regulations is ensured by a mandatory vehicle type approval scheme (Statutory Instrument No. 981, 1984; The Road Traffic Act, 1988) that covers:

- Mass-produced vehicles for sale or use in the UK.
- Low-volume-produced vehicles for sale or use in the UK.
- Single vehicles (produced in the UK or imported) that do not enjoy type approval by virtue of their unique construction or subsequent modification, or were imported from a country without suitable type approval.

In addition to national legislation, many countries place restrictions on type approval according to international agreements. In the UK (as in other European Community (EC)) countries, restrictions are placed in accordance with

- EC Directives.
- Economic Commission for Europe (UN–ECE) Regulations.

There are a significant number of UN–ECE Regulations, mostly concerning individual components or systems. The principal EC Directive for type approval is 70/156/EEC and this has been amended many times since first publication (notably 92/53/EEC). In the UK, the mechanism by which the Secretary of State regulates EC type approval is currently provided by the Motor Vehicles (EC Type Approval) Regulations 1998.

Homologation is the process by which approval for sale or use of a vehicle in a particular country is obtained. This varies from one country to other and may take the form of:

- Type approval (conducted by an independent body).
- Single vehicle approval (conducted by an independent body).
- Self certification by the manufacturer (who certifies that the vehicle complies with all the legislative requirements).

Automotive EC Directives require third party approval which has three components to it:

1. Testing (to particular technical EC Directives).
2. Certification.
3. Product Conformity Assessment (assessing the ability to produce series products in conformity with the specification, performance and labelling/marking requirements of the type approval).

In the United Kingdom, the Vehicle Certification Agency (VCA – an executive agency of the Department of the Environment, Transport and the Regions – DETR) is the only Competent Authority (as defined in 70/156/EEC) for the type approval of vehicles under EC Directives. The VCA is therefore the only body in the UK to issue EC type approval certificates. However, an EC type approval certificate issued by a Competent Authority in another Member State is equally valid in all Member States including the UK. The VCA is also the only test authority in the UK for EC type approval. In other Member States, such as Germany, there are several test authorities but only one Government Authority may issue type approval certificates.

EC Component Type Approval may be obtained for generic components (to be fitted to any vehicle) after evaluation against particular technical EC Directives. EC Separate Technical Unit Type Approval may be obtained for components to be fitted to only one type of vehicle after evaluation against particular technical EC Directives.

EC approval of most new mass-produced road vehicles is based on whole vehicle approval. In such cases, a representative example of the production intent vehicle is tested/evaluated by the Competent Authority (they must at least attend all tests if they do not actually conduct them). The items tested include:

- seat belts
- lights
- tyres
- brakes
- emissions (including noise)
- seats and head restraints
- reward visibility
- vehicle impact test.

Compliance with around 45 technical standards is checked. VCA practises worst case selection in order to reduce the amount of testing needed across a range of product types. (*Source*: www.vca.gov.uk.)

3.1.2 EC noise homologation

Current EC automotive Directives are what are known as old approach whereby a third party tests for compliance with the requirements of particular technical Directives. In new approach Directives (that Member States have so far rejected for automotive applications) more obligation is placed on the manufacturer to ensure that the product complies with the appropriate requirements.

The original technical Directive relating to type approval and noise was 70/157/EEC. This defined limiting noise levels outside the vehicle during a particular form of acceleration test (the test being improved since then and now detailed in ISO 362:1998). An acceleration test was chosen to provide a vehicle operating condition with worst-case noise emissions. The original test was as follows:

- The test site, a section of level road (more than 3 m wide) in an open area of minimum radius 50 m, is constructed in accordance with ISO 10844. ISO 10844 defines the texture and porosity and hence the acoustic effect of the road surface. The road widens at one point to make a 20 m by 20 m test area of special road surface as shown in Figure 3.1.
- A line is drawn down the centreline of the road. The driver follows that line throughout the test.

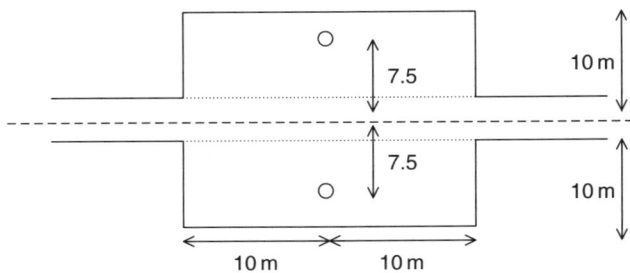

Figure 3.1 The test area defined in ISO 362:1998.

- One microphone position is marked on each side of the road, 7.5 m from the centre point of the test area and at a height of 1.2 m. A type 1 sound level meter is placed at one of the microphone positions.
- The test vehicle approaches the test area at a fixed speed and in a pre-determined gear. For common EC passenger cars of category M1 with 5-speed manually operated transmissions the approach speed is 50 km hr^{-1} and the test is performed in both second and third gears.
- When the front of the vehicle enters the 20-m long test area, the driver depresses the throttle briskly and fully and holds it in that position until the rear of the vehicle leaves the test area.
- The maximum, fast-weighted, A-weighted, noise level recorded whilst the vehicle is within the test area is noted. The results are considered valid when four consecutive results are within 2 dB from each other.
- At least four valid levels are recorded on both sides of the vehicle.
- The results from each side of the vehicle are averaged separately. The intermediate result is the highest of the two averages.
- The final reported result for the 5-speed M1 car is the arithmetic average of the two intermediate results obtained from the tests for both gears.
- The reported level may be reduced by 1 dBA to account for the natural variability of the test.

The homologation noise levels for type approval have reduced significantly since their introduction in 70/157/EEC. The level for M1 passenger cars has been reduced by 8 dBA from 82 dBA to the current level of 74 dBA (92/97/EEC). The reduction has been in stages as shown in Table 3.1.

The current homologation levels correspond to the test detailed in 92/97/EEC. ISO 362:1998 is broadly similar but not identical in all details. In particular, 92/97/EEC requires additional measurements to be made 0.5 m (45° angle of incidence, 0.2 m above the ground) from the exhaust tailpipe of a stationary vehicle in order to facilitate subsequent checks of vehicles in use. The engine is run at three-quarters speed and the

Table 3.1 EC homologation noise levels for type approval

EC Directive Enforced for new vehicles	70/157/EEC 1976	77/212/EEC 1983	81/334/EEC 1984**	84/424/EEC 1990	92/97/EEC 1996
Category of vehicle	Maximum permissible noise level (dBA) at 7.5 m (acceleration test)				
Cars (<9 seats)	82	80	80	77	74*
Minibus <2 tonnes	84	81	81	78	76
Minibus 2–3.5 tonnes	84	81	81	79	77
Bus ≤150 kW	89	82	82	80	78
Bus >150 kW	91	85	85	83	80
Light Truck <2 tonnes	84	81	81	79	76
Light Truck 2–3.5 tonnes	84	81	81	80	77
Truck >3.5 tonnes					
≤75 kW	89	86	86	83	77
75–150 kW	89	86	86	85	78
>150 kW	91	88	88	86	80

* 75 dBA if fitted with a DI diesel engine or if very powerful 'supercar'.
** Imposed more stringent requirements for exhaust and intake systems.

maximum sound pressure level (fast, A-weighted) is recorded. The results from three of these tests must be reported as part of the vehicle certification process. 92/97/EEC also includes a method for assessing compressed air noise.

3.1.3 Track and atmospheric effects

Directive 92/97/EEC introduced a requirement for a standard road surface to be adopted at all test tracks used for drive-pass noise homologation. 92/97/EEC is not completely prescriptive over the details of construction for such a standard track and so many Member States have adopted ISO 10844 as a specification. According to Sandberg (1991), the specification in ISO 10844 was aimed initially at achieving three goals:

1. To make the results achieved at any given track repeatable and reproducible at any other track.
2. To make the track surface highly reflective so that all noise sources on the vehicle (even those hidden from view under the vehicle) make a contribution to the overall drive-pass level and thus this would encourage manufacturers to treat the whole vehicle.
3. Minimise the noise radiated from the tyre/road contact as this is a fairly unavoidable source of noise and if it were to dominate, manufacturers would not be incentivised to treat the noise from other sources on the vehicle.

Before the widespread adoption of ISO 10844 surfaces, Dunne and Yarnold (1993) reported that they knew of 4 dBA variations in the drive-pass noise levels recorded for the same car at different test tracks around Europe.

Walker (1994) reports on an experimental investigation to test the effectiveness of these three goals in practice. Noise measurements were made on an ISO 10844 test track and on another track surfaced with high-drainage asphalt. An omni-directional noise source was positioned on the centreline of the track at heights between 100 mm and 400 mm above the surface. A microphone was positioned 7.5 m away from this at a height of 1.2 m from the surface. The comparison showed that:

- The ISO 10844 surface did indeed behave as a close approximation to a perfectly reflective (low absorption) surface at most frequencies of interest.
- The sound absorption characteristics of the ISO 10844 track are more uniform across the frequency spectrum than those of the high-drainage asphalt surface. This suggests consistent surface characteristics over the 7.5 m acoustic path length and should result in results being more repeatable and reproducible.
- Because of the relative smoothness of the ISO 10844 surface, low-frequency tyre noise was reduced compared with the high-drainage surface, but high-frequency tyre noise was increased.

Walker concluded that with all these taken into account, the ISO 10844 would produce drive-pass noise levels some 2 dB greater than those expected for a given car tested on the high-drainage asphalt surface.

In drafting Directive 92/97/EEC, Dunne and Yarnold (1993) report that the European Commission recognised that other factors such as meteorological conditions may influence the drive-pass test result. The effects of wind speed and direction, humidity and temperature profile above the ground on the outdoor propagation of sound are discussed in Section 2.5.1.

3.1.4 Future developments in noise homologation in the EC

Although noise homologation limits have been reduced significantly since 1970, comprehensive noise testing alongside roads in Germany (Steven, 1995) has shown that noise levels produced by vehicles in normal use have reduced only slightly over the first twelve years of that period.

The main reason for this seems to be that the homologation process has forced vehicle manufacturers to reduce engine and intake/exhaust noise which are strongly engine-load-dependent and thus dominate the low-speed, high-load acceleration test. However, the process has had less impact on tyre noise and on the noise caused by vehicles at speeds much in excess of $50 \, \text{km} \, \text{hr}^{-1}$. The higher speed noise is most important in the environmental impact of vehicles in normal use outside the urban setting.

It is intuitively obvious that once engine, intake and exhaust noise are reduced further, tyre noise will dominate even the low-speed acceleration test. Once that becomes the case (and it already is with many passenger cars fitted with wide tyres), the homologation level cannot be reduced further without reducing tyre noise first.

The EC have responded to this problem by proposing noise homologation levels for tyres (C30/8, 28/1/98). These have not yet been adopted. A new higher speed variant of the acceleration test is proposed, whereby a vehicle coasts through the 20-m test area at a constant approach speed but with the engine turned off. The approach speeds would vary in the range of $60–90 \, \text{km} \, \text{hr}^{-1}$ according to the type of tyre. The noise limits would vary according to the type of tyre and tyre width.

3.1.5 Noise homologation in the US and otherwise outside the EC

Most countries outside the EC have their own systems of vehicle type approval that include restrictions on noise levels. Most tests and limiting levels are based on those of the EC or of the United States.

There are two US noise homologation tests – SAE J986 AUG94 which has the vehicle entering a test area with predetermined vehicle speed and SAE J1030 FEB87 which has the vehicle leaving the test area with predetermined engine speed. Both tests feature vehicle deceleration as well as acceleration.

The acceleration part of SAE J986 AUG94 is broadly similar to that in the EC test except that the microphone is positioned at 15 m from the vehicle pathline (rather than 7.5 m) and the test area is much longer (53 m) with the aim of allowing the vehicle to reach its rated engine speed during the test. In the EC test, the vehicle will seldom reach its rated engine speed.

Generally, a vehicle that achieves noise homologation in the EC will achieve US Federal homologation with comparative ease. Japan and Switzerland have traditionally had homologation restrictions that are more onerous than those of the EC or the US.

3.1.6 The consequences of meeting homologation noise limits

The current homologation noise limits according to 92/97/EEC are very demanding. A great deal of engineering effort was devoted in the early 1990s to reduce noise levels from accelerating passenger cars by the 3 dBA required to meet a level of 74 dBA. Vehicle manufacturers are achieving type approval for their current vehicles, but it is difficult for those with higher powered engines (Sports Utilities, GTi models and the like). Tables 3.2 and 3.3 give an indication of this. Note that most of the ten lowest homologation levels shown are for automatic cars and most of the highest homologation levels are recorded for sports-utility vehicles with large engines and manual transmissions.

It should be appreciated that the total wayside noise level during the acceleration test is made up of contributions from many noise sources (engine, intake, exhaust, tyres, etc.). Noise levels from each of these must be controlled in order to achieve type approval. Some suggested noise targets for each noise source are given in Table 3.4.

It should be noted that due advantage is taken in Table 3.4 of the 1 dBA reduction in reported values due to potential inaccuracy of the noise homologation test.

Table 3.2 The ten lowest homologation levels in the UK

Manufacturer	Model	Description	Trans.	Capacity	Fuel	dBA
Daihatsu	Terios	1.3L Efi 4WD	A4	1298	Petrol	65
Daihatsu	YRV	1.3L Efi turbo	A4	1298	Petrol	66
Renault	Vel Satis	2.0 Turbo	A5	1998	Petrol	66.7
Mitsubishi	Carisma	1.6 Mirage	A4	1597	Petrol	67
Smart	Cabrio Hatchback	Cabrio & Passion	A6	599	Petrol	67
Nissan	Primera	2.0 4/5 door	M6	1998	Petrol	67
Mitsubishi	Galant – sedan	2.5 Elegance	A4	2498	Petrol	67
Nissan	Primera	2.0 Estate	A6	1998	Petrol	67
Nissan	Primera	2.0 4/5 door	A6	1998	Petrol	67
Mitsubishi	Galant – wagon	2.5 Elegance	A4	2498	Petrol	67

Source: www.vca.gov.uk. Accessed in November 2003

Table 3.3 The ten highest homologation levels in the UK

Manufacturer	Model	Description	Trans.	Capacity	Fuel	dBA
Chrysler Jeep	Grand Cherokee	4.7L Overland	A5	4701	Petrol	76
Vauxhall	Frontera	3.2I V6 24v	A4	3165	Petrol	76
Mitsubishi	Colt	1.3 Attivo 2	M5	1300	Petrol	76
Volkswagen	Touareg	2.5	M6	2461	Diesel	76
Mitsubishi	Shogun	3.2 Equippe LWB	M5	3200	Diesel	76
Land Rover	Discovery	2.5 TD5	A4	2495	Diesel	76
BMW	X5	X5 3.0d	M6	2993	Diesel	76
Land Rover	Defender Wagon	2.5 TD5 90	M5	2495	Diesel	76
Land Rover	Discovery	2.5 TD5	M5	2495	Diesel	76
BMW	X5	X5 3.0I	M6	2979	Petrol	76

Source: www.vca.gov.uk. Accessed in November 2003

Table 3.4 Suggested target noise levels for achieving type approval under 9297/EEC

	Target levels at 7.5 m, acceleration test (dBA)		
	Passenger car	Light truck	Heavy truck
Engine	69	72	77
Exhaust	69	70	70
Intake	63	63	65
Tyres	68	69	75
Transmission	60	63	66
Other	60	72	65
Combined level	74.2	77.3	80.1

3.2 Noise source ranking

When a vehicle manufacturer is having difficulty homologating a vehicle for noise, it is important to know the relative contributions made to the pass-by level by the various sources (intake, exhaust, tyres, etc.) so that the most significant sources can be controlled first. This is achieved using noise source ranking.

First, a list of potential noise sources should be compiled, and then the next task should be to rank all of the sources in order of significance. The term 'significance' can have several different meanings according to the case in hand. Significance could mean:

- overall linear sound power level or sound pressure level at a particular point in space;
- overall 'A'-weighted sound power level or sound pressure level at a particular point in space;
- sound pressure or power level in a certain frequency band;
- long-distance propagation;
- subjective rating.

The noise source ranking procedure may be undertaken in a variety of ways. The more popular methods include:

- Isolation – where each component of the machine is run in isolation, where possible, and its noise contribution is measured directly.
- Shielding – where the machine is run several times, and each time a different noise source is encapsulated in a sound-retaining structure. The reduction in noise level achieved yields that particular noise source's contribution to the total noise level. A practical example of this technique is given by Balcombe and Crowther (1993).
- Close microphone techniques – where the machine is run in an anechoic chamber and a microphone is passed in the near field over the surfaces of the machine to locate regions of high pressure level (an approximate indicator only usually used to quickly back-up a hunch as to where the noise is coming from).
- Sound intensity mapping – where the machine is run continuously, and an intensity probe is passed over an notional surface set around the machine. Regions of high intensity give an indication of the position and level of noise sources.
- Spatial Transformation of Sound Field method (STSF). This is a relatively new technique that uses an array of microphones in a plane (or for stationary sound fields,

a single scanning microphone) and the principle of near-field acoustical holography along with the Helmholtz Integral Equation to calculate the three dimensional sound field. Taylor and Bridgewater (1998) give an example of the technique in use. Kim and Lee (1990) give a more detailed description of the technical background.

- Noise from vibration – where the machine is run and measurements of vibration velocity are made on the surfaces of potential noise sources. Radiation efficiencies are calculated or assumed, and the noise contribution from each potential noise source is estimated.
- Modelling – where the noise contribution from each potential source is calculated using empirical or mathematical models.

3.2.1 Noise source ranking using shielding techniques

The most commonly used noise source ranking procedure in the automotive industry is still (arguably) the rather laborious shielding or encapsulation method. This usually consists of first eliminating the various noise sources, using acoustic treatments, and then removing the treatments systematically and measuring the noise contribution from each source as it is uncovered. The shielding technique is used for both vehicle and engine noise. The vehicle noise case shall be discussed further here.

The interior noise contribution from each source is tested using microphones placed at each seat in the vehicle, and the exterior noise contribution is assessed using the relevant standard drive-pass test. A typical noise-source ranking programme might take the following form:

- Perform baseline noise tests on the programme vehicle. Interior noise levels are recorded at each seat position with the vehicle accelerating uniformly in second or third gear, first with wide-open throttle and then part throttle. Exterior noise for the baseline vehicle is assessed using the relevant drive-pass standard.
- Fit a large and effective additional silencer(s) to the exhaust tailpipe (often called an infinite silencer). A series of absorptive silencers may be sufficient. The infinite silencer should be of low backpressure design so as not to restrict the performance of the vehicle. Care must be taken to have the outlet from the additional silencers as near to the original exhaust tailpipe position as possible.
- Lag the standard exhaust system silencers with a layer of glass fibre, and then cover in sheet lead. This is to reduce any contribution from exhaust shell noise (noise radiating from the structure of the exhaust system).
- Fit a large and effective silencer to the intake system. This silencer will often be strapped to the hood (bonnet) of the car, and can be constructed from perspex to allow for driver's visibility (large plastic drinking-water bottles are commonly adopted for use). The silencer system must be of low pressure loss design in order not to restrict the performance of the vehicle. It is vital that the outlet from the silencer is as near to the original snorkel position as possible.
- Encapsulate the engine bay with a thick plywood under tray, ensuring that the edges are well sealed. The engine encapsulation must be performed carefully if engine breathing and cooling problems are to be avoided.
- Encapsulate the drive train (if appropriate) using plywood, or lag with lead.
- Fit treadless tyres (or tyres known to be relatively quiet) to reduce tyre noise.

- Perform identical interior and exterior noise tests on the fully treated vehicle.
- Fit conventional tyres and re-test for tyre noise contribution. Once tested, re-fit the treadless tyres.
- Remove the transmission lagging and re-test for drive-line noise contribution. For speed, it is customary not to re-fit any acoustic treatments after testing.
- Remove the exhaust system lagging and test for exhaust shell noise contribution.
- Remove the engine bay encapsulation and re-test for engine noise contribution.
- Remove the infinite intake silencer and re-test for intake noise contribution.
- Remove the infinite exhaust silencer and re-test for exhaust noise contribution.

The apparent sound pressure level due to each noise source in isolation from the others is derived from the increase in overall sound pressure level as each additional source is uncovered.

$$L_{Si} = 10 \ \log_{10}\left[10^{\frac{L_u}{10}} - 10^{\frac{L_S}{10}} \right] dB \qquad (3.1)$$

where

L_{Si} is the apparent sound pressure level due to source i in isolation.
L_u is the total sound level recorded once source i is uncovered.
L_S is the sound level recorded before source i is uncovered.

This noise source ranking procedure is relatively simple, but involves a large number of tests. The results should be viewed only as being approximate, due to the potential for significant test-to-test variations.

It is important to reduce the number of variables during the tests, so it is wise to use the same test routine for all tests and try to perform the tests on or near the same day. Once the relative noise contribution from each noise source is determined, the degree of additional silencing required is assessed. Only when there is a potential for significant decrease in overall noise levels is the costly process of component development carried out. Some sample vehicle noise source ranking results are shown in Figure 3.2.

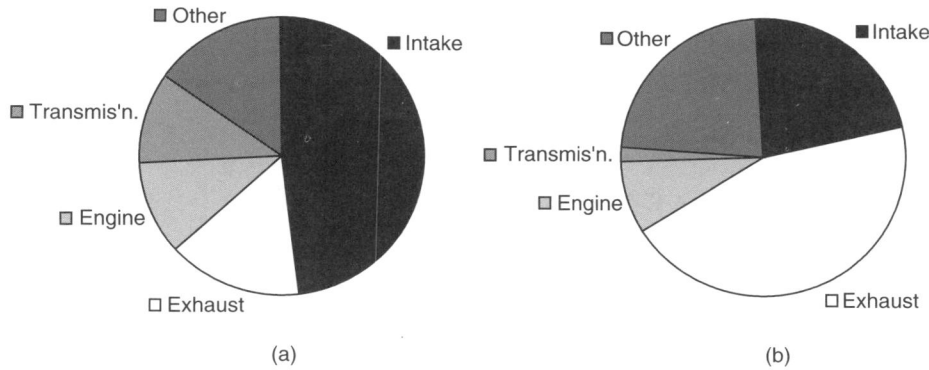

(a)   (b)

Figure 3.2 Noise source ranking results adapted from Balcombe and Crowther (1993). (a) Results at the start of the pass-by test. (b) Results at the end of the pass-by test.

3.3 Air intake systems and exhaust systems: performance and noise effects

3.3.1 Introduction

Intake and exhaust noise must feature prominently in this chapter as they both can be significant sources of exterior vehicle noise. Arguably for the first sixty years of the motor car the exhaust tailpipe was the most significant cause of vehicle noise (even brief exposure to an unsilenced light aircraft provides a clear reminder of that). With the rise in popularity of front-wheel drive passenger cars, permitting the adoption of large-volume exhaust silencers (driven of course by the arrival of noise control legislation), the levels of exhaust tailpipe noise reduced considerably from around 1960. This uncovered hitherto unheard intake noise.

For some twenty years (from around 1980) the intake system was commonly the dominant noise source in family sedans operating at full load. It has only been much more recently that improvements in plastic moulding technologies (and the adoption of port fuel injection removing the need for a metallic intake manifold) have freed intake designers to add several silencing elements to even low-cost intake systems. As a result, the levels of intake noise now match those of the exhaust system and so both engine and tyre noise now commonly dominate the noise emissions from most family sedans.

This chapter is more heavily biased towards intake system design rather than exhaust system design. There are two reasons for this:

1. The careful design of the intake system for a naturally aspirated (NA) engine can raise volumetric efficiency by more than 10% whereas the corresponding improvement caused by detailed exhaust system design is seldom more than 3–4%. Therefore, the intake designer must consider both engine performance and noise control whereas (within back-pressure limits) the exhaust designer is free to concentrate more on noise control.
2. The skills and techniques used for exhaust system design are only slightly modified versions of those acquired for intake system design.

This rather extended section will commence with an overview of intake system requirements and then move on to a detailed assessment of intake design for engine performance followed by intake design for noise control. Only a brief discussion of exhaust system design is offered as this concentrates on the main differences in approach for exhaust system design as compared with intake system design.

3.3.2 Intake noise – objectives

Some intake noise is desirable and can help create an impression of speed or power. Other noise is undesirable and can contribute to excessive interior noise levels or failure of the legislative drive-pass test.

Unwanted intake noise may be classified into two categories:

1. Exterior noise which contributes to the overall noise level during the type approval drive-pass test (see Section 3.1).
2. Interior noise which reduces the comfort of the driver or passengers, or interferes with speech communication (see Chapter 4).

3.3.3 The issues involved in intake system design

It is important to understand the constraints on design which include:

- wave action tuning of the intake system to improve engine performance;
- intake acoustic design;
- intake system static pressure loss;
- intake system mounting;
- packaging space (under-hood);
- turbo or super-charging;
- cost;
- controlling vibration in the system;
- intake snorkel positioning;
- intake water inhalation;
- manufacturing process;
- intake filter durability/servicing;

3.3.4 Intake systems

Intake systems fulfil a number of roles in an overall vehicle design which are:

- channel air to the engine;
- filter particulates;
- enhance engine performance through wave action tuning;
- reduce noise.

3.3.5 The intake system designer

The intake system designer is in a unique position within a vehicle development programme, falling within the interest areas of several design groups. His or her design decisions will directly affect the work of other designers as illustrated in Figure 3.3.

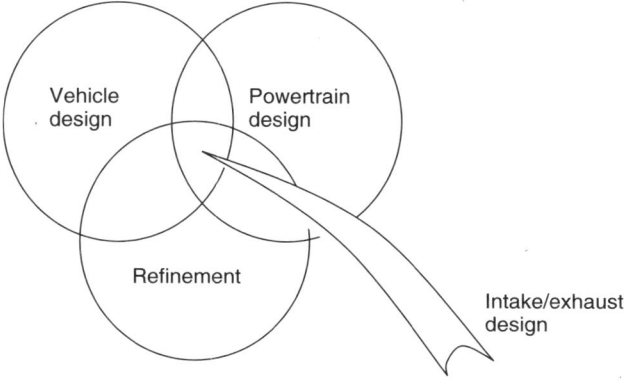

Figure 3.3 The role of the intake system designer.

Changes to the system geometry or layout which are beneficial for noise control may well not be advantageous for power or pollutant emissions.

3.3.6 The development cycle for intake systems

The initial design and further development of a vehicle intake system takes place in step with the design and development cycle for the whole car.

The key activities are:

- *Concept stage*: Claim packaging volumes.
- *Engine prototype stages*: Develop engine breathing and wave action tuning through manifold design. Minimise system pressure loss. Develop air filtration. Benchmark orifice noise. Develop silencers and resonators.
- *Vehicle prototype stages*: Refine packaging. Test for exterior noise levels.
- *Pre-production vehicles*: Refine interior and exterior noise levels.
- *Pre-production assessment*: Address any problems of noise quality.

3.3.7 Principal intake system components

Filter box	housing for a paper filter element. Has a dirty side and a clean side.
Snorkel	dirty-side pipe feeding fresh air to the filter box.
Zip tube	clean-side tube feeding air into the plenum.
Plenum	collecting volume used to join the manifold runners together.
Runners	join each intake port to the intake plenum. These are also commonly known as primary lengths.

There are many issues that need to be considered when designing air intake systems. The following will be discussed here:

- location of the snorkel orifice;
- snorkel and filter box dimensions;
- intake system design for improved engine performance;
- sources of intake (and exhaust) noise;
- flow duct acoustics;
- acoustic performance of a simple intake system;
- conical snorkels;
- protrusion of the snorkel into the filter box;
- Helmholtz resonators;
- sidebranch (quarter wave) resonators;
- effect of flow on sidebranch resonators;
- intake system mounting.

3.3.8 Location of the snorkel orifice

Issues to consider are:

- Location is often chosen for packaging reasons only.
- A long snorkel will give good low-frequency attenuation but will also result in many dips in the attenuation at the resonant frequencies of the snorkel.

- Acoustic flow duct silencing predictions usually relate to an unflanged snorkel. Fitting the snorkel into a hole in the bodywork will act like a flange and alter the acoustic performance of the system.
- Positioning the snorkel orifice near to reflective surfaces may increase interior or exterior noise levels.
- The intake air needs to be relatively cool to preserve intake charge density.
- Possible water ingestion must be considered.
- A snorkel orifice position that is good for interior noise may not be appropriate for the control of drive-pass noise.
- The snorkel orifice position should not excite acoustic cavity modes, whether in the engine bay, the inner wing of the bodyshell or passenger cell.

3.3.9 Snorkel and filter box dimensions

Consider a snorkel position at the top of the front grille, above the radiator. The filter box is to be mounted on the inner wheel arch. This effectively fixes the snorkel length to, for example, 375 mm.

The snorkel cross-sectional area is often dictated by:

- the requirement for minimum mass flow rate required to preserve engine performance;
- packaging constraints.

The dimensions of the filter box are often dictated by:

- The dimensions of a preferred filter element;
- packaging constraints;
- the need for easy access for exchanging the filter element.

3.3.10 Intake and exhaust system design for improved engine performance

Common engine performance indicators include:

- The power curve being the power (usually brake power as measured at the output shaft) in kW plotted against engine speed for a particular engine load (frequently full load/wide-open throttle). Typical gasoline engines (also known as petrol engines or spark-ignition (SI) engines) of the I-4 (four cylinders, in line with each other down the engine), 1.6–2.0-litre class yield a maximum brake power of perhaps 80–90 kW at an engine speed of perhaps 5000–6000 rev min^{-1}.
- The torque curve being the torque (as measured at the output shaft) in Nm plotted against engine speed for a particular engine load. Typical gasoline engines of the I-4, 1.6–2.0-litre class yield a maximum brake torque of perhaps 150–180 Nm at an engine speed in the range of 2500–4000 rev min^{-1}.
- The brake-specific fuel consumption (bsfc) being the mass of fuel consumed in grams per kWh of power produced at a particular engine speed and load. The expected bsfc of a gasoline engine of the I-4, 1.6–2.0-litre class operating at full engine load is 250–300 g/kwhr.

Indicated mean effective pressure (imep) is given by Bosch (1986) as:

$$\text{imep(bar)} = \frac{1200P_I}{V_H N} \quad \text{for the four-stroke engine} \tag{3.2}$$

$$\text{imep(bar)} = \frac{600P_I}{V_H N} \quad \text{for the two-stroke engine} \tag{3.3}$$

where

P_I = indicated power (kW)
V_H = displacement volume of engine (litres) $= A_p s \times 1000$
A_p = piston area (m^2)
s = stroke (m)
N = revolutions of the engine per minute

Typical values for NA gasoline engines are around the 10 bar mark. Imep is given in Heywood (1988) as:

$$imep\,(\text{kPa}) = \frac{P_I \times n_r \times 10^3}{V_H \times n} = \frac{W_{c,I} \times 10^3}{V_d} \tag{3.4}$$

where

P_I = indicated power (kW)
n_r = number of revolutions per power stroke
V_H = displacement volume of engine (litres)
n = revolutions of the engine per second

The *imep* is the constant pressure which acting on the piston area throughout the stroke would produce the indicated work per cycle $W_{c,I}$.

The indicated work per cycle is given by the area of the pressure–volume (PV) diagram of the engine. Only if the area of compression and expansion stroke parts of the PV diagram are considered, the gross indicated work per cycle is found. If all parts of the PV diagram are used, then the net indicated work per cycle is found.

A summary of the engine requirements for excellent performance is offered here:

Excellent engine performance requires a combination of good airflow and good combustion characteristics.

(Advice given by Tim Drake, the author's colleague whilst at Lotus Engineering.)

This is shown quite clearly in the governing equations (Heywood, 1988)

$$P_I = \frac{\eta_f m_a N Q_{hv} \,(F/A)}{n_r} \tag{3.5}$$

$$T_I = \frac{P_I}{2\pi N} \tag{3.6}$$

where

T_I = indicated engine torque (Nm)
P_I = indicated engine power (W)

m_a = mass of air inducted per cycle (kg)
F/A = fuel–air ratio $\dot{m}_f / \dot{m}_a$
$\dot{m}_f$ = mass flow rate of fuel (kg s^{-1})
$\dot{m}_a$ = mass flow rate of air (kg s^{-1})
N = number of revolutions of the crankshaft per second
n_r = number of revolutions of the crankshaft per power stroke
       (one for two-stroke, two for four-stroke)
Q_{hv} = heating value of fuel, typically 42–44 MJ kg^{-1} ($M = 1 \times 10^6$)
η_f = fuel conversion efficiency

$$\eta_f = \frac{W_c}{m_f Q_{hv}} = \frac{\dfrac{Pn_r}{N}}{\left(\dfrac{\dot{m}_f n_r}{N}\right) Q_{hv}} = \frac{P_I}{\dot{m}_f Q_{hv}} \quad \text{(Heywood, 1988)} \tag{3.7}$$

where

m_f = mass of fuel inducted per cycle (kg)

Now introducing the volumetric efficiency η_v for the four-stroke engine only (where there is a distinct induction stroke) (Heywood, 1988)

$$\eta_v = \frac{2\dot{m}_a}{\rho_{a,i} V_d N} \tag{3.8}$$

where

V_d = volume displaced by piston (m^3)
$\rho_{a,i}$ = inlet air density (kg m^{-3})

The following equations are obtained for four-stroke engines (Heywood, 1988):

$$P_I = \frac{\eta_f \eta_v N V_d Q_{hv} \rho_{a,i} (F/A)}{2} \tag{3.9}$$

$$T_I = \frac{\eta_f \eta_v V_d Q_{hv} \rho_{a,i} (F/A)}{4\pi} \tag{3.10}$$

$$imep = \eta_f \eta_v Q_{hv} \rho_{a,i} (F/A) \tag{3.11}$$

The mechanical efficiency η_m allows the conversion of indicated power to brake power P_b (the useful power as measured by a dynamometer)

$$\eta_m = \frac{P_b}{P_{ig}} = 1 - \frac{P_f}{P_{ig}} \tag{3.12}$$

where P_{ig} is the gross indicated power (found from the gross indicated work per cycle – the area under the PV diagram for compression and expansion strokes only) being the sum of brake power and friction power P_f.

$$P_{ig} = P_b + P_f \tag{3.13}$$

Friction power includes the power used to overcome friction in the engine (the rubbing friction) and in pumping the gas in and out of the engine (the pumping loss). Therefore, the mechanical efficiency is a function of throttle setting and engine speed.

Brake mean effective pressure (bmep) is a useful measure of the torque output of an engine. It will be maximum at the engine speed corresponding to peak torque (Heywood, 1988) and will be 10–15% less at peak power.

Typical maximum bmep are (Heywood, 1988):

- NA gasoline engine   8–11 bar
- Turbo gasoline engine   12–17 bar
- NA diesels   7–9 bar
- Turbo-diesels   10–12 bar

The relationship between bmep and brake torque is (Heywood, 1988):

$$bmep \ (\text{kPa}) = \frac{2\pi n_r T(\text{Nm})}{V_d} \tag{3.14}$$

where V_d is in dm^3.

Note: If V_d were in m^3 then the resulting bmep would be in Pascals.

The bmep is commonly used to assess the torque output or *load* on an engine. If bmep is measured to be less than the known maximum bmep expected from an engine, then the engine must be operating at part load. *Bmep* tends to zero at idle (η_m tends to zero), with the useful work produced during the cycle being used up in overcoming pumping losses (Heywood, 1988).

Note that:

$$imep = bmep + fmep + pmep + amep \tag{3.15}$$

where

$imep$ = indicated mep (bar)
$bmep$ = brake mep (bar)
$fmep$ = rubbing friction mep (bar)
$pmep$ = pumping (or gas exchange) mep (bar)
$amep$ = accessory (alternator, pumps etc.) mep (bar)

So, to use Tim Drake's criteria for excellent performance, the above equations show that the designer needs:

- to maximise η_v and $\rho_{a,i}$ in order to give good airflow;
- to maximise η_f and Q_{hv} along with (F/A) whilst minimising bsfc;

$$bsfc = \frac{\dot{m}_f}{P} \ \text{gJ}^{-1} \ \text{or g/kWh (Heywood, 1988)} \tag{3.16}$$

3.3.10.1 Some basic background terms

Each cylinder of an engine with poppet valves (conventional four-stroke engines) will have the following components:

• Intake system, usually comprising a dirty-side duct (or snorkel), supplying air to a filter (housed in a filter box), which in turn supplies air to a manifold.
• Intake manifold, distributing air (possibly air–fuel mixture) to each cylinder. Different configurations of intake manifold exist, most are variants of end-feed and centre-feed designs (Figures 3.4 and 3.5 respectively).
• The intake manifold has two sections – a plenum being a volume commonly equal to the swept volume of the engine, and primary lengths (or runners) being the ducts that connect the plenum to the cylinder head.
• Intake ports, being the lengths of duct concealed within the cylinder head connecting the manifold with the rear of the intake valves.
• Intake valves, the timing and lift of which are commonly controlled mechanically by a rotating camshaft.
• Exhaust valves, the timing and lift of which are commonly controlled mechanically by a rotating camshaft.
• Exhaust ports.
• Exhaust manifold, combining the exhaust flow from different cylinders into a common (or several common) exhaust tailpipe. The combination of flows may be sudden (four pipes combining as one in the four-into-one manifold) or gradual (two sets of two pipes combine, then the two larger pipes subsequently combine, forming the four-into-two-into-one manifold).

3.3.10.2 Valve lift and timing

Commonly, poppet valves have a maximum lift in the order of 10 mm. This is commonly timed to occur (being known as the maximum opening point – MOP) at around 100–110 ATDC (after top dead centre) for the intake valves and 110–120 ABDC (after bottom dead centre) for the exhaust valves.

The period during which a valve is open (often called the cam duration and measured in degrees of revolution of the crankshaft) is commonly measured between the top of

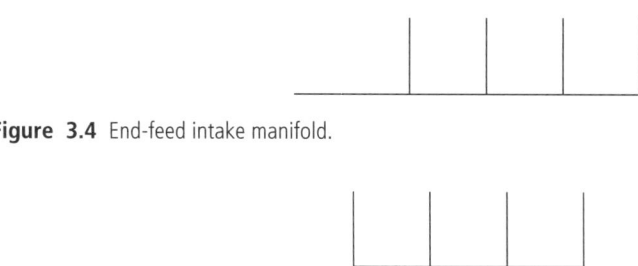

Figure 3.4 End-feed intake manifold.

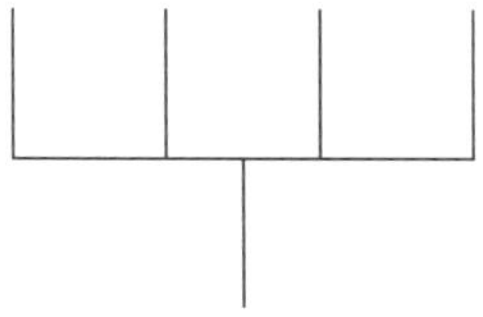

Figure 3.5 Centre-feed intake manifold.

ramp (TOR) of the valve rise and the TOR of the valve fall. The TOR marks a position on the cam profile where there is a transition between a gradual rise and a steeper rise. These ramps are used to smooth the path of the cam follower which rides on the cam and provides the mechanical link between the cam and the valve. The TOR is often used as a basis for valve timings. The four valve timings are:

1. IVO (inlet valve open);
2. IVC (inlet valve close);
3. EVO (exhaust valve open);
4. EVC (exhaust valve close).

Cam durations are commonly in the range of 220°–250° of crankshaft rotation. The intake duration may be different from that of the exhaust. With longer durations, it is possible that EVO will occur before IVC (so both intake and exhaust valves will be open) and a period of valve overlap occurs. This is shown in Figure 3.6.

Valve overlap is often quoted as an area (mmdeg) and is generally less than 5 mmdeg. Peak engine power increases with increasing valve overlap. However, high overlap causes combustion instability (exhaust gases drawn back into the cylinder) particularly at low engine speeds and this is important for stable idle. Combustion instability manifests itself as a cylinder-to-cylinder variation in IMEP. Low valve overlap is beneficial for low-speed torque. Twenty degree overlap is a typical compromise for producting gasoline engines.

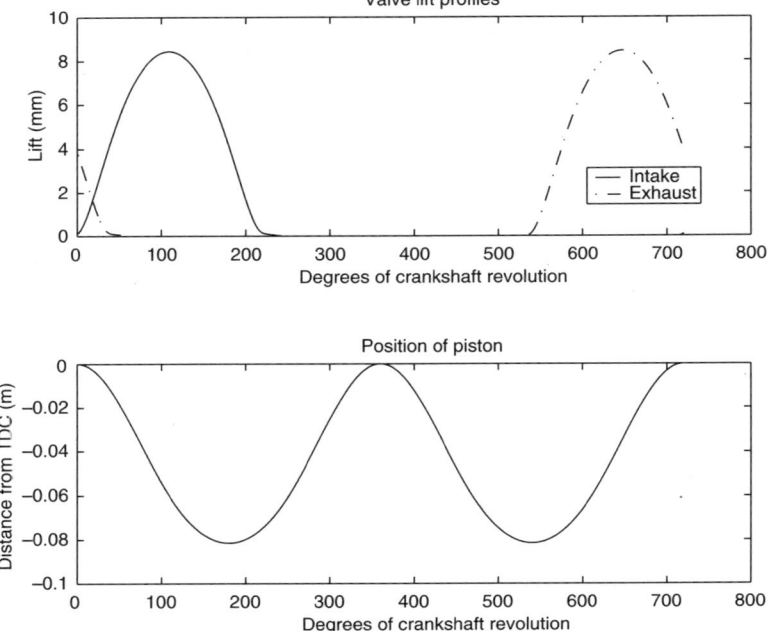

Figure 3.6 Valve lift profiles and the instantaneous position of the piston.

3.3.10.3 On the flow through the intake valve

Knowledge of the gas dynamics at the intake valve is useful to the refinement engineer. A non-conservative form of the conservation of momentum equation for a fluid in three dimensions is (Hirsch, 1988)

$$\rho \frac{d\bar{u}}{dt} = -\nabla p \bar{\bar{I}} + \nabla \cdot \bar{\bar{\tau}} + \rho \bar{f}_e \tag{3.17}$$

where

$$\text{grad } p = \nabla p = \frac{\partial p}{\partial x}\bar{i} + \frac{\partial p}{\partial y}\bar{j} + \frac{\partial p}{\partial z}\bar{k}$$

$\bar{i}, \bar{j}, \bar{k}$ are unit vectors
$\bar{\bar{I}}$ is the unit tensor
$\bar{\bar{\tau}}$ is the viscous shear stress tensor
$\bar{f}_e$ is the external force vector
p is pressure
ρ is density
$\bar{u}$ is the velocity vector

The total inertial term on the left-hand side of equation (3.17) can be re-written as the sum of linear, kinetic and rotational forces

$$\rho \frac{d\bar{u}}{dt} = \rho \left[\frac{\partial \bar{u}}{\partial t} + \nabla \left(\frac{u^2}{2} \right) - \left(\bar{u} \times \bar{\xi} \right) \right] \tag{3.18}$$

where $\bar{\xi}$ is known as the vorticity vector.

Simplifying the analysis to consider only one dimension x, and neglecting viscosity effects, external forces and vorticity, equation (3.17) reduces to the familiar non-linear inviscid Euler equation (see Appendix 4G for a derivation)

$$\frac{\partial u}{\partial t} + \frac{\partial}{\partial x}\left(\frac{u^2}{2} \right) = -\frac{1}{\rho}\frac{\partial p}{\partial x} \tag{3.19}$$

From the second law of thermodynamics (for instance Zemansky and Dittman (1997)):

$$T ds = dh - v dp \tag{3.20}$$

or

$$T\frac{ds}{dx} = \frac{dh}{dx} - v\frac{dp}{dx} \tag{3.21}$$

where T is the temperature, s is specific entropy, h is specific enthalpy and v is specific volume. This can be re-written as

$$T\frac{ds}{dx} = \frac{dh}{dx} - \frac{1}{\rho}\frac{dp}{dx} \tag{3.22}$$

Substituting equation (3.22) into equation (3.19) gives

$$\frac{\partial u}{\partial t} + \frac{\partial}{\partial x}\left(\frac{u^2}{2}\right) = T\mathrm{d}s - \mathrm{d}h \tag{3.23}$$

Now the total (or stagnation) enthalpy H is given as

$$H = h + \frac{u^2}{2} \tag{3.24}$$

and substituting the differential of equation (3.24) into equation (3.23)

$$\frac{\partial u}{\partial t} = T\mathrm{d}s - \mathrm{d}H \tag{3.25}$$

Now for the assumption of homentropic flow, equation (3.25) reduces to

$$\frac{\partial u}{\partial t} + \mathrm{d}H = 0 \tag{3.26}$$

Now along a streamline the stagnation enthalpy is constant, so

$$\frac{\partial u}{\partial t} + H = H_0 = \text{constant} \tag{3.27}$$

A scalar potential function ϕ can be declared so that

$$u = \overline{\nabla}\phi \tag{3.28}$$

$$u = \frac{\partial \phi}{\partial x}\text{in a 1-D model} \tag{3.29}$$

and thereby create what is known as a potential flow model given by

$$\frac{\partial \phi}{\partial t} + H = H_0 = \text{constant} \tag{3.30}$$

For an ideal gas, where a is the speed of sound $(\mathrm{m\,s}^{-1})$ and γ is the ratio of specific heats $c_\mathrm{p}/c_\mathrm{v}$

$$h = \frac{a^2}{\gamma - 1} \tag{3.31}$$

and so using equation (3.24), equation (3.30) can be written for flow along a streamline (or a Fanno line) as

$$\frac{\partial \phi}{\partial t} + \frac{a^2}{\gamma - 1} + \frac{u^2}{2} = \frac{a_0^2}{\gamma - 1} \tag{3.32}$$

Now as

$$\frac{\partial \phi}{\partial t} = \frac{\partial^2 u}{\partial x \partial t} \tag{3.33}$$

most workers omit that term so that the 1-D, non-conservative, inviscid, irrotational, homentropic momentum equation, along a streamline (!) becomes

$$a_0^2 = a^2 + \frac{\gamma - 1}{2} u^2 \tag{3.34}$$

Benson (1982) famously calls this an energy equation (although this is probably a misnomer).

From the derivation of equation (3.34) a simple and very well-known intake valve flow model can be constructed that assumes a flow from a large reservoir (representing the manifold) into one of the cylinders via a single orifice of negligible length. This was first derived by Tsu (1947). Stagnation conditions (subscript 0,1 meaning zero flow velocity in zone 1) are assumed in the cylinder.

For inflow to the cylinder via the intake valve, the intake manifold is assumed to constitute a sufficiently large volume for constant pressure conditions to occur, and the conditions in the manifold are given subscript '2'. For outflow (reverse flow through the intake valve) the subscripts are reversed.

From a principle of continuity of mass, assuming quasi steady flow it can be written that:

$$\dot{m} = \rho_2 u_2 A_m \tag{3.35}$$

where

$\dot{m}$ = mass flow rate (kg s^{-1})
A_m = open flow area of the valve (given in Appendix 3A at the end of this chapter, after [Heywood 1988])

The following isentropic relationships apply for an ideal gas

$$\rho_2 = \rho_1 \left(\frac{p_2}{p_1} \right)^{1/\gamma} \tag{3.36}$$

$$T_2 = T_1 \left(\frac{p_2}{p_1} \right)^{\gamma - 1/\gamma} \tag{3.37}$$

Now also for an ideal gas, where R is the specific gas constant

$$p = \rho RT \tag{3.38}$$

and

$$a = \sqrt{\gamma RT} \tag{3.39}$$

so

$$\rho = \frac{\gamma P}{a^2} \tag{3.40}$$

Substituting equation (3.40) into equation (3.36) and thus assuming isentropic expansion of the gas as it enters the cylinder

$$\rho_2 = \frac{\gamma p_{01}}{a_{01}^2} \left(\frac{p_2}{p_{01}} \right)^{1/\gamma} \tag{3.41}$$

Re-writing equation (3.34) with the appropriate subscripts

$$a_{01}^2 = a_2^2 + \frac{\gamma - 1}{2} u_2^2 \tag{3.42}$$

Re-arranging equation (3.42)

$$u_2^2 = \left(a_{01}^2 - a_2^2\right) \frac{2}{\gamma - 1} \tag{3.43}$$

Substituting equation (3.39) into equation (3.37) gives, with the appropriate subscripts,

$$T_2 = \frac{a_{01}^2}{\gamma R} \left(\frac{p_2}{p_{01}}\right)^{\gamma - 1/\gamma} \tag{3.44}$$

Use of equation (3.39) once more to replace T_2 yields

$$a_2^2 = a_{01}^2 \left(\frac{p_2}{p_{01}}\right)^{\gamma - 1/\gamma} \tag{3.45}$$

Substituting equation (3.45) into equation (3.43)

$$u_2^2 = \left(a_{01}^2 - a_{01}^2 \left(\frac{p_2}{p_{01}}\right)^{\gamma - 1/\gamma}\right) \frac{2}{\gamma - 1}$$

$$u_2^2 = \frac{2a_{01}^2}{\gamma - 1} \left(1 - \left(\frac{p_2}{p_{01}}\right)^{\gamma - 1/\gamma}\right) \tag{3.46}$$

Now substituting both equation (3.46) and equation (3.41) into (3.35) gives

$$\dot{m} = \frac{\gamma p_{01}}{a_{01}^2} \left(\frac{p_2}{p_{01}}\right)^{1/\gamma} \left[\frac{2a_{01}^2}{\gamma - 1}\right]^{1/2} \left[1 - \left(\frac{p_2}{p_{01}}\right)^{\gamma - 1/\gamma}\right]^{1/2} A_m$$

$$\dot{m} = \frac{p_{01} A_m}{a_{01}} \left[\left(\frac{2\gamma^2}{\gamma - 1}\right) \left(\frac{p_2}{p_{01}}\right)^{2/\gamma} \left[1 - \left(\frac{p_2}{p_{01}}\right)^{\gamma - 1/\gamma}\right]\right]^{1/2} \tag{3.47}$$

Equation (3.47) presents the mass flow rate as a function of the open valve area and the pressure ratio across the valve for subsonic flow through the orifice (valve). However, equation (3.47) will tend to predict much higher flow rates than are encountered with practical engines, due to the large number of simplifying assumptions that went into its derivation. One way to correct this effect is to introduce a discharge coefficient c_d where

$$c_d = \frac{A_e}{A_r} \tag{3.48}$$

and A_e is the effective area and A_r is some suitable reference area. The effective area (Annand and Roe, 1974) is the outlet area of an imaginary frictionless nozzle which would pass the required flow when drawing from a large constant pressure reservoir and discharging into another reservoir. The reference area can be the cross-sectional area of any suitable part of the real flow path such as the curtain area under the open valve.

Measured discharge coefficients can be used to calculate the effective area for a given reference area, and hence equation (3.47) becomes

$$\dot{m} = \frac{p_{01} A_e}{a_{01}} \left[\left(\frac{2\gamma^2}{\gamma - 1} \right) \left(\frac{p_2}{p_{01}} \right)^{2/\gamma} \left[1 - \left(\frac{p_2}{p_{01}} \right)^{\gamma - 1/\gamma} \right] \right]^{1/2} \tag{3.49}$$

The most well-known discharge coefficients for inflow through the intake valve are given in Annand and Roe (1974) for a reference area equal to the curtain area under the open valve

$$A_r = \pi D L_v \tag{3.50}$$

where D is the valve head diameter and L_v is the valve lift. These discharge coefficients are reproduced in Figure 3.7

At low valve lifts, the flow past the valve remains attached to both the valve head and the valve seat. At intermediate valve lifts, the flow is separated on one side but not on the other, producing a sudden drop in discharge coefficient that subsequently recovers with further valve lift. At high valve lifts, the flow is detached from both sides, and the so-called free-jet is formed.

The discharge coefficients for outflow through the inlet valve (reverse flow) are generally higher (around 0.7 up to $L_v/D = 0.2$, then falling to 0.5 at $L_v/D = 0.4$).

Flow loss coefficients (rather than discharge coefficients) are commonly used in commercial engine simulations (AVL, 2000). These are defined as the ratio between the actual mass flow and the loss-free isentropic mass flow for the same stagnation pressure and the same pressure ratio (AVL, 2000). The difference between a discharge coefficient and a flow loss coefficient is important. The discharge coefficient applies to flow between stagnant reservoirs passing through a frictionless nozzle. The flow loss coefficient applies to steady or pulsating flow through the cylinder head.

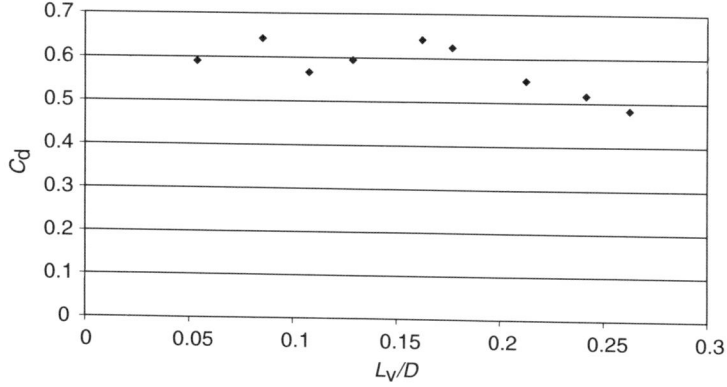

Figure 3.7 Discharge coefficients with respect to reference area given by equation (3.50) for the case of inflow through the intake valve (Annand and Roe, 1974).

Flow loss coefficients (such as those shown in Figures 3.8 and 3.9) are often measured using steady flow on a bench (Blair and Drouin, 1996). Sometimes, they are measured using pulsed flow (for instance Fukutani and Watanabe (1982)) in order to improve the realism of the model represented by equation (3.49).

An example for an intake port is given in Figure 3.8 and an example for an exhaust port in Figure 3.9.

For the flow loss coefficients shown in Figures 3.8 and 3.9

$$A_e = \text{coefficient} \times \frac{d_{vi}^2 \times \pi}{4} \tag{3.51}$$

where d_{vi} is the inner valve seat diameter (reference diameter corresponding to D in Figures 3.8 and 3.9).

Equation (3.49) reduces to

$$\dot{m} = \frac{A_e \gamma p_{01}}{a_{01}} \left(\frac{2}{\gamma+1}\right)^{\gamma+1/2(\gamma-1)} \tag{3.52}$$

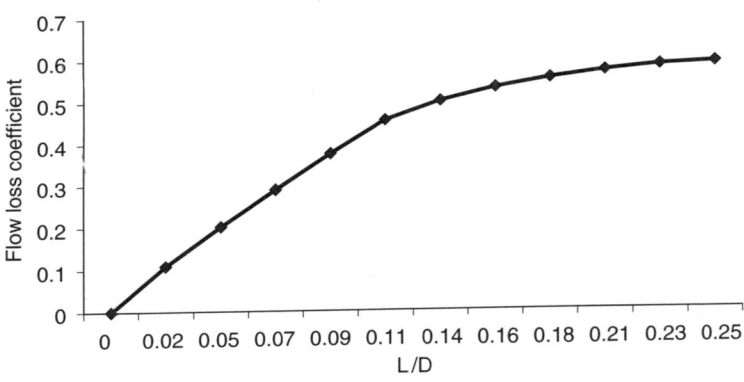

Figure 3.8 Intake port flow loss coefficient (AVL, 2000).

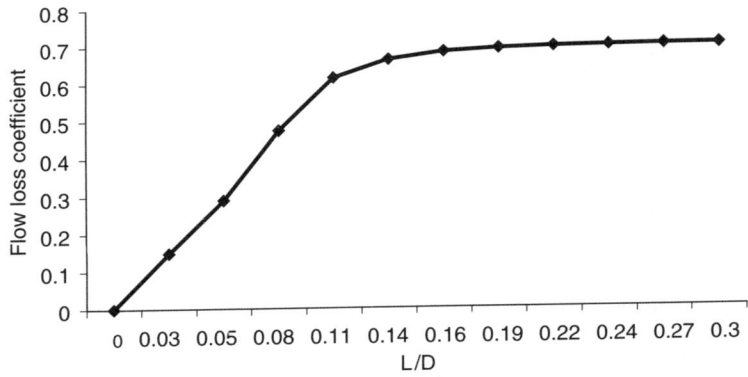

Figure 3.9 Exhaust port flow loss coefficient (AVL, 2000).

for sonic flow through the orifice (valve) which is also known as choked flow as this represents the maximum rate of mass transfer possible through a single orifice of negligible length. Note that this is no longer a function of pressure ratio across the valve, but is solely a function of the conditions in the manifold and of the effective open area of the valve. The sonic condition is reached when

$$\frac{p_{01}}{p_2} \leq \left(\frac{2}{\gamma+1}\right)^{\gamma/\gamma-1} \tag{3.53}$$

which is a pressure ratio of around 0.53.

3.3.10.4 A note on intake manifold design

The volumetric efficiency of a four-stroke engine can be improved through dynamic tuning whereby it is contrived to maximise the pressure in the intake port at around IVC (Ohata and Ishita, 1982). This of course maximises the pressure ratio in equation (3.49) and hence the mass flow rate, and hence the volumetric efficiency (see equation (3.8)).

Achieving the dynamic tuning effect is strongly dependent on intake manifold design, and in particular the length of the primary runners (Winterbone and Pearson, 1999). Generally, longer runner lengths are used to maximise low-speed torque, but often with a penalty of reducing high-speed power.

There is a connection between ignition timing and runner length. Fifteen degree changes in advance may be required at a given speed when changing runner length (Harrison, 2003).

In addition to optimising spark advance, one really needs to optimise the valve timing as well. IVC is important but good intake manifold design is more important. A good cam design cannot make up for a poor manifold.

The balance in flows between runners is important. With one restrictive intake runner, that cylinder will tend to run rich at high engine speeds. As a result, the other cylinders will run lean. Intake port velocity is typically 70–90 m s^{-1}.

3.3.10.5 The motion of the piston during the intake stroke

Earlier discussions show that volumetric efficiency is strongly influenced by the pressure ratio across the inlet valves at IVC. Consider what is happening in the cylinder around IVC. The position of the piston relative to its TDC position can be calculated using equation (6.170) in Section 6.4.3.

Consider one cylinder of an engine with the following specification:

- 1.6 litre I4 four-stroke;
- stroke 81.5 mm;
- compression ratio 10.0;
- conrod length = 129 mm;
- piston pin offset = 1.0 mm.

From this data, and using equation (6.170) it is easy to calculate the instantaneous volume in that cylinder as it changes with the rotation of the crankshaft. The results of such a calculation are shown in Figure 3.10 where the volume has been normalised by the maximum volume (swept volume + clearance volume) and expressed as a percentage.

Inspecting Figure 3.10 then at TDC, the volume is 10% of the maximum volume. This of course corresponds to the clearance volume in an engine with a compression ratio of 10. At the intake stroke BDC (540°) the volume is at its maximum (100%). The swept volume is the difference between the volume at BDC and the volume at TDC.

Inspection of the data used to generate Figure 3.10 reveals that for this typical case, the cylinder volume is in the range of 99–100% of the maximum for 28° of crankshaft rotation in the range of 526–554°.

The volume lies in the range of 98–100% for the 40° of crankshaft rotation between 520° and 560°. So, for the 30–40° around BDC the volume in the cylinder is changing by less than 1–2%.

Typically, MOP for the intake valve is 100–110° ATDC (see Section 3.3.10.2). Take an intake cam with MOP at 102° ATDC (462°) and 230° duration. That places IVC at 37° ABDC (577°). For the case shown in Figure 3.10 the cylinder volume at 577° is 93.6% of the maximum value at BDC. By the time intake MOP is reached, the cylinder volume is 70% of its maximum value at the BDC position.

A very short duration camshaft (MOP 102° duration 184°) will position IVC at a point when the cylinder volume is >99° of that at BDC. A MOP 102° duration 196° intake camshaft will position IVC at a point when the cylinder volume is >98% of that at BDC. With more conventional camshafts (duration >220°) the point when the cylinder volume is >98% of its maximum will occur 20–50° before IVC.

It is the pressure ratio across the valve during this part of the intake lift profile 20–50° before IVC, that most strongly influences volumetric efficiency. Winterbone and Pearson (1999) suggest that a significant period is some 50° before IVC.

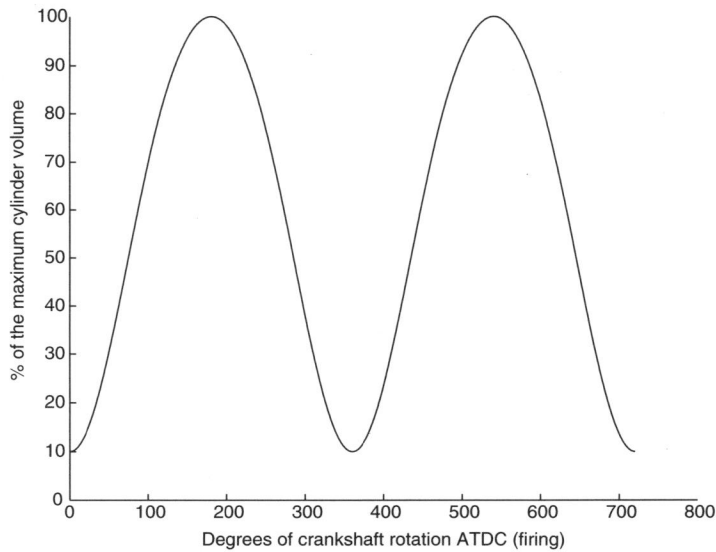

Figure 3.10 Time history of the cylinder volume.

Now the piston velocity around IVC will be calculated. This is easily done by dividing the distance that the piston travels in 1° crank by the time period for 1° crank. Figure 3.11 is the result for 4000 rev min^{-1} using the same input data as for Figure 3.10.

At BDC (540°) the piston velocity is zero; therefore there is a 90° phase lag relative to displacement. The acceleration will be at its maximum value at BDC (180° phase lag relative to displacement). At 577° the piston velocity in the case shown in Figure 3.11 is 7.6 ms^{-1} compared with a maximum of 17.9 ms^{-1}.

The maximum piston velocity in this case occurs at 646° (at this point the piston is slightly more than halfway back to TDC – 106/180 through the compression stroke) and also at 434° (the piston is 74/180 through the intake stroke).

The mean piston speed for this case is

$$2 \times \frac{\text{rev min}^{-1}}{60} \times \text{stroke} = 10.9 \text{ m s}^{-1}$$

In Section 3.3.12.12 it is shown that a typical gasoline engine producing 50 kW consumes of the order of 0.068 m^3 of air per second. If the engine here has 8 intake ports of diameter 30 mm then the average flow velocity through the ports is 12.0 m s^{-1} averaged over 720°. However, there is flow through the port for only say 230° of the 720° so the average velocity in the port is more likely to be 37.6 m s^{-1}. Peak flow velocities are of the order of 70–90 m s^{-1} for the case of a single intake valve per cylinder.

Note that there are several algorithms to describe the motion of the piston:

- Equation (6.170) in Section 6.4.3;
- Bosch (1986);
- Heywood (1988).

The results are subtly different for each one.

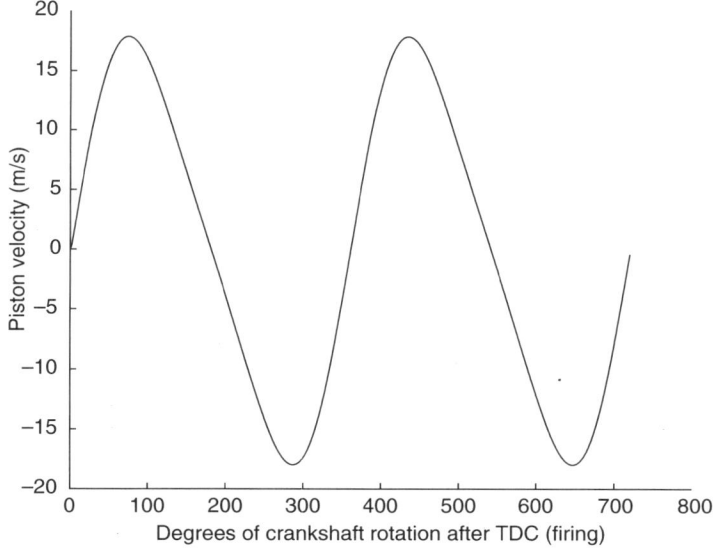

Figure 3.11 Piston velocity, 4000 rev min^{-1}.

3.3.10.6 *The pressure waves in the intake system*

How can the volumetric efficiency be maximised by increasing the pressure in the intake port at around IVC? Figure 3.12 shows the intake port pressure trace for a single cylinder racing engine fitted with a simple intake pipe which is a conical pipe some 150 mm long (the intake primary). The engine is at 3/4 of its rated speed, full load (wide-open throttle) and it is running 'on tune' (Dunkley, 1999).

In Figure 3.12:

- 0° corresponds to ignition TDC.
- IVO occurs a little before 360°
- IVC occurs around 600° (>40° ABDC).

There are three distinct features on Figure 3.12:

1. A period of pressure depression after IVO and lasting some 120° until the intake MOP.
2. A clear pressure peak around IVC. In this case the wave action in the port has a pressure amplitude of 0.6 bar (60 kPa, 190 dB re 20 μPa) at around IVC.
3. After IVC the pressure decays with (in this case) three successive oscillations before the intake valve opens once more.

The origin of the pressure peak at IVC can be explained thus:

- Just after IVO (350°) the intake valve opens, the piston accelerates downwards and air is drawn into the cylinder. If the intake primary length is sufficiently long (as it is in this case) this causes a pressure depression in the port just behind the valve.
- The pressure depression travels away from the valve at the speed of sound. It is reflected at the open end of the intake primary (with a phase shift – 180° for an

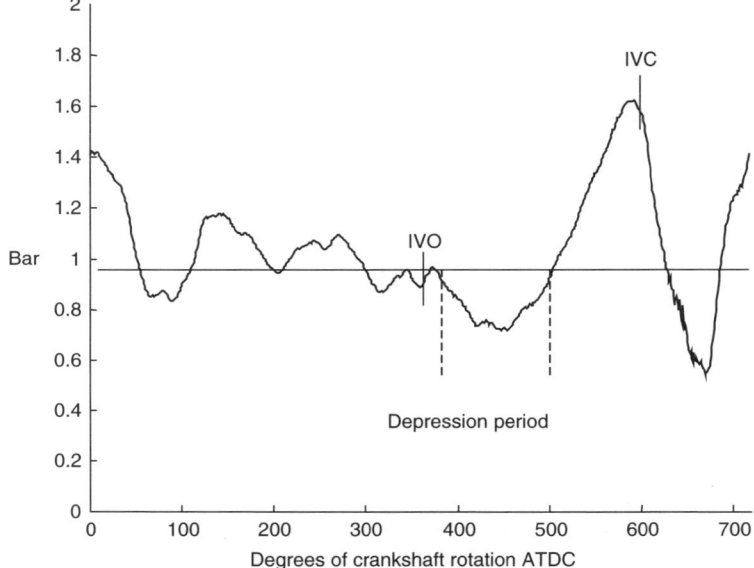

Figure 3.12 Intake port pressure trace, 3/4 rated speed, full load (Dunkley, 1999).

unflanged open pipe) and the reflected pressure pulse travels back towards the valve as a positive pressure (pressure peak). The 180° phase shift is a result of the open pipe end being a pressure release surface. The net dynamic pressure at the open end must be equal to the prevailing static pressure just beyond the pipe end. The only way to produce this net cancellation of the dynamic pressure is to have two waves travelling in opposite directions, in anti-phase, thus cancelling each other out.

- By choosing a suitable primary length (and the dimensions of the rest of the intake system if fitted) one can phase the pressure peak to arrive back at the valve at around IVC. This is known to be happening in the case of Figure 3.12.
- In the case shown in Figure 3.12, the arrival of a positive pulse at the valve might account for 20 kPa of dynamic pressure at IVC but one can note a level of 60 kPa. The additional 40 kPa is the result of the so-called inertia effects or ram. During the pressure depression after IVO the flow through the valve builds up a certain momentum. As the valve closes towards IVC this momentum is transferred to static pressure as the air is brought to a halt. Providing that the acceleration of the air after IVO was efficient (a function of timing the IVO) then the ram effect produces a net gain in volumetric efficiency (usually only at high speeds) (Harrison and Dunkley, 2004).

Note that in the exhaust a similar phenomenon occurs except that a positive pressure peak is generated at the exhaust valve which is subsequently reflected back as a pressure depression that aids scavenging.

Consider Figure 3.13 where the same engine is operating at a much lower speed, full load and 'off tune' (Dunkley, 1999). Comparing this pressure trace with that of the 'on tune' operating condition in Figure 3.12:

- When, 'off tune' (in this case), the pressure depression period is much shorter than when 'on tune'. In this case this is due to flow reversal in the intake valve caused by the particular timing of IVO.
- There is a pressure depression (in this case) at IVC which will reduce volumetric efficiency. The magnitude of this depression is 1.5 bar, which is much less than the peak magnitude seen in Figure 3.12.

So, in summary, inspection of Figures 3.12 and 3.13 has identified two physical mechanisms by which the intake port pressure may be altered around IVC:

1. *Wave effects*   whereby a travelling pressure pulse is timed to be positive at the valve near IVC at the 'on tune' engine speed. Away from this speed, a pressure depression might occur at IVC thus reducing volumetric efficiency.
2. *Inertia effects*   where by flow momentum is transformed to pressure around IVC (usually most significant at higher engine speeds).

Group demonstration – acoustic resonance in a pipe

A loudspeaker, a signal generator, an amplifier, a long piece of pipe and a sound level meter are used in this demonstration. Acoustic resonances in the pipe appear as peaks in the noise level at the end of the pipe. These resonances are similar to those found in a straight-pipe intake system when the valve is closed that produce the oscillations seen in Figure 3.12.

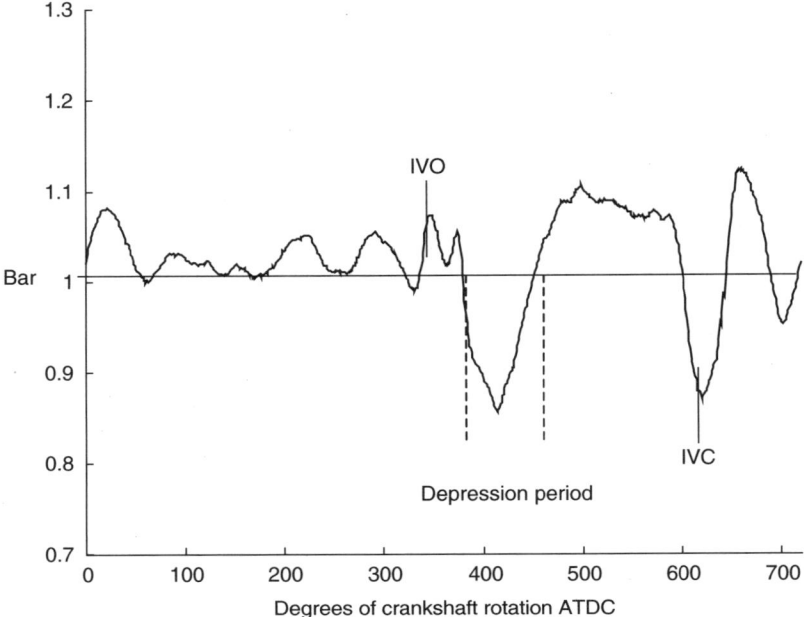

Figure 3.13 Intake port pressure trace, 35% rated speed, full load (Dunkley, 1999).

In practical production engines, there will be more than one cylinder. A different intake port pressure trace results from the wave effects due to the filling of adjacent cylinders. Figure 3.14 shows such a trace for a four cylinder engine.

Ohata and Ishida (1982) measured the intake pressure trace in PORT # 2 of a four-cylinder engine with only one cylinder operating at any given time. When they summed the four pressure traces they found that the sum matched the pressure trace recorded in PORT # 2 when all four-cylinders were activated.

The Ohata and Ishida experiment supports the notion that the wave action in the intake system is linear as the additivity part of the theory of superposition seems to hold (see Chapter 4, Appendix 4A).

3.3.10.7 *On turbo and supercharging the engine to improve performance*

Torque is a useful description of engine performance (see Section 3.3.10). Low-speed torque strongly affects driveability. High-speed power affects maximum speed and high-speed torque affects elasticity in the gear selection.

Torque and power are simply related by engine speed N (rev/s)

$$T = \frac{P}{2\pi N}$$

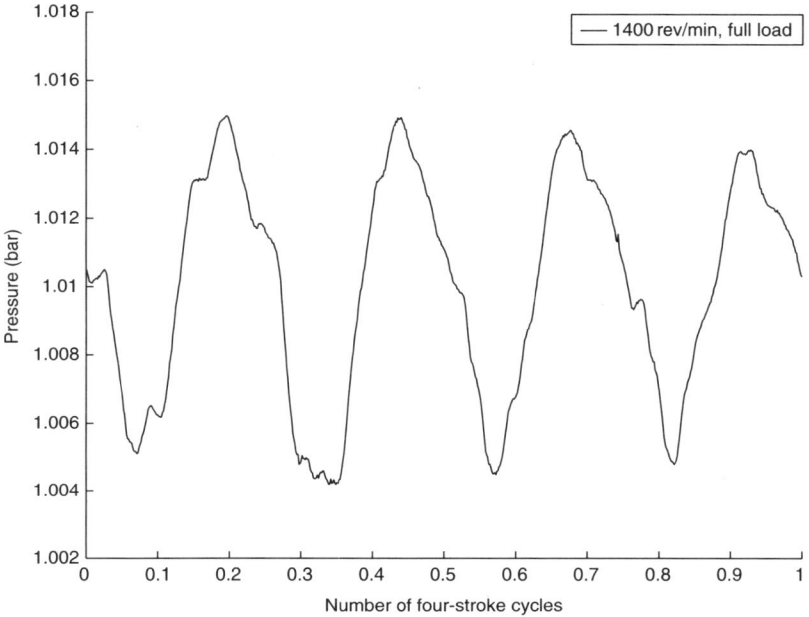

Figure 3.14 Intake pressure trace for a four-cylinder engine.

The governing torque equation is

$$T_{\mathrm{I}} = \frac{\eta_{\mathrm{f}}\,\eta_{\mathrm{v}}\,V_{\mathrm{d}}Q_{\mathrm{hv}}\rho_{\mathrm{a,i}}(F/A)}{4\pi}$$

where

T_{I} = indicated engine torque (Nm)
F/A = fuel air ratio $\dot{m}_{\mathrm{f}}/\dot{m}_{\mathrm{a}}$
Q_{hv} = heating value of fuel, typically 42–44 MJ kg^{-1} (M = 1×10^6)
η_{V} = volumetric efficiency
η_{f} = fuel conversion efficiency
V_{d} = swept volume (m^3)

Assuming those variables connected with fuelling and combustion (η_{f}, Q_{HV}, F/A) are to be kept constant, inspection of the torque equation reveals the following mechanisms for increasing torque:

- increase η_{v};
- increase V_{d};
- increase $\rho_{\mathrm{a,i}}$.

Increasing engine capacity has always been an option for increasing torque (witness the massive aero engines fitted to early twentieth-century racing cars). More recently, design skills have been applied with modern multi-point fuel injection technologies to raise η_{v} by careful tuning of the intake and exhaust manifolds and appropriate valve lift timing (and multi-valve engines).

Raising torque by increasing the intake air density is the subject here. It requires additional work on the intake gas beyond the pumping work found in the NA engine. This work may be supplied by either:

- *Supercharging*   where a mechanical drive from the engine powers a positive displacement device (a sweep or a pump basically) or
- *Turbocharging*   where a turbine runs off the engine's exhaust gas and is used to power a compressor (a high-speed impeller).

Bmep is a useful measure of torque output. Typical maximum bmep are:

NA gasoline	8–11 bar
Turbo gasoline	12–17 bar
NA diesel	7–9 bar
Turbo diesel	10–12 bar

Based on these data, it appears that the turbocharger is being used to raise inlet air densities by a factor of ≈ 1.5 in the gasoline engine and by $\approx 1.3–1.4$ in the diesel engine.

3.3.10.8 The rationale for turbocharging

Turbochargers are now routinely fitted to diesel engines. The reasons for this are:

- Specific output (power/swept volume) of the NA diesel is poor, leading to large, heavy expensive engines (due to the need to run lean to avoid excessive smoke (Heywood, 1988)).
- Power output of the diesel is smoke-limited. This restriction is relaxed if more air mass is added to a given cylinder size by turbocharging.
- There are no knock problems to overcome even when turbocharging diesel engines (unlike gasoline engines).
- Diesel engines are more costly, allowing the additional costs of a turbocharger to be absorbed (Watson and Janota, 1982).

The best-known reference work on turbocharging (Watson and Janota, 1982) is mostly devoted to diesel engines. The rationale for turbocharging the gasoline engine is, to be honest, rather less compelling than for the diesel engine.

If the designer needs/wants to boost the torque output from a restricted swept volume then they might choose to turbocharge. However, there are two limits to this. The first is that the increased air density at inlet, when combined with the correct ratio of fuel will produce higher cylinder pressures and temperatures thus increasing the tendency to knock (Heywood, 1988) – this can be controlled by reducing the compression ratio and accepting the loss of thermal efficiency caused. Even with the loss of efficiency, the work achieved per cycle is greater in the turbocharged engine than achieved by the NA engine because more fuel can be burned whilst preserving the desired air–fuel ratio. The second limitation to turbocharging the gasoline engine is speed. The turbocharger is ideally matched at one engine operating point (speed/load combination) and achieving adequate performance at other operating points is difficult.

The turbocharged diesel engine may be more fuel efficient than its NA counterpart (unlikely to be the case in gasoline engines). This is because the turbocharged engine produces more torque at a given engine speed (Watson and Janota, 1982). As friction losses in the engine are speed-dependent, these form a smaller proportion of the turbocharged IMEP than that of the NA IMEP. Thus the overall efficiency of the turbocharged engine is greater.

Watson and Janota have an interesting perspective on turbocharging (Watson and Janota, 1982).

> Turbocharging is a specific method of supercharging. An attempt is made to use the energy of the hot exhaust gas of the engine to drive the supercharging compressor. The user is not getting something for nothing, but is merely using energy that would normally go to waste; however it is clear that it is no longer necessary to debit the power requirement of the compressor from the indicated power of the engine.

3.3.10.9 On turbocharger noise

The addition of a turbocharger to an engine is known to reduce the levels of intake noise. The silencing mechanism is believed to be simple with the compressor housing behaving as a small, reactive silencer element. The presence of the spinning rotor will have little effect except to add some flow-induced noise. Reductions in narrow-band intake noise of the order of 3–8 dB should be expected depending on the frequency content of the intake noise.

3.3.11 Sources of intake (and exhaust) noise

The noise due to the operation of the intake and exhaust systems can be classified as follows:

Primary noise sources   being the unsteady mass flow through the valves, which causes pressure fluctuations in the manifold and these propagate to the intake orifice (or exhaust tailpipe) and are radiated as noise. It should be noted that the mechanism that causes primary (or engine breathing) noise is the same mechanism that is usefully harnessed to improve the volumetric efficiency of the NA engine by wave action tuning (see Section 3.3.10.6).

Harrison and Stanev (2004) propose the following hypothesis to explain the fluctuating pressure time history found in the intake manifold:

> The early stages of the intake process are governed by the instantaneous values of the piston velocity and the open area under the valve. Thereafter, resonant wave action dominates the process. The depth of the early depression caused by the moving piston governs the intensity of the wave action that follows. A pressure ratio across the valve that is favourable to inflow is maintained and maximised when the open period of the valve is such to allow at least, but no more than, one complete oscillation of the pressure at its resonant frequency to occur whilst the valve is open.

Harrison and Dunkley (2004) also identified the role for intake flow momentum in the breathing performance of higher-speed engines.

Secondary noise sources   being noise created by the motion of the flow through the intake and exhaust systems. This self-induced noise is commonly called flow noise.

Shell noise   being the structure-radiated sound from the intake or exhaust tailpipes as excited by either primary or secondary noise sources.

The first two classifications are the subject of this section and are illustrated in Figure 3.15.

When analysing the sound recorded at the intake orifice (or exhaust tailpipe), it is notoriously difficult to distinguish between sound that is due to primary noise sources and that due to secondary noise sources. There is a common misconception that whilst primary (engine breathing) noise is tonal and dominated by low-frequency components of the fundamental cycle frequency (which is true), flow noise is chaotic, broadband and mostly of high frequency (which is not true). In reality, flow noise in the intake or exhaust system is likely to be tonal, with both low- and high-frequency components.

The fact that engine breathing noise and flow noise can be confused and misdiagnosed is hardly surprising as they both result in similar effects: a level of sound power in the duct that propagates away from some source towards the open end of the pipe and along the way that sound power flux can be filtered, attenuated or amplified.

Both engine-breathing noise and flow noise are classes of aerodynamic noise. Before concentrating on the differences between the two, it is useful to briefly consider how they are similar. Howe (1975) provides the most convenient framework for this. Howe offered a contribution to the theory of sound generation by flow turbulence and vorticity by using the specific stagnation enthalpy as the fluctuating acoustic quantity rather than vorticity (Powell, 1964) or fluctuating momentum (Lighthill, 1952; 1954) as had been used before. This is particularly convenient for the case of IC engine intake and exhaust noise as the engine-breathing process is well modelled by the component of enthalpy that is associated with inviscid irrotational flow (see Section 3.3.10.3). By contrast,

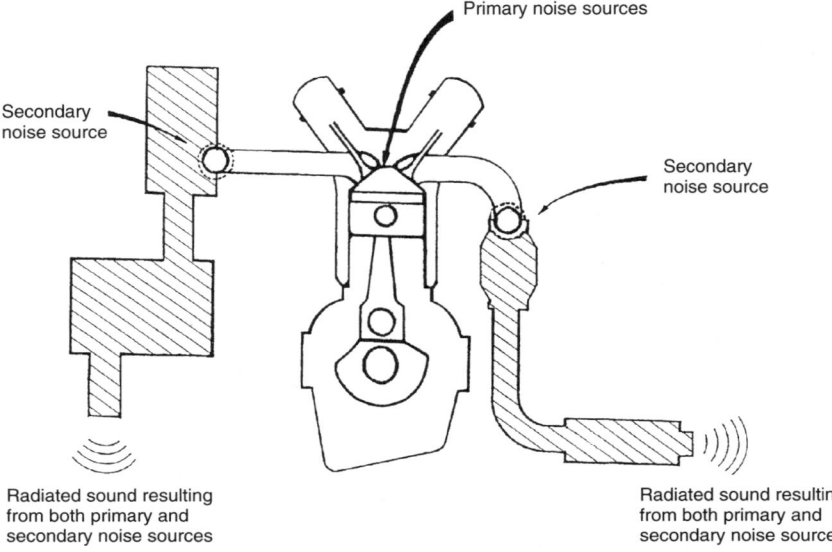

Figure 3.15 Primary and secondary sources of intake and exhaust noise.

the flow noise can be considered as having its origin in the component of enthalpy associated with:

- rotational flow;
- irreversibilities in the flow;
- viscous forces;
- local heat input or loss.

Intuitively the formation of vortices in the flow are the most likely manifestation of this component of enthalpy. These might be formed:

- at sudden expansions of the flow such as at the entrance to reactive silencer chambers;
- at points of flow separation such as at tight radius bends;
- at free jets such as at the exhaust tailpipe.

Although in one way Howe's model unifies both engine-breathing noise and flow noise as being related to enthalpy, it also leads to a useful point of distinction: engine-breathing noise is associated most with volume velocity (or mass flux) sources whereas flow noise is associated mostly with stresses at a boundary or aerodynamic forces or fluctuating pressures.

Davies (1996) presents two acoustic source/filter models for use in flow duct acoustics: one for excitation by fluctuating mass or volume velocity (Figure 3.16a) and one for excitation by fluctuating pressure or aerodynamic force (Figure 3.16b).

Davies shows that for a volume source the acoustic power W_m of the source is given by:

$$W_m = 0.5 \mathrm{Re}\, \{p_1^* V_s\} = 0.5\, |V_s^2|\, \mathrm{Re}\, \{Z_1/(1+Z_1/Z_e)\}\, /S_s \qquad (3.54)$$

and for a fluctuating force

$$W_D = 0.5 \mathrm{Re}\, \{f_1^* u_s\} = 0.5\, |f_s^2|\, \mathrm{Re}\, \left\{ \frac{1}{Z_e S_s} \right\} \qquad (3.55)$$

where S_s is the associated surface area of the source.

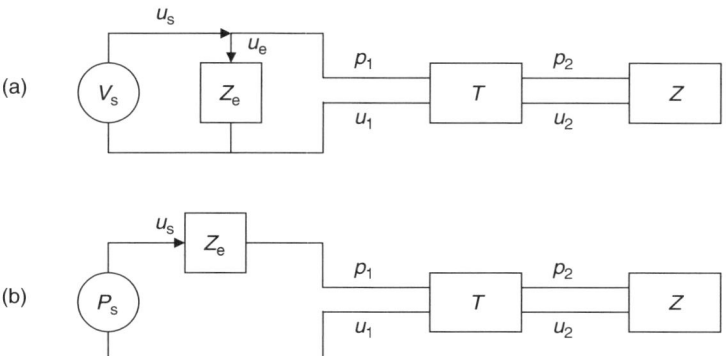

Figure 3.16 Acoustic circuits for flow ducts. (a) Excitation by a fluctuating mass or volume velocity. (b) Excitation by fluctuating pressure or aerodynamic force (after Davies (1996)).

These two equations imply that the sound power of the sources is a function of the termination impedance Z caused by the reflection coefficient at the open end and the transfer element T (being the acoustics of the flow duct network).

Harrison and Stanev (2004) used a volume velocity source located at the intake valve and linear models for one-dimensional acoustic propagation in flow ducts to successfully calculate the engine-breathing noise component in an IC engine inlet flow. The model can be used to identify the effects of:

- engine speed;
- valve timing, lift and open period;
- intake system acoustic resonances

in the prediction of engine-breathing noise.

In a similar way, Davies and Holland (1999) used a fluctuating pressure source positioned at the outlet of a reactive silencer chamber to predict a common component of flow noise (see Figure 3.17).

In a later publication, Davies and Holland (2001) describe the physical process at work in the pressure source. When the flow first enters the silencer chamber it leaves the downstream facing edge and separates, forming a thin shear layer or vortex sheet. Such sheets are very unstable and quickly develop waves that roll up to form a train of vortices with well-ordered spacings. When the vortices impact on the downstream face of the silencer, a fluctuation in pressure occurs. Acoustic energy propagating upstream from this source can affect the formation of vortices and a feedback mechanism is created. The feedback is strongly influenced by resonances downstream of the vortex generating expansion. Davies (1981) identified the influence of exhaust tailpipe and chamber resonances in the spectrum of flow noise generated in a simple reactive silencer element (see Figure 3.18).

In addition, Davies (1981) highlighted the possibility that simple reactive silencer elements could act as amplifiers of sound rather than attenuators due to the feedback processes that generate flow noise. The use of a length of perforated pipe to bridge the gap between inlet and outlet of a simple silencer is effective at eliminating this amplification by suppressing the formation of vortices at the inlet. Providing the porosity of the perforate pipe is greater than 15% it will have a negligible effect on the attenuation of engine-breathing noise afforded by the silencer.

The above discussion paints a picture of volume velocity sources of engine-breathing noise located at the intake/exhaust valves superimposed with fluctuating pressure sources of flow noise distributed down the intake (and exhaust) system. A rational way to separate the contributions made to intake orifice (or exhaust tailpipe) noise by the two classes of

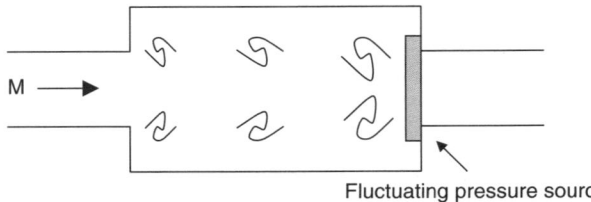

Fluctuating pressure source

Figure 3.17 Prediction of a common flow noise source (after (Davies and Holland, 1999)). M denotes a flow with a finite Mach number.

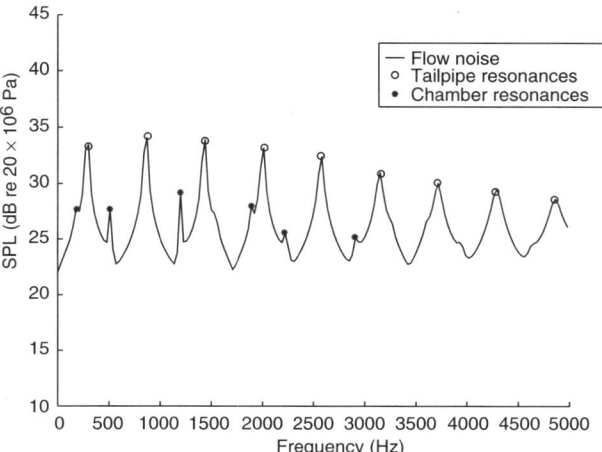

Figure 3.18 Predicted levels of flow noise from a simple silencer element using the method of Davies and Holland (1999). The exhaust tailpipe and chamber resonances have been marked as in Davies (1981).

sound source would be to measure the sound energy flux at several locations down the length of the intake (or exhaust) system. Near to the valves, the engine-breathing noise sources should dominate and the level due to this source can be predicted elsewhere in the system. Any local differences between predicted levels of engine-breathing noise energy flux and measured energy flux must be due to sources of flow noise.

Morfey (1971) showed that for a non-uniform flow, the acoustic intensity I which is the wave energy per unit area is given by

$$I = (1+M^2)\langle pu \rangle + M\left[\left(\frac{\langle p^2 \rangle}{\rho_0 c_0}\right) + \rho_0 c_0 \langle u^2 \rangle\right] \tag{3.56}$$

where $\langle \; \rangle$ denotes a time average and M is the Mach number. The first term corresponds to the sound intensity associated with the wave motion itself and the second with that due to the convection of acoustic energy density by the mean flow.

With plane wave propagation this becomes (Davies, 1988)

$$I = \frac{1}{\rho_0 c_0}\left[(1+M)^2 |p^+|^2 - (1-M)^2 |p^-|^2\right] \tag{3.57}$$

Holland et al. (2002) demonstrate the use of an experimental method of measuring sound intensity flux down the exhaust systems of running IC engines. This seems the most promising way of separating engine-breathing noise from flow noise. Rather, simpler experimental methods have been used elsewhere (Selemet et al., 1999; Sievewright, 2000) although the distinction between primary and secondary noise sources is not possible with these (Kunz and Garcia, 1995).

The narrow band noise spectra shown in Figures 3.19 and 3.20 give an indication of the typical level and spectral content of both intake and exhaust noise. The tonal quality of the noise, arising from the cyclic operation of the engine is obvious.

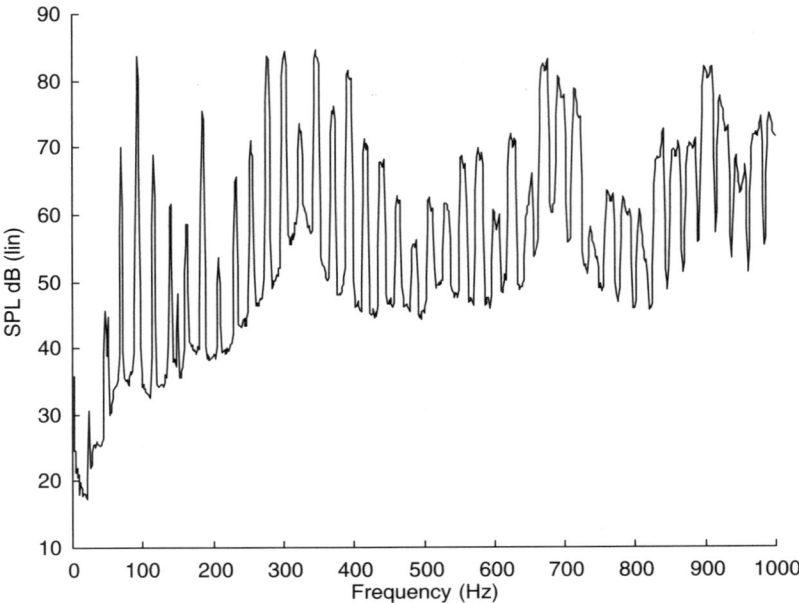

Figure 3.19 Sound pressure levels recorded 100 mm from the intake orifice of a 1.6-litre four-cylinder gasoline engine running full load at 2800 rev min^{-1}.

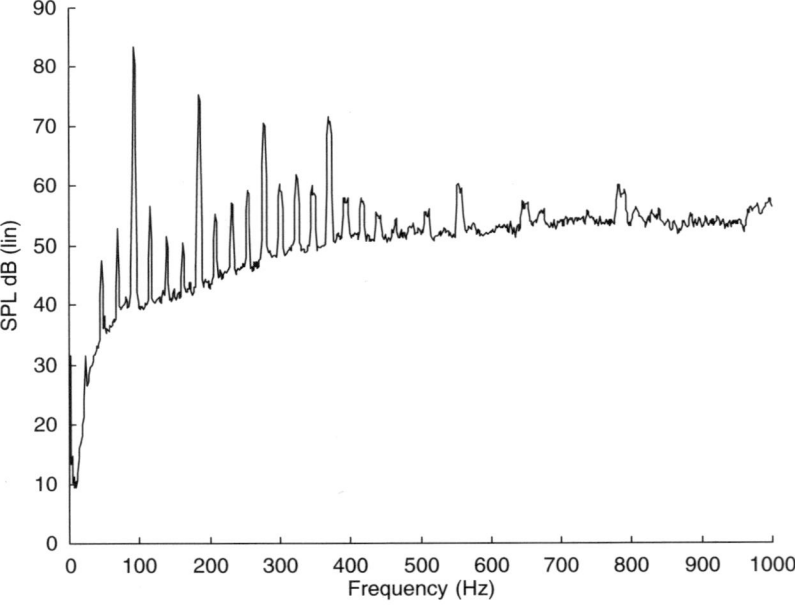

Figure 3.20 Sound pressure levels recorded 500 mm from the exhaust tailpipe of a 1.6-litre four-cylinder gasoline engine running full load at 2800 rev min^{-1}.

3.3.12 Flow duct acoustics

More space in this book is devoted to the control of intake noise than to the control of exhaust noise. Once learnt for the intake system, the design methods and supporting theory of acoustics can be readily transferred to the design and development of the exhaust system. The additional complicating factors for the exhaust system are:

- higher flow velocities;
- higher amplitude sound;
- steep temperature gradients;
- increased levels of flow-generated noise.

3.3.12.1 Basic design concepts

Basic intake and exhaust systems are made up of expansions, contractions and pipe protrusions. In the absence of temperature gradients or flow, these elements behave in a predictable manner as shown in Figure 3.21.

Inspection of Figure 3.21 leads to the following basic rules for the design of flow duct silencers (intake or exhaust):

- The sudden expansion of the gas at an area discontinuity strongly reflects acoustic waves back towards their source (the engine) and results in the attenuation of that part of the acoustic wave that finally radiates from the end of the system (snorkel or exhaust tailpipe noise).

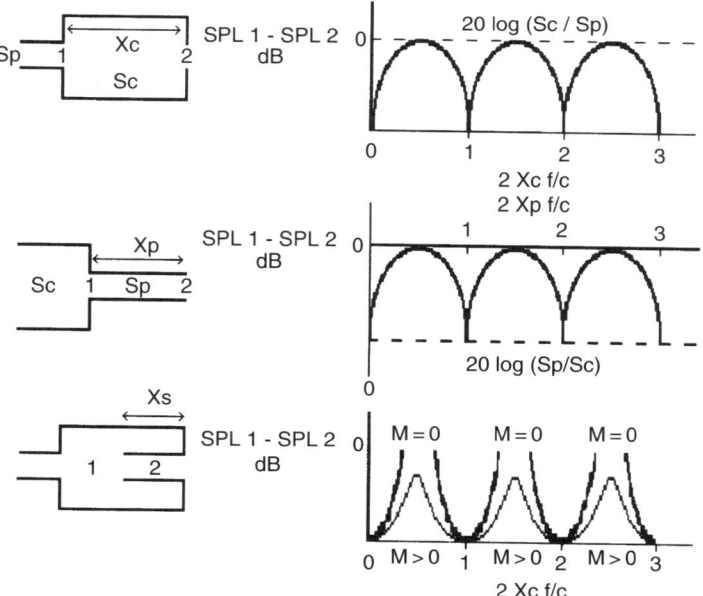

Figure 3.21 The performance of silencer elements in the absence of flow or temperature gradients.

- Lengths of duct (pipes, expansion chambers and the like) that are open at both ends have acoustic resonances that reduce the attenuation achieved at certain predictable resonant frequencies.
- Lengths of pipe that are open at one end and closed at the other act as resonators that increase the attenuation achieved at certain predictable resonant frequencies.

The silencing elements of expansions, contractions and sidebranches can be used to construct rather complex silencing units as shown in Figure 3.22.

3.3.12.2 Important design and development parameters

It is important to get the basic silencer expansion ratios and exhaust tailpipe (or snorkel) lengths right at the start of the design. It is not good practice to select a silencer by volume. Rather, a designer should determine the pipe diameter required to channel the flow efficiently and then, selecting an appropriate expansion ratio for the silencer, determine the cross-sectional area of the silencer. The length of the silencer will depend on its position in the system (packaging considerations, length of exhaust tailpipe required,

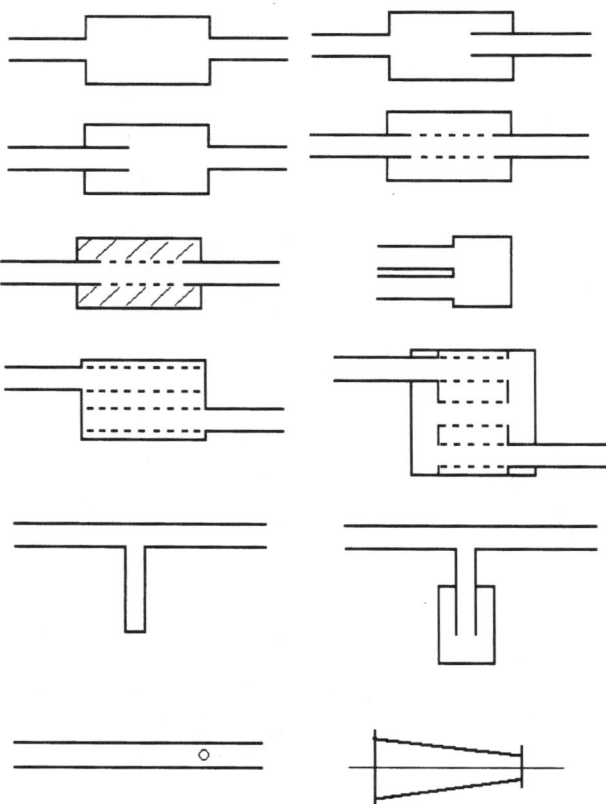

Figure 3.22 Simple and complex silencing units used for intake and exhaust systems.

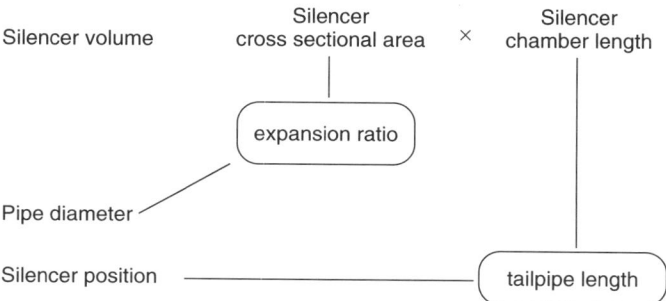

Figure 3.23 Important parameter in flow duct acoustics.

etc.) and on the avoidance of critical resonant frequencies. These considerations taken together determine the location and volume of the silencer in a rational way as illustrated in Figure 3.23.

3.3.12.3 Design and development strategies

There is a spectrum of choice for design and development strategies as illustrated in Figure 3.24. Empirical methods are at one extreme of the spectrum, relying on know-how and on the extensive testing of prototype silencers. The experienced silencer designer will develop their own 'favourite' silencers. However, acoustic theory allows them to tune such silencers to a particular engine or car without too much prototype testing. The theoretical tuning of the complex silencers shown in Figure 3.22 defies simple hand calculations, and a computer model is required.

3.3.12.4 Traditional intake and exhaust system design and development techniques

Cut-and-try methods have been very successful in the past. They allow a designer, time and opportunity to balance the need for low noise emission against the desire for a particular quality of noise. However, there are a number of disadvantages to this approach namely:

- these are often time consuming;
- they may be costly due to the need for a large number of prototypes for testing;

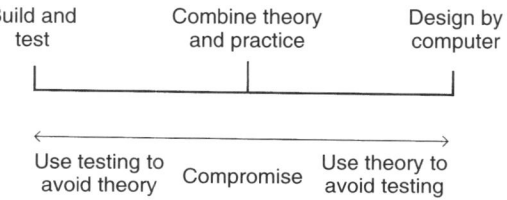

Figure 3.24 Spectrum of choice for intake design strategy.

- it is often difficult to be sure that an optimum solution has been obtained;
- the quality of the final product is highly dependent on the talents and experience of the development engineer.

Some rules of thumb for acoustic design of silencers:

Maximum attenuation	$20 \log_{10}$ (expansion ratio) dB
Intake filter box volume	3–5 times the engine swept volume
Exhaust silencer volume	5–10 times the engine swept volume
Intake snorkel length	300–400 mm (usually bell mounted)
Exhaust tailpipe length	less than 500 mm.
<120 Hz intake noise control	Helmholtz resonator onto the snorkel
120–250 Hz intake noise control	sidebranch resonator onto the snorkel
250–500 Hz intake noise control	tuning holes, conical snorkel
>500 Hz intake noise control	not usually a problem
>500 Hz exhaust noise control	pack silencers with porous materials such as basalt or wire wool

3.3.12.5 *Predicting the acoustic behaviour of complex flow duct systems*

Adopting a suitable acoustic prediction method during the development programme can help reduce programme time and cost by minimising the number of prototypes that need to be manufactured and tested.

Acoustic prediction models fall into two groups:

1. Time domain predictions – where a prediction of the gas dynamic behaviour in the intake or exhaust manifold is extended through the entire intake system.
2. Frequency domain predictions – where the acoustic characteristics of the intake system are predicted in terms of their variation with frequency.

Without predicting radiated noise the acoustic performance of a system is assessed relative to a baseline, such as a straight pipe, or another reference system.

Frequency domain computer programs are popular due to their fast run-times and their ability to adequately model complex system geometries. Most frequency domain methods make the following assumptions:

- linear acoustic theory remains realistic;
- one-dimensional models remain realistic;
- all wave propagation is planar;
- all disturbances are isentropic.

3.3.12.6 *Models for acoustic propagation in flow ducts*

Acoustic motion in flow ducts is very complex. The overall motion of the fluid being excited by an acoustic source can however be described as the sum of a variety of modes of propagation.

These modes can be described as patterns of motion such as axial oscillation, sloshing and spiralling. All these modes are generally present in a flow duct acoustic field, but it is known that each mode will only propagate efficiently at frequencies below its own specific cut-off frequency.

This frequency is generally expressed in terms of Helmholtz number ka, where in this section

$$k = \text{ wavenumber} = \frac{\omega}{c} \tag{3.58}$$

a = duct radius and ω is the radial frequency in rads^{-1} and c is the speed of sound for the fluid in ms^{-1}.

The most efficient mode of propagation is in the form of plane waves. Here the pressure and velocity of the disturbance are constant across a given plane along the duct. It is assumed that below the first cut-off frequency given by

$$ka = 1.84$$

only plane waves propagate any great distance, while other modes decay rapidly away from their source and are referred to as evanescent waves.

For vehicle systems the first cut-off frequency is in the 2000+ Hz range, which for a four-stroke engine would represent the 24th engine order at 5000 rev min^{-1}. It is therefore a reasonable and convenient simplification to consider only plane wave propagation for vehicle intake and exhaust systems.

3.3.12.7 Limits of linear acoustic theory

Acoustic or linear theory remains appropriate as long as the waves travel along the uniform sections of duct between discontinuities without significant change in their shape. Typically, this is the case for pressure amplitudes in the region of 0.01–0.001 bar, with the limit falling with increasing frequency. Different investigations into the limits of the linear assumption in practical flow ducts are reported in Davies and Holland (2004) and Payri et al. (2000).

The plane wave restriction is useful as:

- it allows for a simpler means for including the effects of flow into the analysis;
- it allows measurements to be made at duct walls.

3.3.12.8 Acoustic plane waves in ducts

Standing waves occur as a result of the interference between waves travelling out of the source and waves reflected by each discontinuity in area or more generally each discontinuity in acoustic impedance. The concept of one forward-travelling wave and one backward-travelling wave is illustrated in Figure 3.25. The background theory that supports this concept is given in detail in Section 2.1.4.

$$p(x, t) = p^+(x, t) + p^-(x, t) \tag{3.59}$$

Figure 3.25 Sound field in a duct.

The magnitudes of p^+ and p^- remain effectively invariant between discontinuities, while the relative phase will vary in an organised manner.

However, $p(x, t)$ will vary along the duct due to the presence of standing waves. These so-called standing waves do not actually stand at all but are the result of the interference between waves travelling in opposite directions. Traversing a microphone down the pipe will show the standing wave pattern which may be viewed as the mode shape of a particular resonance.

3.3.12.9 *Acoustic sources and acoustic loads*

The acoustic performance of a flow duct element depends on the acoustic source and the acoustic load acting on it as illustrated in Figure 3.26 (Davies and Harrison, 1997).

Consider an exhaust system. The acoustic performance of the first silencer will depend on the acoustic load imposed on it. That load is the combined acoustic impedance of the remainder of the exhaust system downstream of the first silencer. Therefore, the acoustic performance of the first silencer depends on where in the system it is placed (hence a silencer that performs in a certain way on the flow bench, may behave differently on the vehicle). This is illustrated in Figure 3.27 (Z is the acoustic impedance, the ratio of acoustic pressure to acoustic volume velocity).

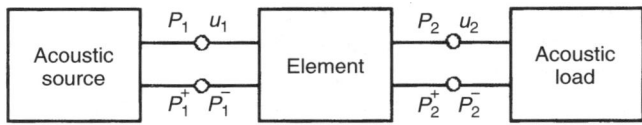

Figure 3.26 The interaction between an acoustic source and an acoustic load (Davies and Harrison, 1997).

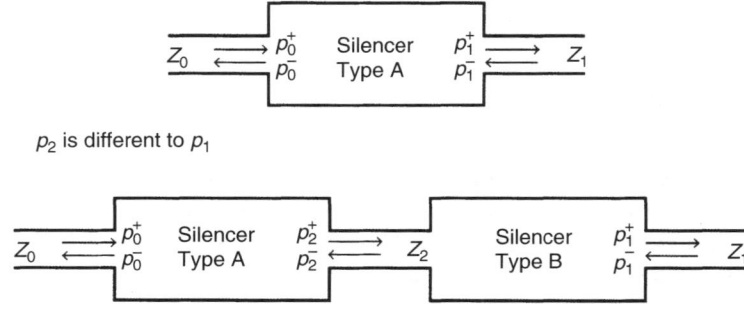

Figure 3.27 The interacting sound fields in a twin-silencer exhaust system.

3.3.12.10 The acoustics of an unflanged pipe with flow

The final acoustic termination in an intake system is at the snorkel orifice. The final acoustic termination for the exhaust system is the exhaust tailpipe. The inflow of air or outflow of exhaust gas produces a final acoustic load on the system that is well understood and takes the form of a reflection coefficient r:

$$r = \frac{p^-}{p^+} \tag{3.60}$$

$$\frac{Z}{\rho_0 c} = \frac{1+r}{1-r} \tag{3.61}$$

and

$$r = Re^{i\theta} = -Re^{-i2kl} \tag{3.62}$$

where R is the modulus, θ is the phase, k the wavenumber and l is an end correction (Davies et al., 1980; Davies, 1987). Ideally, on the assumption that the pressure waves remain plane, $r \rightarrow -1$, $R \rightarrow 1$ and $\theta \rightarrow \pi$. In reality, the plane wave assumption remains realistic only when the dimension a (pipe radius) remains a small fraction of a wavelength; that is, for the limiting case when the Helmholtz number ka approaches zero (the plane wave cut-off occurs at $ka = 1.84$).

For zero outflow, the modulus of the reflection coefficient is unity at low Helmholtz number and reduces with increasing Helmholtz number (the higher-frequency sound transmitted is reflected less than the low-frequency sound – i.e. the high-frequency sound is radiated more than the low-frequency sound).

The effect of mean outflow is to increase R throughout the frequency range. The effect of mean inflow is to decrease R throughout the frequency range.

3.3.12.11 The general effects of temperature and flow speed

In addition to its effect on the reflection coefficient at an unflanged pipe termination, the effect of a steady mean flow on isentropic acoustic plane wave propagation in a duct of constant area is to modify the appropriate wavenumber k. With flow, the distance between the nodes of a standing wave is reduced by a factor $(1 - M^2)$ where M is the Mach number of the flow. In other words, the effective duct lengths are reduced.

The presence of a mean flow alters the values of the wavenumber by a factor $1/1 \pm M$ (Davies, 1988). The Mach number in the duct varies with the mass flow rate of gas (or the engine speed and load) and the gas density (or the gas composition and temperature).

The wavenumbers are also affected by changes in temperature through variations in the speed of sound (see also the effect of sound-absorbing materials – Section 4.9) as

$$\beta = \text{complex wavenumber} = \left(\frac{\omega}{c}\right) - i\alpha \tag{3.63}$$

where α is a visco-thermal attenuation coefficient which for plane wave propagation in a circular pipe of radius a is given by

$$\alpha = \left(\frac{1}{ac}\right)\left(\frac{\nu\omega}{2}\right)^{1/2}\left[1+(\gamma-1)\left(\frac{1}{P_r}\right)^{1/2}\right] \quad (3.64)$$

where ν is the local kinematic viscocity and P_r is the Prandtl number for the gas in the pipe $P_r = \frac{\mu c_P}{k}$ where μ is the shear viscosity and k is the thermal conductivity (Davies, 1988).

The speed of sound varies with the relation

$$c = \sqrt{\gamma RT} \quad (3.65)$$

where

$R = $ specific gas constant,
$T = $ absolute temperature (K)

As an added complication the values of γ and R also vary with gas composition and temperature, but for air in an intake, values of $\gamma = 1.4$ and $R = 290$ are reasonable, whilst for exhaust gas $\gamma = 1.35$ and $R = 285$ might be chosen.

For intake systems, it is reasonable to assume a sound speed in the order of $343\,\mathrm{m\,s^{-1}}$. The final effect of flow on the acoustic performance of silencers is complex, and defies simple calculations.

3.3.12.12 Calculating Mach numbers

Regrettably, the car fitted with an internal combustion engine is a rather inefficient means of turning chemical energy into tractive effort.

In fact, as will be demonstrated here, under urban driving conditions, only 20% (or less) of the chemical energy in gasoline is made available for tractive effort at the wheels. For the case of the engine in isolation, Heywood calls this the thermal conversion efficiency (Heywood, 1988)

$$\eta_{tc} = \frac{W_c}{\eta_c m_f Q_{HV}} \quad (3.66)$$

where

$W_c \quad = $ work output from the cycle
$\eta_c \quad = $ combustion efficiency
$m_f \quad = $ mass of fuel
$Q_{HV} = $ heating value of fuel

To obtain the thermal conversion efficiency for a complete vehicle, the chain of efficiencies has to be analysed, first for the engine at full load (Bosch, 1986) and then subsequently for part load operation, and finally for the engine installed in the vehicle.

The chain of efficiencies for the engine itself, operating at full load throttle is given by

$$\eta_{engine} = \eta_i \times \eta_d \times \eta_c \times \eta_m \quad (3.67)$$

where

η_i = ideal fuel–air cycle efficiency, which Taylor (1985) shows to be 46% for stoichiometric octane/air mixture and a compression ratio of 10.

η_d = diagram factor – the ratio of work output from the real cycle to work output from
an ideal cycle. Taylor (1985) suggests 78% for stoichiometric octane/air mixture
and a compression ratio of 10.

η_m = mechanical efficiency. Lumley (1999) suggests 85% for low engine speeds.

η_c = combustion efficiency. Heywood suggests 98% for stoichometric mixtures
(Heywood, 1988).

Thus the product of the engine efficiency chain for full load operation:

$$\text{Engine efficiency} = 0.46 \times 0.78 \times 0.85 \times 0.98 = 30\%$$

This figure is a little above the energy balance of 25–28% given for the gasoline engine
by Heywood (1988). It is commonly held that 1/3 of the chemical energy ends up as
useful work, 1/3 goes to heat up the cooling water and 1/3 goes down the exhaust tailpipe.

It is useful to try out some numbers now (Harrison, 2003). The ratio of energy put in
to the system compared with energy converted to useful work is $100/30 = 3.333$

Now, for gasoline

$$Q_{LHV} = 43.1 \, \text{MJ} / \text{kg}^{-1}$$

Thus the fuel required per second for 1 kW output from the cycle is

$$= (1000 \times 3.333)/43.1 \times 10^6 = 0.000077 \, \text{kg} / \text{s}^{-1} = 278 \, \text{g}/\text{kWh}$$

In 1988, Heywood expected the values of bsfc to be around 270 g/k Wh for gasoline
engines and as low as 200 g/kWh for diesel engines (Heywood, 1988). By the mid-1990s,
port fuel injection resulted in engines operating in the 250–300 g/kWh range with bsfc
at around 250 g/kWh (full load) at the medium engine speed ranges.

For a 50 kW output, and a bsfc of 278 g/kWh, fuel flows would be $0.278 \times 50 = 14 \, \text{kg}/\text{hr}^{-1}$. With the density of typical gasoline being 733 kg m^{-3} this equates to 19 litres
of fuel burned per hour at 50 kW (4.2 gallons). At stoichiometric air–fuel mixtures, this
requires $14 \times 14.7 = 206 \, \text{kg}/\text{hr}^{-1}$ of air.

With the density of air at standard conditions being 1.19 kg m^{-3} this is equal to 173 m^3
of air an hour or 0.048 m^3 per second. With throttle diameters being commonly around
60 mm, the local flow speed through the wide-open throttle body would be around 17 ms^{-1}
(0.05 Mach).

The method adopted here allows the intake designer to calculate intake Mach numbers
using practical engine performance data (brake power and bsfc) with the need for only a
limited set of additional data, all of which is freely available in the literature.

3.3.12.13 Describing the acoustic performance of flow duct systems

Referring to Figure 3.28 and to Section 3.3.12.11, the acoustic field in the intake pipe
is the sum of the positive and negative wave components p_1^+ and p_1^- respectively and is
described by the equation

$$P_{inlet} = p_1^+ e^{i(\omega t - \beta_1^+ x)} + p_1^- e^{i(\omega t + \beta_1^- x)} \tag{3.68}$$

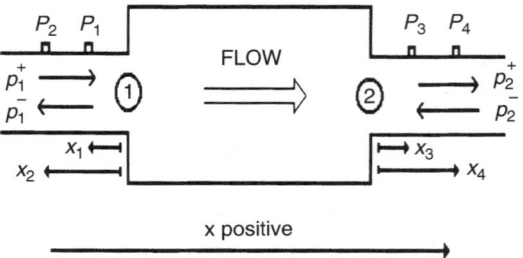

Figure 3.28 Inlet and outlet acoustic fields in a silencer element.

The acoustic field in the outlet pipe is:

$$P_{\text{outlet}} = p_2^+ e^{i(\omega t - \beta_2^+ x)} + p_2^- e^{i(\omega t + \beta_2^- x)} \tag{3.69}$$

The acoustic performance of the duct section can be characterised in terms of ratios of the complex wave components:

$$\text{Attenuation (dB)} = 20 \log_{10} \left(\frac{p_1^+}{p_2^+} \right) \tag{3.70}$$

$$\text{Reflection at plane (1)} = \frac{p_1^-}{p_1^+} \tag{3.71}$$

$$\frac{Z}{\rho_0 c} = \frac{1+r}{1-r} \tag{3.72}$$

Note that a dip in system attenuation will result in a peak in radiated noise. Therefore, much of the designer's work is in filling in the dips in attenuation at critical frequencies by the alteration of pipe lengths and the addition of resonators.

There are various methods available for bench testing the acoustic performance of flow duct systems:

- Pressure ratio measurements in duct.
- Radiated sound measurements.
- Wave decomposition techniques (Davies et al., 1999) – Measure sound pressures at four known positions and solve the four simultaneous equations that result:

$$P_1 = p_{\text{up}}^+ e^{i(\omega t - \beta_{\text{up}}^+ x_1)} + p_{\text{up}}^- e^{i(\omega t + \beta_{\text{up}}^- x_1)} \tag{3.73}$$

$$P_2 = p_{\text{up}}^+ e^{i(\omega t - \beta_{\text{up}}^+ x_2)} + p_{\text{up}}^- e^{i(\omega t + \beta_{\text{up}}^- x_2)} \tag{3.74}$$

$$P_3 = p_{\text{down}}^+ e^{i(\omega t - \beta_{\text{down}}^+ x_3)} + p_{\text{down}}^- e^{i(\omega t + \beta_{\text{down}}^- x_3)} \tag{3.75}$$

$$P_4 = p_{\text{down}}^+ e^{i(\omega t - \beta_{\text{down}}^+ x_4)} + p_{\text{down}}^- e^{i(\omega t + \beta_{\text{down}}^- x_4)} \tag{3.76}$$

3.3.13 Intake noise control: a case study

Frequently, the fundamental dimensions of an intake system (such as minimum pipe diameters and the choice of air filter element) are set in advance of noise control engineering by the engine development team. In this case study, those considerations lead to an expansion ratio of 18.33:1.00 between the snorkel and the filter box. The peak attenuation in this filter box can be estimated using:

$$\text{attenuation dB} = 20 \ \log_{10} (\text{expansion ratio}) = 20 \ \log_{10} (18.33) = 25.27 \, \text{dB} \qquad (3.77)$$

The peak attenuation will occur at frequencies away from resonances in the snorkel and in the filter box. These resonant frequencies can be estimated using the equation below:

$$\frac{2xf}{c} \approx n \quad n = 1, 2, 3, 4, \ldots \qquad (3.78)$$

where

$x = $ length
$f = $ frequency (Hz)
$c = $ speed of sound (ms^{-1})

The attenuation (predicted using a software known as APINEX) afforded by this simple intake system is shown in Figure 3.29.

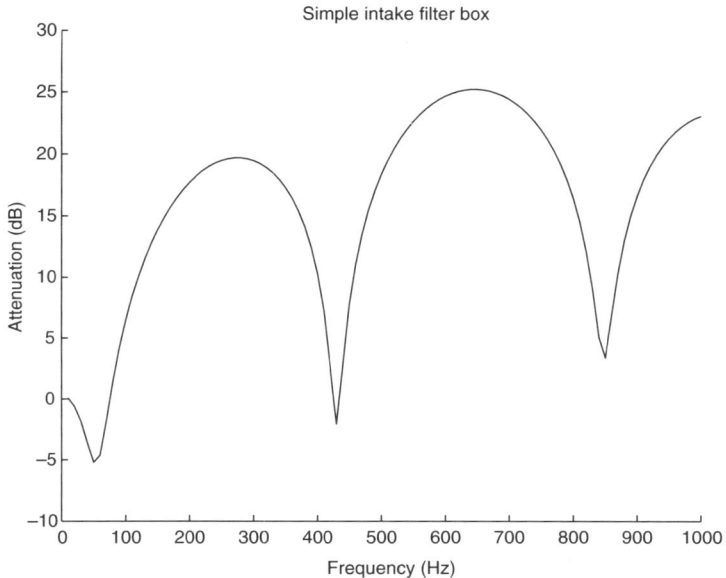

Figure 3.29 Attenuation of a simple intake filter box calculated using APINEX software.

Inspection of Figure 3.29 reveals:

- peak attenuation = 25 dB at 650 Hz;
- two dips in the attenuation relating to the snorkel length when $n = 1$ and 2 (filter box depth is too short for any chamber resonances <1000 Hz).

The attenuation characteristics of the expansion portion of the filter box are shown in Figure 3.30. The attenuation characteristics of the contraction section of the filter box are shown in Figure 3.31.

The effect of the intake snorkel is to have negative attenuation near its resonant frequencies. Contrasting the last two figures illustrates the point that it is the expansion within the filter box which provides the basic attenuation, while the snorkel and the filter box lengths tune the dips in the attenuation characteristics. Figures 3.30 and 3.31 (and others in this section) were constructed using sophisticated flow duct acoustic modelling tools (Davies, 1988; Harrison et al., 2004). These results compare favourably with common understanding of how reactive silencers work (see Figure 3.21).

3.3.13.1 Conical snorkels

A conical snorkel can be used to reduce the depth of the troughs in the attenuation due to the tuning effect of the snorkel length as shown in Figure 3.32. The typical conical snorkel is tapered such that the snorkel orifice is reduced in area. Care must be taken not to seriously limit engine performance by throttling the intake flow as a result.

Note: It is generally assumed that all intake snorkels, tapered/conical or not have a slight flare at the intake orifice in order to improve the initial inlet flow.

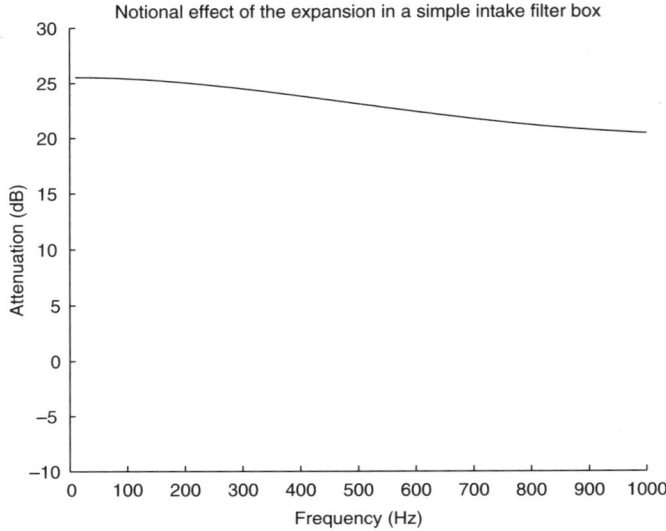

Figure 3.30 Attenuation of the expansion section of a simple intake filter box calculated using APINEX software.

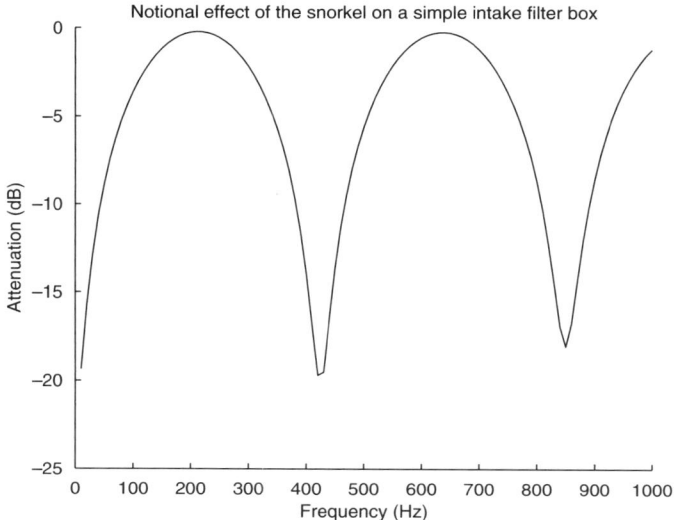

Figure 3.31 Attenuation of the contraction section of a simple intake filter box calculated using APINEX software.

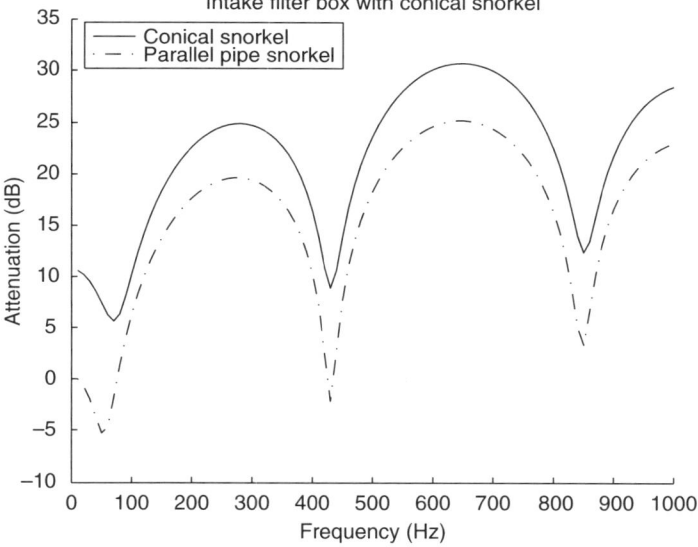

Figure 3.32 Effect of adopting a conical snorkel. Calculated using APINEX software.

3.3.13.2 *Protrusions of the snorkel into the filter box*

Protrusions of the snorkel into the base of the filter box act like sidebranch resonators. The effect of a 120-mm protrusion to the snorkel along the bottom of the filter box is shown in Figure 3.33.

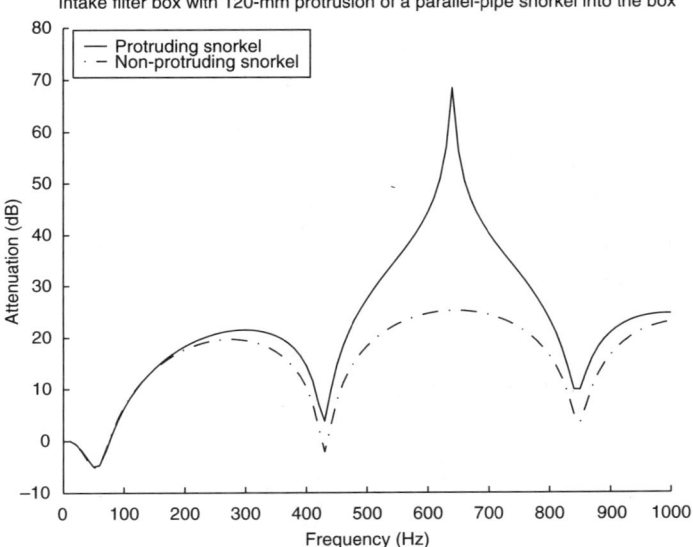

Figure 3.33 Effect of protruding the snorkel into the filter box. Calculated using APINEX software.

The frequencies at which the system attenuation is improved through the action of the sidebranches may be calculated for the zero flow case from:

$$\frac{4xf}{c} = n \quad n = 1, 3, 5, 7, 9, \ldots \tag{3.79}$$

where x is the sidebranch length and f is the resonant frequency.

In practical cases, the attenuation shown in Figure 3.33 would be limited to around 40 dB by the effects of flow noise.

3.3.13.3 Helmholtz resonators

These feature a 1–3-litre volume placed as a sidebranch to the snorkel. Connection to the snorkel is by a tuned length of pipe. These devices provide useful low frequency attenuation with little extra static pressure loss. The effect of adding of a Helmholtz resonator, tuned to around 90 Hz to the snorkel (as illustrated in Figure 3.34) is shown in Figure 3.35.

The Helmholtz resonator is a reliable (and often the only practical) way of increasing intake system attenuation at frequencies below 120 Hz.

3.3.13.4 Sidebranch resonators

A simple 300-mm sidebranch resonator (illustrated in Figure 3.36) is fitted instead of the Helmholtz resonator, with the effect shown in Figure 3.37. Figure 3.38 shows a

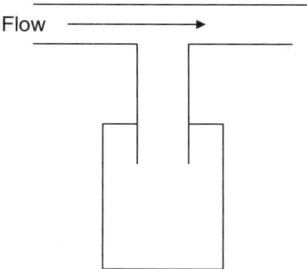

Figure 3.34 The Helmholtz resonator.

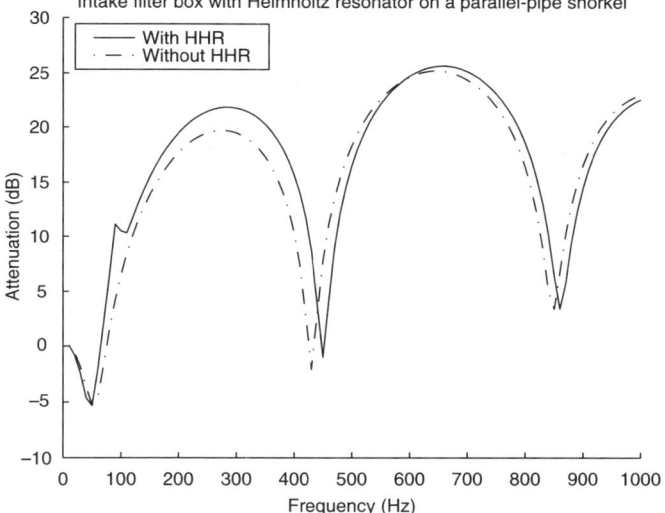

Figure 3.35 Effect of an added Helmholtz resonator calculated using APINEX software.

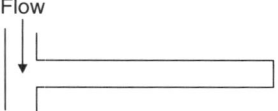

Figure 3.36 The sidebranch resonator.

photograph of a typical production example of this. The frequencies at which the system attenuation is improved through the action of the sidebranches may be calculated for the zero flow case from:

$$\frac{4xf}{c} = n \quad n = 1, 3, 5, 7, 9, \ldots \tag{3.80}$$

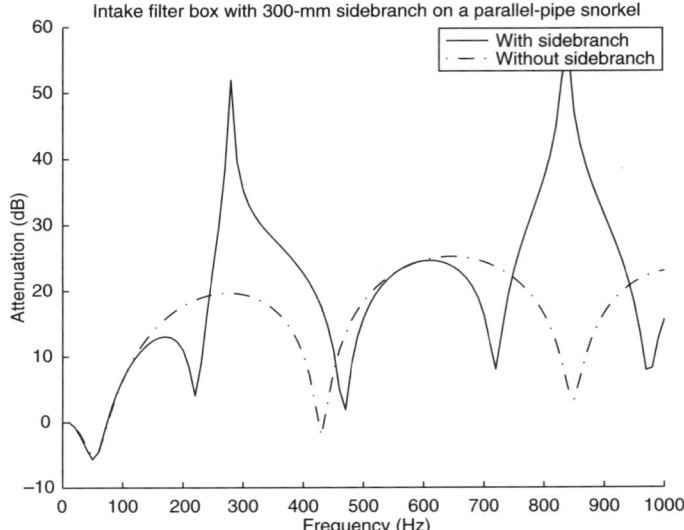

Figure 3.37 The effect of adding a sidebranch resonator calculated using APINEX software.

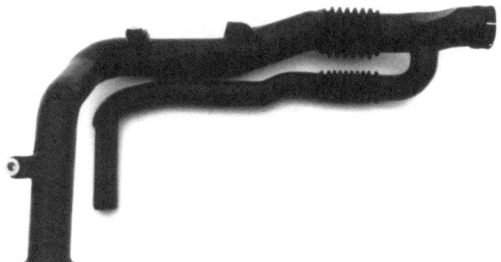

Figure 3.38 Photograph of a production snorkel with sidebranch mounted.

where x is the sidebranch length and f is the resonant frequency. In practical cases, the attenuation shown in Figure 3.37 would be limited to around 40 dB by the effects of flow noise.

3.3.13.5 The effects of the manifold and the zip tube

The effect of adding a simple manifold and a zip tube to the filter box is shown in Figure 3.39. The primary runner lengths of the manifold's branches act as sidebranch resonators when they are terminated by closed intake valves. Therefore, in the absence of cylinder-to-cylinder overlap in intake valve opening, at any given time only one valve will be open and the remainder of the branches act as sidebranch resonators to the branch with the open valve. This results in a strong peak in the attenuation spectrum that is related to the primary runner length.

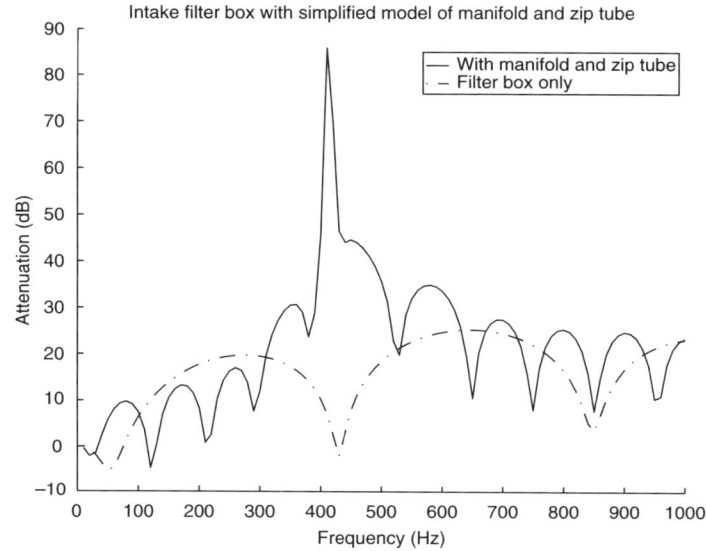

Figure 3.39 The influence of the intake manifold and zip tube calculated using APINEX software.

Away from this strong peak, there are additional troughs in the attenuation caused by resonances of the zip tube. The frequencies at which these troughs occur may be controlled by varying the length of the zip tube as shown in Figure 3.40.

Figure 3.41 shows a line illustration of an intake system where the design has been optimised using linear acoustic theory.

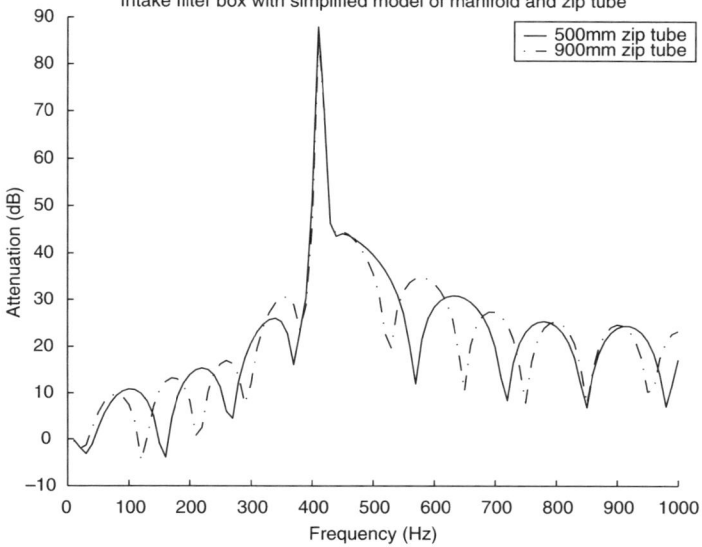

Figure 3.40 Influence of zip tube length calculated using APINEX software.

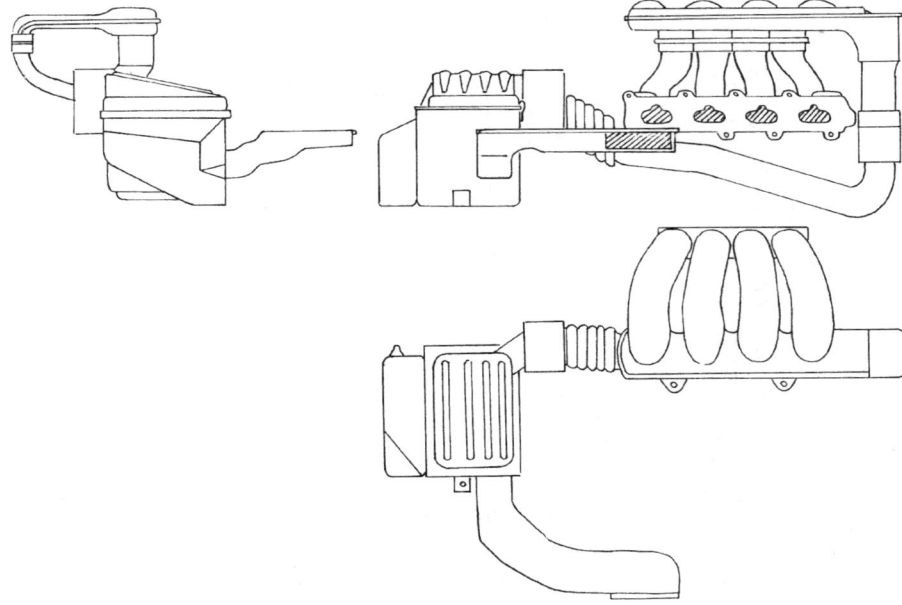

Figure 3.41 An intake system where the design has been optimised using linear acoustic theory.

3.3.13.6 Intake system mounting

The mounting of the intake system must give effective attenuation of:

- transmitted engine vibration;
- shell noise.

Care should be taken to avoid noise 'breaking out' through the walls of flexible hoses. If the filter box is mounted on the vehicle body, the mounting system must be flexible enough to allow for engine movement. The filter box should be fixed to a low-mobility piece of bodywork. Filter box mounts should be sufficiently compliant to give adequate vibration isolation. In terms of vehicle refinement, interior noise concerns should generally influence the mounting of the filter box.

3.3.14 Exhaust noise control

The earlier discussion of intake system design applies equally to the exhaust system with the following notable exceptions:

- The influence of the flow is much greater in the exhaust system due to the higher flow speeds.
- Temperature gradients are very steep, particularly in the exhaust manifold (300°C drop in gas temperature between the exhaust port and the outlet of the catalytic converter is typical).

- Durability/corrosion resistance is often a limiting factor in the design of silencers.
- Backpressure has a stronger influence on engine performance than the wave action in the exhaust manifold.
- Perforate tubes are commonly used (porosity less than 15%) to provide gradual expansion of the gas into the silencer, achieving noise reduction with a lesser backpressure penalty. Several parallel flow paths may be provided, each linked via perforated tube (Davies et al., 1997).
- Many silencer chambers can be provided within a single silencer shell by dividing the silencer using baffle plates. The flow path through the silencer may be convoluted with many reversals of flow direction. In this way, the flow path length is commonly much longer than the physical length of the exhaust system. This is sharp contrast to common practice with intake systems. One example is given in the line illustration shown in Figure 3.42.
- Noise radiated from the outer surfaces of the silencers may be significant (shell noise). This is commonly controlled using two-layer skins of thin-sheet steel for the silencer construction.
- Noise generated by the flow itself may be significant. This flow noise is generated within the silencers themselves, and it is common for a silencer to actually amplify sound as a result of flow acoustic coupling (Davies and Harrison, 1997).
- There is less use of sidebranch and Helmholtz resonators compared with common practice with intake systems.

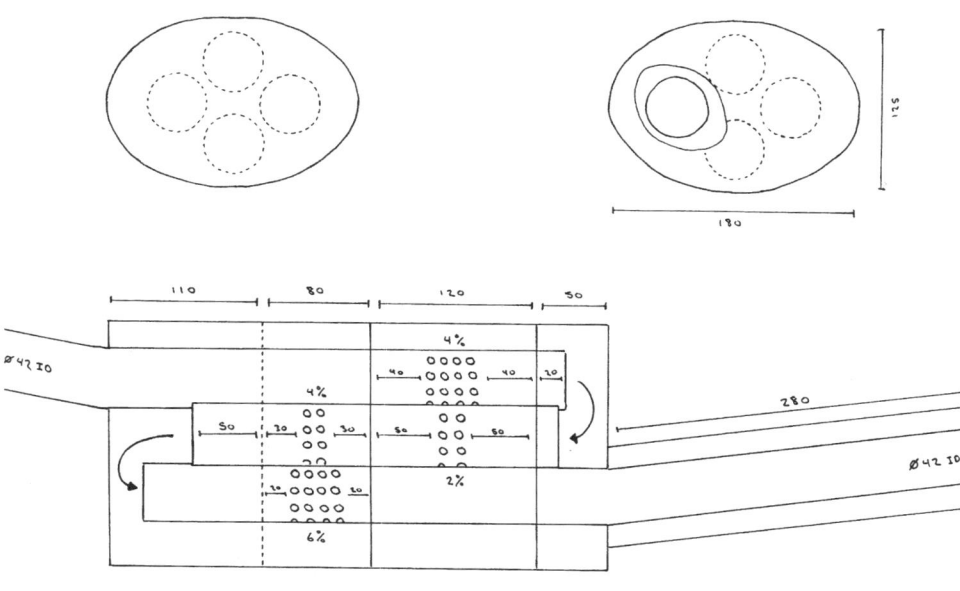

Figure 3.42 The internals of a compact exhaust silencer element showing many distinct chambers and a flow path that is longer than the physical length of the element. Dimensions are in millimetres.

- Silencers are often packed with sound absorbing materials such as basalt, wire wool or glass fibre strands. This improves attenuation at high frequencies, but often at the expense of durability due to the retention of water within the materials.
- The mounting of the exhaust system below the vehicle is important to avoid the creation of paths for structure-borne noise to enter the vehicle.

3.3.14.1 On the flow through the exhaust valve

The same principles apply to flow through the exhaust valve as for the case of the inlet valve. However, the model that results in equation (3.49) is too simplistic to realistically model the high temperature, high speed flow through the exhaust valve. An alternative model that allows for sonic flow through the valve and/or through the port is given in both Benson (1982) and Winterbone and Pearson (1999). That model requires an iterative procedure to account for the entropy loss across the valve.

It is clear that, for practical engines, the performance may be optimised (or 'tuned') by varying:

- Valve timing and lift to influence intake and exhaust flow rates and promote effective scavenging.
- Inlet air density (by super or turbo charging) to maximise the air charge (and hence fuel charge).
- Port design to influence in-cylinder gas motion to improve combustion and reduce flow losses (often these are mutually exclusive).
- Varying the exhaust design to reduce backpressure.
- Varying the intake geometry (and usually to a lesser extent the exhaust manifold geometry) to maximise the pressure ratio across the valve at IVC.

3.3.14.2 Calculating exhaust temperatures

Just as for intake system design, flow duct acoustic modelling is useful for exhaust system design. However, in order to determine the important changes in the speed of sound down the length of an exhaust system, it is first necessary to calculate the gas temperature gradient. Common flow duct acoustic models require the user to declare exhaust manifold, and exhaust tailpipe gas (or ambient) temperatures and the temperature gradient is calculated accordingly. The exhaust manifold temperature may not be known early in a development cycle but it is possible to make realistic estimates of in-cylinder gas pressure and temperature at certain points in the cycle using suitable simplifying assumptions. These points are:

- firing TDC (top dead centre);
- peak pressure/temperature (assumed crank angle in the range of 10–30° ATDC);
- EVO (exhaust valve open);
- IVO (intake valve open);
- IVC (intake valve close);
- Ignition point.

At the heart of the approach, the temperature rise across the engine ∂T_E is estimated using an enthalpy balance (Weaving, 1990):

$$\partial h = Q_{LHV} m_f (1 - \eta_{th} - Q_c) \tag{3.81}$$

where

∂h = enthalpy rise across the engine (J)
Q_{LHV} = lower heating value of the fuel (assume 43.1×10^6 J/kg for gasoline)
m_f = mass of fuel trapped in the cylinder at IVC
η_{th} = thermal efficiency
Q_c = fraction of heat input lost to the cooling (in the range of 0.2–0.35)

$$\partial T_E = \frac{\partial h}{C_{Pe} m_f (1 + AFR)} \tag{3.82}$$

where

C_{Pe} = specific heat capacity of the exhaust gas (J/kgK)
AFR = air fuel ratio

It is clear from equations (3.81) and (3.82) that m_f cancels in the calculation of the temperature rise. However, as it is of general interest to the engine designer, it can be found from the bsfc, the brake power and the engine speed. Thus

$$\text{kg/s fuel flow} = (bsfc/1000) \times \frac{1}{3600} \times P_b \tag{3.83}$$

where

$bsfc$ = brake specific fuel consumption (g/kWh)
P_b = brake power (per cylinder – kW)

$$m_f = \frac{\text{kg/s} \times 2}{\text{rpm}/60} \tag{3.84}$$

The thermal efficiency can be calculated from

$$\eta_{th} = 100/Q_{LHV} \times \left(bsfc \times 10^{-3}/(1000 \times 3600) \right) \tag{3.85}$$

A value for the specific heat capacity of exhaust gas at elevated temperature is needed. This can be found from a polynomial fit of published data for CO_2 (Rogers and Mayhew, 1980) assuming that CO_2 is the main constituent of exhaust gas:

$$C_{Pe(CO_2)} = -3.6459 \times 10^{-11} T^4 + 3.0779 \times 10^{-7} T^3 - 9.7959 \times 10^{-4} T^2$$

$$+ 1.4606 T + 485.6034 \tag{3.86}$$

The elevated temperature $T(K)$ can be estimated from the following empirical relationships for peak-burnt gas temperature, using excess air ratio λ and ambient temperature T_{amb} (Benson and Whitehouse, 1979).

If $\lambda > 1$ (lean mixture)

$$T = T_{amb} + \frac{2500}{\lambda} \tag{3.87}$$

If $\lambda < 1$ (rich mixture)

$$T = T_{amb} + \frac{2500}{\lambda} - 700 \left(\frac{1}{\lambda} - 1 \right) \tag{3.88}$$

So, from a knowledge of bsfc (typically in the range of 250–280 g/kWh for a gasoline engine operating at full load) and an estimate of the heat loss coefficient to the coolant, the temperature rise across the engine can be calculated.
Therefore
Temperature at EVC

$$T_{exh} = \partial T_e + T_{amb}$$

where T_{amb} is the ambient temperature (K).
Now, the ideal thermal efficiency for the gasoline engine is given by

$$\eta_{ideal} = 1 - \frac{1}{r^{\gamma-1}} \tag{3.89}$$

where r is the compression ratio and γ is the ratio of specific heats. With knowledge of the compression ratio, and the estimate of the actual thermal efficiency from equation (3.85), equation (3.89) can be used in an iterative procedure to find the value of γ for which $\eta_{ideal} = \eta_{th} + Q_c$. That value of γ can be used to estimate the peak cylinder temperature assuming isentropic expansion during the expansion stroke:
At 10–30° ATDC

$$T_{peak} = T_{exh} r^{\gamma-1} \tag{3.90}$$

The pressure of the reactants at TDC can be estimated using the following polytropic relationship for the compression stroke
At TDC

$$P_{TDC} = P_{amb} r^k \tag{3.91}$$

where the polytropic index k is in the range of 1–1.2.
The reactant temperature at TDC can be obtained assuming isentropic compression of air ($\gamma = 1.4$)
At TDC

$$T_{TDC} = T_{amb} \left(\frac{P_{TDC}}{P_{amb}} \right)^{\gamma-1/\gamma} \tag{3.92}$$

The peak pressure can be obtained assuming an ideal gas, and that the combustion chamber volume at the crank-angle corresponding to peak pressure is the same as at TDC, and that the gas constant remains invariant with changing temperature:

At 10–30° ATDC

$$P_{\text{peak}} = P_{\text{TDC}} \left(\frac{T_{\text{peak}}}{T_{\text{TDC}}} \right)$$

(3.93)

The exhaust pressure can be estimated by assuming isentropic expansion and using the value for γ for which $\eta_{\text{ideal}} = \eta_{\text{th}} + Q_{\text{c}}$

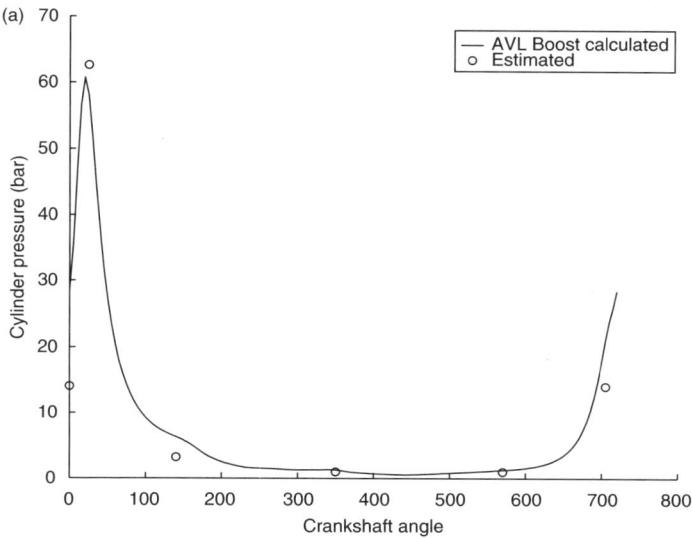

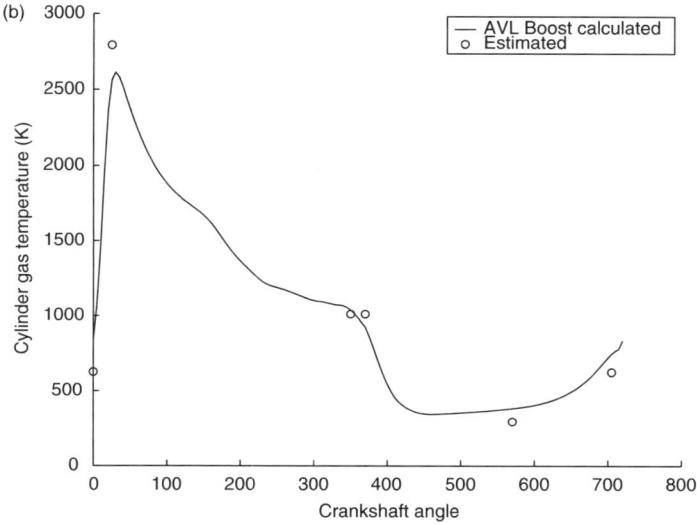

Figure 3.43 Estimates of in-cylinder gas pressure and temperature made using simplified assumptions detailed in section 3.3.14.2. Comparison is made with the results obtained from a full engine simulation of a 2.0-litre gasoline engine at 5000 rev min^{-1} using AVL *Boost*.

At EVO

$$P_{exh} \approx P_{peak} \left(\cfrac{1}{r \times \cfrac{evo}{180}} \right)^{\gamma} \qquad (3.94)$$

In summary, the pressure estimates are:

- at TDC, P_{TDC}
- at 10–30° ATDC, P_{peak}
- at EVO, P_{exh}
- at IVO, P_{amb}
- at IVC, P_{amb}
- at ignition, P_{TDC}.

And the temperature estimates are:

- at TDC, T_{TDC}
- at 10–30° ATDC, T_{peak}
- at IVO, T_{exh}
- at EVC, T_{exh}
- at IVC, T_{amb}
- at ignition, T_{TDC}.

In order to validate this approximate approach, the case of a 2.0-litre gasoline engine with a compression ratio of 10.5:1 is considered. The results of a full engine simulation (AVL Boost), using the integral of the first law of thermodynamics, a heat release model and Woschini's heat transfer relationships for the cylinder (Woschini,1967) give a bsfc of 271 g/kWh, and the cylinder gas pressure and temperature curves shown in Figure 3.43. The results from the simplified approach are seen to compare favourably.

3.4 Tyre noise

This section will deal with exterior noise resulting from the contact between the tyres and the road. This is often labelled tyre noise.

In a separate section (Section 4.5) the following is considered. Interior noise resulting from the contact between the tyres and the road, being transmitted to the interior by both airborne and structure-borne paths. This is often labelled road noise.

It is important to make a distinction between tyre and road noise for the reason that the motivation for controlling tyre noise is usually a desire to pass the drive-pass noise test for type approval (see Section 3.1) whereas the motivation for controlling road noise is usually to maximise passenger comfort and preserve the quality of speech communication (see Section 4.1).

3.4.1 Sources of airborne tyre noise

Airborne tyre noise has dominated the wayside noise levels caused by vehicles travelling at higher speeds for years, and more recently has begun to affect the low-speed acceleration tests used for type approval. As a result, a proposed EC Directive aims to reduce the problem by setting noise limits for different tyre types [C30/8, 28/1/98].

There is some debate over the sources of airborne tyre noise. The two noise-generating mechanisms given most attention are:

1. Noise generated when air is pumped in and out of tyre tread and road cavities during the contact process – the so-called air-pumping noise.
2. Noise generated by vibrations in the tyre caused by the contact process.

The most plausible explanation for the doubt over noise-generating mechanisms is that both may prove significant depending on:

- tyre construction and tread pattern;
- road surface;
- speed of the tyre.

The air-pumping mechanism has been shown to be significant for tyres with deep cross-grooves (known as cross-bar or cross-lug tyres) (Wilken et al., 1976). The effect of a single cross-groove cut into a treadless tyre was studied. Filling the groove with foam helped identify that the air-pumping mechanism is reinforced by acoustic resonance of the groove near its quarter wavelength frequency. Opening the closed end of the groove to circumferential grooves helped control this resonance.

The common observation that many treadless tyres are as noisy as tyres with treads suggests that tyre vibration also cause noise in addition to air pumping. With most modern tyre tread patterns that are not block like, the tyre vibration is commonly the dominant noise-generating mechanism.

Comparisons made between noise measurements near to and far from a rotating tyre suggest that (reported in Nilsson (1976)):

- most of the noise originates near to the contact patch;
- the sound intensity is greatest at the entrance and exit surfaces of the contact patch;
- the exit of the contact patch is important for tonal components of tyre noise;
- the tyre sidewall is not a significant radiator of sound.

As a result of the work described above, subsequent investigations have concentrated on measuring vibration and noise levels within the tread of the tyre (Jennewein and Bergmann, 1985).

The tonal tyre noise originates from regularities in the tyre construction. The random tyre noise originates first by radial excitation due to roughness in the road but also from random tangential movements of the tread pattern (Nilsson, 1976).

Tonal tyre noise is more speed-dependent than random tyre noise. Random tyre noise is strongly affected by the characteristics of the road surface. A simple empirical relationship between noise levels at 7.5 m and the tyre noise caused by coasting vehicles is presented in (Nilsson, 1976).

$$L_A = C + 10 \ \log_{10} (V^n) \ \text{dB} \tag{3.95}$$

where

L_A = sound pressure level at 7.5 m dBA due solely to tyre noise
V = vehicle coasting speed ($km\,h^{-1}$)

    Rib tyre on wet road $C = 47$, $n = 1.7$
    Smooth tyre on wet road $C = 23$, $n = 2.7$
    Smooth tyre on dry road $C = 10$, $n = 3.4$
    Regular tyre on dry road $C = 18$, $n = 3.0$

Measurements of vibration acceleration made on the tread show it to be greatest during the contact process (Jennewein and Bergmann, 1985). Removal of the influence of accelerations due solely to the flattening of the tyre contour yields the following:

- Radial acceleration of the tread bottom (particularly in the run-in section of the contact patch) is most important at frequencies below 1000 Hz.
- Tangential vibration of the tread blocks is most important at frequencies above 1000 Hz (particularly in the run-out section of the contact patch).

Noise measurements made with tiny microphones placed in the tread grooves (Jennewein and Bergmann, 1985) show:

- As a tread block strikes the ground, a groove that did have both ends open has one end sealed now. This forms a one quarter wavelength resonator with a resonant frequency commonly in the 1250 Hz third octave band. The resonance is excited by tread vibrations.
- As the tread block lifts at the trailing edge it forms a new resonator with the volume of air trapped in the groove behind. This second resonant frequency is typically in the range of 1500–2500 Hz.
- Different block arrangements create different types of resonator.
- There is a high amplification of sound (20 dB or more) within the resonators formed in the contact patch due to the leading and trailing edges of the contact patch acting as acoustic horns.

The various acoustic resonances are clearly seen in the noise spectrum measured at the contact patch and remain evident in the wayside noise.

3.4.2 The influence of the road surface on airborne tyre noise

It is commonly known that the characteristics of the road surface affect the wayside noise levels. As a result, a reference road surface is provided for use in type approval tests (ISO 10844: 1994) as described in Section 3.1.3.
    The three road characteristics influencing tyre noise are:

1. surface roughness;
2. the ability to shed surface water;
3. sound absorption.

There are some published data for the surface roughness of different roads (Cebon, 1999). These are in the form of displacement spectra, with the highest wavenumber components (leading to the highest-frequency noise) having the lowest amplitude. The simplest descriptor of the spectral density is

$$S_u(\kappa) = S_0 |\kappa|^{-n} \, \text{m}^3/\text{cycle} \tag{3.96}$$

where κ is the wavenumber (cycles/m) and $n = 2.5$ and S_0 varies according to the type of road as shown below

Motorway $\qquad S_0 = 3-50 \times 10^{-8} \, (\text{m}^{0.5}\text{cycle}^{1.5})$
Principal road $\quad S_0 = 3-800 \times 10^{-8} \, (\text{m}^{0.5}\text{cycle}^{1.5})$
Minor road $\qquad S_0 = 50-3000 \times 10^{-8} \, (\text{m}^{0.5}\text{cycle}^{1.5})$

The frequency associated with each wavenumber is the product of wavenumber and forward velocity (m s^{-1}).

Two common ways to texture a road surface to improve safety in wet conditions are:

- to roll stone chippings into hot asphalt (HRA);
- to produce closely spaced ridges that run transverse across a concrete road using a wire brush.

The second method of texturing a road surface will produce wayside noise levels 5 dB greater than the first method (Wright, 1999).

One alternative road surface, thought to be 3.5 dBA quieter than HRA (DoT, Calculation of road traffic noise, 1988) is the use of porous asphalt (PA). This is a thick (100 mm) road with a porous layer of larger stones sitting on an impermeable base course of asphalt. It is free flowing for surface water. Unfortunately, it is costly, gives poor wear and tends to freeze. Alternatives to PA are now available (Wright, 1999) which are thinner and more hard wearing. They are generally 2–5 dBA quieter than HRA and cheaper than PA.

It is important to remember road safety when considering tyre noise. Although noise levels increase when a vehicle is accelerating rather than coasting (Donavan, 1993) and when tyre load increases, fortunately there does not seem to be any correlation between wet friction and noise (Sandberg et al.,1999).

3.4.3 Measuring airborne tyre noise

Many test schedules have been developed for measuring airborne tyre noise. Some important ones include:

- Wayside noise from a coasting vehicle (C30/8, 28/1/98, J57 SAE 1994).
- Sound intensity measured alongside a rolling tyre mounted on a vehicle in use on the road (Donavan, 1993; Bolton et al., 1995).
- Sound pressure measured alongside a rolling tyre mounted on a special trailer. Various trailers are described in Sandberg (1998). One example, taken from Sandberg (1998) is sketched in Figure 3.44.
- Measuring tyre noise on a rolling road rig in the laboratory (Pope and Reynolds, 1976; Sandberg and Ejsmont, 1993)

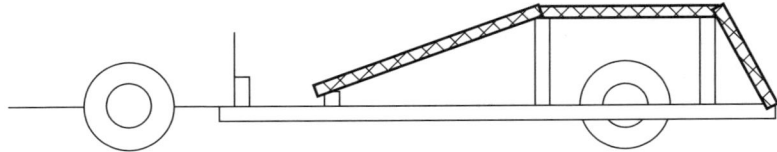

Figure 3.44 Measuring tyre noise using an Austrian trailer: after Sandberg (1998).

3.4.4 Controlling airborne tyre noise by design

Some tyre design guidance has emerged for reducing airborne tyre noise:

- Reduce the tread modulus (Muthukrishnan, 1990).
- Use a softer rubber on the tread to reduce the impact of the block at the leading edge of the contact patch (Jennewein and Bergmann, 1985).
- Avoiding tension in the tread blocks to decrease tangential vibration at the trailing edge of the contact patch (Jennewein and Bergmann, 1985).
- Avoid transverse grooves (Jennewein and Bergmann, 1985) to reduce the effect of acoustic cavity resonances. Short grooves with one end open are preferred.
- Provide frequency the modulation in the tyre by selectively arranging tread elements of various sizes – so-called pitch sequencing (Williams, 1995).

References

70/156/EEC, Council Directive of 6 February 1970 on the approximation of the laws of the Member States relating to the type-approval of motor-vehicles and their trailers, *Official Journal of the European Communities*, No. L 42/1, 23.2., 1970

70/157/EEC, Council Directive of 6 February 1970 on the approximation of the laws of the Member States relating to the permissible sound level and the exhaust system of motor-vehicles, *Official Journal of the European Communities*, No. L 42/16, 23.2., 1970

Statutory Instrument No. 981, 1984, The Motor Vehicles (Type Approval) (Great Britain) Regulations, HMSO, London, 1984

Statutory Instrument No. 1078, 1986, The Road Vehicles (Construction and Use) Regulations 1986, HMSO, London, 1986

SAE J1030 FEB87, Maximum sound level for passenger cars and light trucks, 1996 SAE Handbook, Society of Automotive Engineers, 1996

The Road Traffic Act, HMSO, London, 1988

Department of Transport Welsh Office, Calculation of road traffic noise, HMSO, London, 1988

92/97/EEC, Council Directive of 10 November 1992 amending Directive 70/157/EEC on the approximation of the laws of the Member States relating to the permissible sound level and the exhaust system of motor-vehicles, *Official Journal of the European Communities*, No. L 371/35, 19.2., 1992

ISO 10844: 1994, Acoustics – Test surface for road vehicle noise measurements, International Organization for Standardization, 1994

J57, Sound level of highway truck tires, Society of Automotive Engineers, June 1994

SAE J986 AUG94, Sound level for passenger cars and light trucks, 1996 SAE Handbook, Society of Automotive Engineers, 1996

Statutory Instrument No. 2051, 1998, The Motor Vehicle (EC Type Approval) Regulations 1998, HMSO, London, 1998

C30/8, 28/1/98, Proposal for a European Parliament and Council Directive amending Council Directive 92/23/EEC relating to tyres for motor vehicles and their trailers and to their fitting, *Official Journal of the European Communities*, 1998

BS ISO 362: 1998, Acoustics – Measurement of noise emitted by accelerating road vehicles – Engineering method, British Standards Institution, 1998

Annand, W.J.D., Roe, G.E., Gas flow in the internal combustion engine – power, performance, emission control and silencing, GT Foulis, 1974

GmbH, A.V.L., Boost user's manual – version 3, April 2000

Balcombe, D.R., Crowther, P., Practical development problems in achieving 74 dBA for cars, Proc. IOA, Vol. 15: Part 1, pp. 49–58, 1993

Benson, R.S., Whitehouse, N.D., Internal combustion engines – Vols 1 and 2, Pergamon Press, 1979

Benson, R.S., The thermodynamics and gas dynamics of internal combustion engines – Vols. 1, Clarendon Press, Oxford, 1982

Blair, G.P., Drouin, F.M.M., Relationship between discharge coefficients and accuracy of engine simulation, Society of Automotive Engineers, Paper No. 962527, 1996

Bolton, J.S., Hall, H.R., Schumacher, R.F., Stott, J., Correlation of tire intensity levels and passby sound pressure levels, SAE Paper No. 951355, Proceedings of the 1995 noise and vibration conference – Vol. 2, Society of Automotive Engineers, 1995

Bosch, Automotive handbook – Second edition, Robert Bosch GmbH, 1986

Cebon, D., Handbook of vehicle–road interaction, Swets & Zeitlinger, 1999

Davies, P.O.A.L., Flow-acoustic coupling in ducts, *Journal of Sound and Vibration*, 77(2), 191–209, 1981

Davies, P.O.A.L., Plane wave reflection at flow intakes, *Journal of Sound and Vibration*, 115(3), 560–564, 1987

Davies, P.O.A.L., Practical flow duct acoustics, *Journal of Sound and Vibration*, 124(1), 91–115, 1988

Davies, P.O.A.L., Aeroacoutics and time varying systems, *Journal of Sound and Vibration*, 190(3), 345–362, 1996

Davies, P.O.A.L., Harrison, M.F., Predictive acoustic modelling applied to the control of intake/exhaust noise of internal combustion engines, *Journal of Sound and Vibration*, 202, 249–274, 1997

Davies, P.O.A.L., Holland, K.R., I.C. engine intake and exhaust noise assessment, *Journal of Sound and Vibration*, 223(3), 425–444, 1999

Davies, P.O.A.L., Holland, K.R., The observed aeroacoustic behaviour of some flow-excited expansion chambers, *Journal of Sound and Vibration*, 239(4), 695–708, 2001

Davies, P.O.A.L., Holland, K.R., The measurement and prediction of sound waves of arbitrary amplitude in practical flow ducts, *Journal of Sound and Vibration*, in press 2004

Davies, P.O.A.L., Harrison, M.F., Collins, H.J., Acoustic modelling of multiple path silencers with experimental validations, *Journal of Sound and Vibration*, 200(2) 195–225, 1997

Davies, P.O.A.L. Bento Coelho, J.L., Battacharya, M., Reflection coefficients for an unflanged pipe with flow, *Journal of Sound and Vibration*, 72(4), 543–546, 1980

Donavan, P.R., Tire-pavement interaction noise measurement under vehicle operating conditions of cruise and acceleration, SAE Paper No. 931276, Included in *Tires and Handling – PT-59* (Ellis Johnson (ed.)), Society of Automotive Engineers, 1993

Dunkley, A., The acoustics of IC engine manifolds, Cranfield University – MSc Thesis, 1999

Dunne, J.M., Yarnold, I.C., Vehicle noise legislation – an overview, Proc. IOA, Vol. 15: Part 1, pp. 1–8, 1993

Fukutani, I., Watanabe, E., An analysis of the volumetric efficiency characteristics of 4-stroke cycle engines using the mean inlet Mach number Mim, SAE Paper No. 790484, 1979

Fukutani, I., Watanabe, E., Air flow through poppet inlet valves – analysis of static and dynamic flow coefficients, Society of Automotive Engineers, Paper No. 820154, 1982

Harrison, M.F., Notes to accompany the course on Piston Engines – Thermofluids, MSc Automotive Product Engineering, Cranfield University, 2003

Harrison, M.F., Stanev, P.T., A linear acoustic model for intake wave dynamics in IC engines, *Journal of Sound and Vibration*, 269(1), 361–387, 2004

Harrison, M.F., Dunkley, A., The acoustics of racing engine intake systems, *Journal of Sound and Vibration*, 271, 959–984, 2004

Harrison, M.F., De Soto Beodo, I., Rubio Unzueta, P.L., A linear acoustic model for multi-cylinder I.C. engine intake manifolds, including the effects of the intake throttle, *Journal of Sound and Vibration*, in press, 2004

Heywood, J.B., Internal combustion engine fundamentals, McGraw Hill Book Company, 1988

Hirsch, C., Numerical computation of internal and external flows – Vol.1, John Wiley & Sons, 1988

Holland, K.R., Davies, P.O.A.L., van der Walt, D.C., Sound power flux measurements in strongly excited flow ducts, *Journal of the Acoustical Society of America*, 112(6), 2863–2871, 2002

Howe, M.S., Contributions to the theory of aerodynamic noise, with applications to excess jet engine noise and the theory of the flute, *Journal of Fluid Mechanics*, 71, 625–673, 1975

Jennewein, M., Bergmann, M., Investigations concerning tyre/road noise sources and possibilities of noise reduction, Proceedings of the Institution of Mechanical Engineers, Vol. 199, D3, 199–205, 1985

Kim, G.T., Lee, B.H., 3-D sound source reconstruction and field prediction using the Helmholtz Integral Equation, *Journal of Sound and Vibration*, 136(2), 245–261, 1990

Kunz, F., Garcia, P., Simulation and measurement of hot exhaust gas flow noise with a cold air flow bench, SAE Paper No. 950546, 1995

Lighthill, M.J., On sound generated aerodynamically: I, general theory, Proceedings of the Royal Society of London, A211, 564–587, 1952

Lighthill, M.J., On sound generated aerodynamically: II, turbulence as a source of sound, Proceedings of the Royal Society of London, 222, 1–32, 1954

Lumley, J.L., Engines – an introduction, Cambridge University Press, 1999

Morfey, C.L., Acoustic energy in non-uniform flows, *Journal of Sound and Vibration*, 14(2), 159–170, 1971

Muthukrishnan, M., Effects of material properties on tire noise, SAE Paper No. 900762, Included in *Tires and Handling – PT-59* (Ellis Johnson (ed.)), Society of Automotive Engineers, 1990

Nilsson, N.A., On generating mechanisms for external tire noise, SAE Paper No. 762026, Proceedings of the SAE highway tire noise symposium, 10–12 November 1976, Society of Automotive Engineers, 1976

Ohata, A., Ishida, Y., Dynamic inlet pressure and volumetric efficiency of four cycle four cylinder engine, Society of Automotive Engineers, Paper No. 820407, 1982

Payri, F., Torregrosa, A.J., Payri, R., Evaluation through pressure and mass velocity distributions of the linear acoustical description of I.C. engine exhaust systems, Applied Acoustics, 60, 489–504, 2000

Pope, J., Reynolds, W.C., Tire noise generation: the roles of tire and road, SAE Paper No. 762023, Proceedings of the SAE highway tire noise symposium, 10–12 November 1976, Society of Automotive Engineers, 1976

Powell, A., Theory of vortex sound, *Journal of the Acoustical Society of America*, 36(1), 177–195, 1964

Rogers, G.F.C., Mayhew, Y.R., Thermodynamic and transport properties of fluids – Third edition, Basil Blackwell, 1980

Sandberg, U., Standardisation of a test track surface for use during vehicle noise testing, SAE Paper No. 911048, 1991

Sandberg, U., Ejsmont, J.A., The art of measuring noise from vehicle tyres, SAE Paper No. 931275, Included in *Tires and Handling – PT-59* (Ellis Johnson (ed.)), Society of Automotive Engineers, 1993

Sandberg, U., Noise trailers of the world – tools for tire/road noise measurements with the close-proximity method, Proceedings of the 1998 National Conference on noise control engineering, 5–8 April 1998

Sandberg, U., Ejsmont, J.A., Mioduszewski, P., Taryma, S., Noise – the challenge, Tire Technology International, 99, 98–101, 1999

Selemet, A., Kurniawan, D., Knotts, D., Novak, J.M., Study of whistles with a generic sidebranch, SAE Paper No. 1999-01-1814, 1999

Sievewright, G., Air flow noise of plastic air intake manifolds, SAE Paper No. 2000-01-0028, 2000

Steven, H., Effects of noise limits for powered vehicles on their emissions in real operation, SAE Paper No. 951257, 1995

Taylor, C.F., The internal combustion engine in theory and practice – 1, Second edition, revised, MIT Press, 1985

Taylor, N.C., Bridgewater, A.P., Airborne road noise source definition, IMechE, C521/011, 1998

Tsung-chi Tsu Theory of the inlet and exhaust processes of internal combustion engines, NACA TN 1446, 1947

Walker, A.W., The effect of the ISO surface on noise passby levels, Reproduced in: *Noise and the automobile*, selected papers from Autotech 93, Birmingham, 16–19 November 1993, Published by Mechanical Engineering Publications Ltd, London, 1994

Watson, N., Janota, M.S., Turbocharging the internal combustion engine, Macmillan Press Ltd, 1982

Weaving, J. (ed.) Internal combustion engineering – science and technology, Elsevier Applied Science, 1990

Wilken, I.D., Oswald, L.J., Hickling, R., Research on individual noise source mechanisms of truck tyres: aeroacoustic sources, SAE Paper No. 762022, Proceedings of the SAE highway tire noise symposium, 10–12 November 1976, Society of Automotive Engineers, 1976

Williams, T.A., Tire tread pattern noise reduction through the application of pitch sequencing, SAE Paper No. 951352, Proceedings of the 1995 noise and vibration conference – Vol. 2, Society of Automotive Engineers, 1995

Winterbone, D.E., Pearson, R.J., Design techniques for engine manifolds – wave action methods for IC engines, Professional Engineering Publishing, 1999

Woschini, G., A universally applicable equation for the instantaneous heat transfer coefficient in the internal combustion engine, Society of Automotive Engineers, Paper No. 670931, 1967

Wright, M., Thin and quiet? An update on new quiet road surface products, Acoustics Bulletin, September/October 1999, Institute of Acoustics, 1999

Zemansky, M.W., Dittman, R.H., Heat and thermodynamics (an intermediate textbook) – Seventh edition, McGraw Hill International Editions, 1997

Appendix 3A: Valve and port geometry

At low valve lifts, the open area of the valve is given by (Heywood, 1988)

$$A_\text{m} = \pi L_\text{v} \cos \beta \left(D_\text{v} - 2w + \frac{L_\text{v}}{2} \sin 2\beta \right) \tag{A3.1}$$

for $\dfrac{w}{\sin \beta \cos \beta} > L_\text{v} > 0$

At intermediate valve lift,

$$A_\text{m} = \pi D_\text{m} \left[(L_\text{v} - w \tan \beta)^2 + w^2 \right]^{1/2} \tag{A3.2}$$

for $\left[\left(\dfrac{D_\text{p}^2 - D_\text{s}^2}{4 D_\text{m}} \right) - w^2 \right]^{1/2} + w \tan \beta \geq L_\text{v} > \dfrac{w}{\sin \beta \cos \beta}$

where D_p is the port diameter.

When the valve lift is sufficiently large, the minimum flow area actually becomes the port area minus the valve stem area

$$A_\text{m} = \frac{\pi}{4} \left(D_\text{p}^2 - D_\text{s}^2 \right) \tag{A3.3}$$

for $L_\text{v} > \left[\left(\dfrac{D_\text{p}^2 - D_\text{s}^2}{4 D_\text{m}} \right) - w^2 \right]^{1/2} + w \tan \beta$

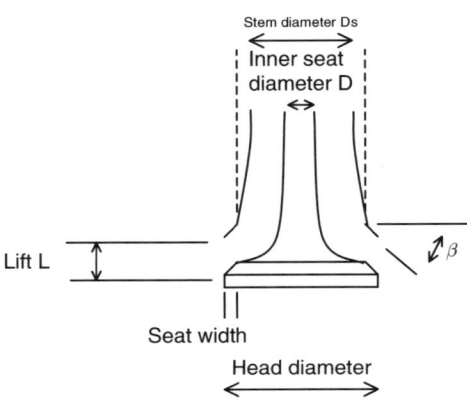

Figure 3.A.1 Valve geometry.

Interior noise: assessment and control

4.1 Subjective and objective methods of assessment

4.1.1 Background

Vehicle interior noise is a combination of:

- engine noise;
- road noise;
- intake noise;
- exhaust noise;
- aerodynamic noise;
- noise from components and ancillaries;
- brake noise;
- squeaks, rattles and 'tizzes'.

Apart from squeaks, rattles and tizzes that occur inside the passenger compartment, noise or vibration usually originates from outside, interacting with the vehicle structure in some way (and possibly hence with other noise sources) and then producing radiated sound inside the compartment. This process is illustrated in Figure 4.1.
The interaction with the structure can be either as:

- An airborne noise path – airborne noise from outside the passenger compartment leaking in to cause airborne noise inside.
- A structure-borne noise path – vibration from outside causing the surfaces of the passenger compartment to vibrate and radiate noise.

Direct airborne noise paths are found where there is a lack of sealing between the interior and the exterior environments (around door seals, grommets in the bulkhead, etc.). Indirect airborne noise paths are found when airborne noise outside impinges on the surfaces of the passenger compartment, causing them to vibrate and radiate noise inside.

The interaction between the noise source and the structure has a filtering effect on the final interior noise level (see Figure 4.1). For instance, in the case of indirect airborne paths, the transmission of sound to the passenger compartment will be greatest at low frequencies due to the action of the mass law for transmission loss (TL) (see Section 4.10.5). In the case of structure-borne paths, the use of resilient mounts will

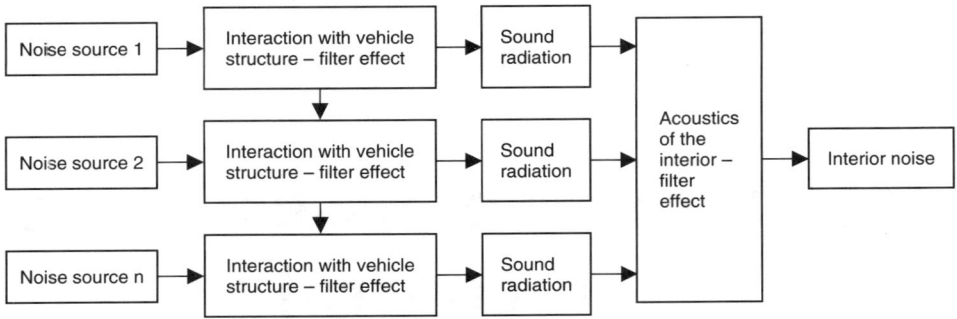

Figure 4.1 Schematic showing the origin of interior noise.

isolate the passenger compartment at higher frequencies and so transmission will also be greatest at low frequencies (see Section 6.3).

From the above observations, one would expect a strong low frequency component to interior noise levels. This is found in practice (see for instance Steel et al. [2000]) in spite of the filtering effect due to the fact that steel panels are rather poor radiators of sound at low frequencies (see Section 4.3.7).

One of the reasons for this is due to acoustic modes within the passenger compartment enhancing low-frequency noise levels. Section 2.5.1 discusses the fact that the so-called acoustic modes are set up at the natural frequencies of the space, frequencies given by (for a rectangular space)

$$f_{a,b,c} = \frac{c}{2\pi}\sqrt{\left(\frac{a\pi}{x}\right)^2 + \left(\frac{b\pi}{y}\right)^2 + \left(\frac{c\pi}{z}\right)^2} \qquad (4.1)$$

where

c = speed of sound in air $(\mathrm{m\,s^{-1}})$
a, b, c = integer indices $1, 2, 3, \ldots$
x, y, z = acoustic dimensions of the space (m)

For the modern generation of small European car, the longest acoustic dimension of the passenger compartment is that between the footwell and the rear screen. This dimension is typically a little more than the wheelbase of the vehicle, perhaps 2.5 m. The lowest frequency acoustic mode of the passenger compartment (the 1, 0, 0 mode, $a = 1$, $b = 0$, $c = 0$) therefore has pressure maxima (anti-nodes) at the footwells and at the rear screen and pressure minima at the midpoint of the wheelbase. Assuming $x = 2.5$ m this would occur at 69 Hz. Note that the driver's head position is usually just aft of the midpoint of the wheelbase and so the driver seldom enjoys the benefits of being precisely at the nodal position.

The 1, 0, 0 mode is commonly expected in the frequency range of 65–75 Hz. Our estimate of 69 Hz equates to the firing frequency of a four-cylinder, four-stroke engine operating at 2070 rev min^{-1}. European four-cylinder cars commonly exhibit low-frequency noise peaks (known as interior booms or sometimes as body booms) at engine speeds in the 2000–2500 rev min^{-1} range as a result of exciting the 1, 0, 0 mode.

The next longitudinal mode (2, 0, 0) would be expected at around 138 Hz. It often occurs at a lower frequency though in road cars.

The first transverse mode (0, 1, 0) can be expected at 123 Hz if the transverse acoustic dimension is assumed equal to the typical track of a small European car (1.4 m). The (0, 2, 0) mode would be expected at 246 Hz.

The first vertical mode (0, 0, 1) can be expected at 143 Hz if the vertical dimension is assumed to be the typical body height of a small European car (1.2 m). The (0, 0, 2) mode would be expected at 286 Hz.

The firing frequency of a four-cylinder, four-stroke engine lies in the range of 33–200 Hz for the corresponding speed range of 1000–6000 rev min^{-1}. In this range, using the rather over-simplified analysis used above, one might reasonably expect the following body booms:

- (1, 0, 0) at around 70 Hz/2100 rev min^{-1}
- (0, 1, 0) at around 120 Hz/3600 rev min^{-1}
- (2, 0, 0) at around 140 Hz/4200 rev min^{-1}
- (0, 0, 1) at around 140 Hz/4200 rev min^{-1}
- (1, 1, 0) at around 140 Hz/4200 rev min^{-1}
- (2, 1, 0) at around 185 Hz/5550 rev min^{-1}

The boom around 3600 rev min^{-1} is often the most annoying. This is because the lower speed boom is usually only transient as the vehicle will accelerate quickly through it. Also the high speed booms are seldom a problem as most drivers will not run their engines for long periods at 4000+ rev min^{-1} and will select a higher gear when they first hear such booms. However, drivers on high-speed roads may well find themselves having to cruise around 3600 rev min^{-1} in top gear with an annoying boom ever present in the passenger compartment. As the occupants tend to sit near to the sides of the passenger compartment, all find themselves at anti-nodes of the (0, 1, 0) boom and all suffer pressure maxima.

The passengers in the rear seats of most cars suffer from pressure maxima at most modes – longitudinal, lateral, vertical (and other non-orthogonal modes). This is not the case for the occupants of the front seats. For this reason, rear seat noise quality is often a greater issue for concern than the levels and quality of noise at the driving position.

4.1.2 On the balance between airborne and structure-borne noise

The vibration isolating effects of resilient components (engine mounts, subframe mounts, under-carpet treatments) tends to limit the significance of structure-borne noise to frequencies below around 500 Hz. At higher frequencies, noise received via airborne noise paths generally dominates the interior noise levels.

At frequencies below 500 Hz, airborne noise may remain a significant contributor to overall interior noise levels, particularly if the sealing of the passenger compartment (door seals, window seals, grommets in the bulkhead) is not perfect.

4.1.3 On the measurement of interior noise

A procedure for measuring vehicle interior noise is specified in BS 6086 – 1981 (ISO 5128 – 1980).

BS 6086 calls for:

- The measurement of sound pressure level, fast weighted, 'A'-weighted – 1/3 octave analysis if possible (see Sections 2.3.4 and 2.4.2 for details of these).
- The measurement of noise levels at more than one location, and at least at the driver's ear position and one point at the rear of the vehicle.
- Microphones to be horizontal and pointing with their direction of maximum sensitivity in the direction that an occupant would normally be looking.
- Microphones to be no closer than 0.15 m to walls or upholstery.
- Microphones should be mounted in such a way as to be unaffected by vibrations of the vehicle.
- Tests shall be carried out with the vehicle being stationary (at idle and at full engine speed, held for 5 seconds), at various steady speed in the range of 60–120 km hr^{-1}, and with full-load acceleration from 45% of the maximum power speed to 90% of the maximum power speed with the transmission in the highest position without exceeding 120 km hr^{-1}. Many practitioners make full load accelerations from near idle to near maximum engine speed in both second and third gears as their standard interior noise acceleration tests.

Most organisations have their own test regimes, and in many cases these vary from BS 6086. For this reason it is vitally important to note the microphone positions and orientations used in any test and the exact test conditions used when reporting test results.

The results of transient tests are commonly presented in the form of order diagrams or waterfall plots (see Section 2.4.2). The results of steady state/speed tests are commonly presented in the form of narrow band or third octave band spectra (see Section 2.4.2).

4.1.4 On the subjective assessment of interior noise

BS 6086 presents a method for the objective assessment of interior noise. These assessments are commonly used to guide the process of vehicle refinement. Notwithstanding this, it is clear that the subjective impression of noise – both noise level (as assessed under BS 6086) and noise quality – is a factor in the process of selling cars.

The subjective assessment of vehicle interior noise can be used for different purposes (Van der Auweraer et al., 1997):

- to judge sound quality;
- to diagnose the causes of poor sound quality.

The judgement of sound quality is best performed by a panel of potential customers. A diagnosis is best made by a panel of refinement specialists. Different methodologies can be used for assessment or diagnosis and those designed for assessment will be discussed first.

Subjective assessment of interior noise quality can be performed in two ways:

1. The driving method where panel members answer questions relating to their experience of driving a vehicle (and perhaps some competitor vehicles).
2. The listening room method where panel members assess the quality of recorded sounds in a dedicated listening room or over headphones.

The driving method is the most complex and costly choice as, in order to be statistically significant, it involves many people undertaking extended test drives (safely) along predetermined test routes. If it is undertaken, one of two strategies can be followed. The first is to get drivers to rank sounds that they hear in order of preference. So, if only one vehicle is being assessed, the panel member could rank the following sound quality descriptors according to which they best described their experience:

Quiet
Luxurious
Powerful
Sporty
Pleasant
Unpleasant
Noisy
Commonplace
Unusual

If competitor vehicles are being tested a relative ranking can be produced, for example: car A is quieter than car B; car B is more sporty than car A, etc. In an alternative method of ranking, panel members answer paired questions like: *which best describes the noise level – quiet or noisy?*

Instead of ranking tests, panel members can be asked to rate (usually on a scale of 1–10) each descriptor. A score of 0 on the 'quiet' descriptor would indicate a noisy vehicle and a score of 10 would indicate a very quiet vehicle. This magnitude estimation often proves difficult for untrained members of the public. An alternative is to set a scale for paired questions such as:

Is the vehicle

Quiet?   Noisy?
Extremely, very, somewhat, neither, somewhat, very, extremely.

and a semantic differential system of questioning is obtained (Otto et al., 1999).

The listening room method is different. Here an artificial head (and torso) is placed in a passenger seat of the vehicle and binaural recordings of the interior noise are made. Thereafter, ranking, paired ranking, magnitude estimation or semantic differential questioning can be undertaken for a statistically significant number of panel members in a room (see Johnson (1995) for example). Each panel member can be tested separately, or juries of ten persons can be assembled. Otto et al. (1999) describe the process well.

Apart from safety and cost, one significant advantage that the listening room method has over the driving method is that the recorded sounds can be manipulated digitally and the altered sounds used to establish customer preferences without the cost of constructing prototype vehicles. Russell et al. (1988, 1992) describe an early system with more sophisticated versions being demonstrated by Maunder (1996) and Naylor and Willats (2000). Such advanced digital techniques can be used by panels of refinement specialists to diagnose the causes of poor sound quality. Otherwise, magnitude estimations for the driving method are commonly used when the panel has the opportunity to drive a candidate vehicle. Most involve a panel of people driving and riding in the car(s) along a predetermined test route on public roads and rating the following noise (and vibration) attributes:

- wind noise
- road noise
- engine noise
- idle refinement
- cruising refinement
- transmission noise
- general shakes and vibrations
- squeaks, rattles and tizzes
- ride quality
- driveability
- noise that is a 'feature' (sporty exhaust notes, etc.).

The ratings are made to a common scale from 1 to 10 as shown in Table 4.1.

- A rating of less than 4 is unacceptable for any attribute.
- A rating of 5 or 6 is borderline.
- A rating of 7 or more on any attribute is acceptable.

Most new cars are launched with a subjective rating of 7 or 8 on most attributes. Russell et al. (1992) supply supplementary attributes for rating noise from particular sources such as diesel engines (called dimensions in the reference):

- overall level;
- low-frequency content and 'boom';
- impulsiveness;
- harmonic content;
- strong tones;
- irregularity;
- high frequency content.

Table 4.1 Common subjective rating scheme

1	2	3	4	5	6	7	8	9	10
Not acceptable			Objectionable	Requires improvement	Medium	Light	Very light	Trace	No trace

4.2 Noise path analysis

4.2.1 Background

Interior noise levels can be controlled under certain circumstances by adding sound absorbing material to the passenger compartment. The advantage in controlling sound by absorption in the vehicle interior is that it will take effect on noise in a certain frequency range regardless of origin or noise path providing that the receiver is at some distance from the source. The disadvantage is that its effect is usually rather small unless the vehicle had little sound absorbing trim in the first place.

As an alternative, a noise path analysis may be used to determine the contributions to interior noise levels made by noise using different paths between the source(s) and the vehicle interior. Depending on the dominant noise path(s) identified, the following noise control options are available:

Structure-borne noise from the engine

- Improve the vibration isolation provided by engine mounts.
- Reduce the vibration produced by the engine.
- Add damping treatments to resonant portions of the firewall and floor.

Airborne noise from the engine

- Improve the TL of the firewall and/or floor by adding a barrier layer (Wentzel and Saha, 1995), commonly a mat of EVA, PVC or natural rubber (surface density $1–7 \, \text{kg m}^{-2}$) glued to a decoupling layer of chip foam or something similar (volumetric density in the range of $20–60 \, \text{kg m}^{-3}$) or a fibrous matting (density in the range of $60–80 \, \text{kg m}^{-3}$).
- Plug all gaps in the firewall caused by ill-fitting grommets, etc.
- Add sound absorption treatment to the underside of the hood (bonnet).

Structure-borne noise from the road

- Change tyres.
- Change suspension bushes.
- Change subframe bushes (if an isolated subframe is used).
- Add damping treatments to resonant portions of the firewall and floor.

Airborne noise from the tyres

- Change tyres.
- Improve the TL of the firewall and/or floor by adding a barrier layer along with a decoupling layer.
- Improve the door seals if necessary.

Structure-borne noise from the exhaust

- Improve the vibration isolation afforded by the mounts by using more compliant mounts, fixed to high impedance points on the chassis and nodal points on the exhaust system.
- Improve the TL of the trunk and/or rear floor by adding a barrier layer along with a decoupling layer.

- Add damping treatments to resonant portions of the trunk and floor.
- Add a flexible coupling between the exit of the catalyst and the remainder of the system.

Airborne noise from the exhaust

- Improve the TL of the trunk and/or rear floor by adding a barrier layer along with a decoupling layer.
- Improve the door seals if necessary.

Structure-borne noise from the intake

- Mount the body-side elements of the intake system (filter box and snorkel usually) on resilient mounts.

Airborne noise from the intake

- Improve the TL of the firewall and/or floor by adding a barrier layer along with a decoupling layer.
- Improve the door seals if necessary.
- Add sound absorption treatment to the underside of the hood.

Aerodynamic Noise

- Re-contour wing mirrors, aerials, door-handles, etc.
- Improve the door seals if necessary.

Engine component noise

- Reduce at source by adopting a quieter component.
- Improve the TL of the firewall and/or floor by adding a barrier layer along with a decoupling layer.
- Add sound absorption treatment to the underside of the hood.

4.2.2 Coherence methods for noise path analysis

In the first instance, readers will need to understand the significance of the terms coherence and frequency response function. To do this, they are directed to:

- a review of some background materials on systems in Appendix 4A;
- an explanation of the convolution integral in Appendix 4B;
- explanations of the covariance function, correlation and coherence given in Appendix 4C;
- the derivation of the frequency response function given in Appendix 4D;
- Sinha (1991), Fahy and Walker (1998) and texts similar to Weltner et al. (1986) for further reading.

This section studies one noise source identification method (or noise path analysis method) based on measured coherence – that proposed by Halvorsen and Bendat (1975).

The analysis starts with their linear single input, single output (two pole – Appendix 4A) problem as illustrated in Figure 4.2.

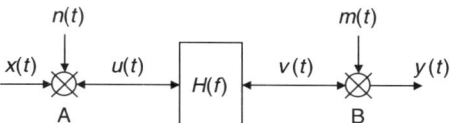

Figure 4.2 Halvorsen and Bendat's single input/single output problem (1975).

If data were acquired at point A in Figure 4.2, then

$$x(t) = u(t) + n(t) \qquad (4.2)$$

would be recorded, where u is the wanted input and n is some unwanted but unavoidable noise.

Equally, if data were acquired at point B in Figure 4.2, then

$$y(t) = v(t) + m(t) \qquad (4.3)$$

would be recorded, where v is the wanted input and m is some unwanted but unavoidable noise.

Now one can write down the following relationships (where $G(f)$ is a one-sided spectrum)

$$G_{xx}(f) = G_{uu}(f) + G_{nn}(f) \qquad (4.4)$$
$$G_{yy}(f) = G_{vv}(f) + G_{mm}(f) \qquad (4.5)$$
$$G_{xy} = G_{uv}(f) \qquad (4.6)$$

because $n(t)$ and $m(t)$ are assumed to be uncorrelated with $u(t)$ and $v(t)$

$$G_{vv}(f) = |H(f)|^2 G_{uu}(f) \quad \text{from Appendix 4D} \qquad (4.7)$$
$$G_{uv}(f) = H(f) G_{uu}(f) \quad \text{from Appendix 4D} \qquad (4.8)$$
$$\gamma_{xy}^2(f) = \frac{|G_{xy}(f)|^2}{G_{xx}(f) G_{yy}(f)} \quad \text{from Appendix 4C} \qquad (4.9)$$

The coherence function given by equation (4.9) is the measured coherence not the true coherence.

If the measured coherence is multiplied by the measured output power spectrum, the coherent output power spectrum is obtained:

$$\gamma_{xy}^2(f) G_{yy}(f) = \frac{|G_{xy}(f)|^2}{G_{xx}(f)} \qquad (4.10)$$

Substituting equation (4.6) into equation (4.10)

$$\gamma_{xy}^2(f) G_{yy}(f) = \frac{|G_{uv}(f)|^2}{G_{xx}(f)} \qquad (4.11)$$

Substituting equation (4.4) into equation (4.11)

$$\gamma_{xy}^2(f)G_{yy}(f) = \frac{|G_{uv}(f)|^2}{G_{uu}(f) + G_{nn}(f)} \tag{4.12}$$

Substituting equation (4.8) into equation (4.12)

$$\gamma_{xy}^2(f)G_{yy}(f) = \frac{|H(f)G_{uu}(f)|^2}{G_{uu}(f) + G_{nn}(f)} = \frac{|H(f)|^2 G_{uu}^2(f)}{G_{uu}(f) + G_{nn}(f)} \tag{4.13}$$

Substituting equation (4.7) into equation (4.13)

$$\gamma_{xy}^2(f)G_{yy}(f) = \frac{G_{vv}(f) \cdot G_{uu}(f)}{G_{uu}(f) + G_{nn}(f)} = \frac{G_{vv}(f)}{1 + \left(\dfrac{G_{nn}(f)}{G_{uu}(f)}\right)} \tag{4.14}$$

So, the measured coherent output power spectrum will yield a good measure of the true system output spectrum $G_{vv}(f)$ providing the input signal-to-noise ratio is high. This method is not commonly used for automotive noise path analysis for the following reasons:

- $n(t)$ and $u(t)$ are often correlated when they result from the same source – like the engine. For example, the vibration experienced at the engine mount under study will contain contributions (the noise $n(t)$) provided by the same engine via other nearby engine mounts.
- $m(t)$ and $v(t)$ are often correlated. For example, the sound in the cabin due to transmission of vibration power through one engine mount is partially correlated with the sound due to power transmitted through the other mounts.
- The noise paths are often non-linear, particularly when transmission of vibration power via rubber components is concerned.
- Delays between $u(t)$ and $v(t)$ result in low estimates of coherence due to a lack of properly synchronous sampling. The effects of delays can be minimised by using sample data lengths that are much longer than the longest delay.

See Piersol (1978) and Verhulst and Verheij (1979) for further solutions to these problems.

4.2.3 Standard methods for noise path analysis

There are standard measurement methods for noise path analysis in automotive vehicles that overcome (to some degree at least) the limitations discussed for the coherent output power method.

Both the LMS [1998] and the I-DEAS [MSX 1998] measurement systems commonly used in the automotive industry offer methods for noise path analysis. These are broadly similar, allowing the user to choose between:

- the complex stiffness method (I-DEAS call this the 'force vector method'); and
- the matrix inversion method (I-DEAS call this the 'full matrix method').

Both methods are based on the same principle: that the received sound pressure level (or vibration acceleration) during operational conditions is the superposition of partial results, each describing the contribution of individual transfer paths.

$$r(f) = \sum_{i=1}^{n} \frac{R(f)}{S_i(f)} \cdot S_i(f) \tag{4.15}$$

where

$r(f)$ = received power spectral density
$R(f)/S_i(f)$ = frequency response function between the received power spectral density and the input power spectral density applied to transfer path i
$S_i(f)$ = input power spectral density of operational force or operational volume velocity applied to transfer path i.

The complex stiffness method (force-vector method) is suitable for occasions where the source of input power is connected to the receiver via sprung or compliant mounts and there is a reasonable differential movement across each mount. For example, the complex stiffness method is commonly applied to the engine mounting problem thus:

1. Identify all probable vibration paths between the source (the engine) and the receiver (a microphone positioned in the passenger compartment).
2. Organise a phase reference signal for use in subsequent measurements. This might be an electrical signal taken from the tachometer or the conditioned output from an accelerometer attached to the engine block.
3. On the test track, or the rolling-road dynamometer, measure acceleration levels in the three axes (x, y, z) on the engine side of the engine mounts. Analyse these levels either as power spectral densities or as 'order-levels' (see Section 2.4.2.5). Order levels may be determined either by synchronous sampling relative to the reference phase signal or they may be extracted from the time-averaged power spectral density (in which case the magnitude of the order level is taken as the average of three spectral lines, whereas the phase information is taken from the middle spectral line).
4. Repeat (3) for the body side of each engine mount. Ideally steps (3) and (4) should be undertaken simultaneously.
5. Remove the engine from the vehicle and measure the input accelerances in the three axes (x, y, z) on the body side of the mounts due to force excitation applied in each of the three axes in turn. The force excitation is applied with either a shaker or using a hammer fitted with a force transducer at its tip. The accelerance between the acceleration $(m\,s^{-2})$ at point i due to a force applied at point j is given by

$$\text{accelerance} = \frac{a_{ij}}{F_j} \tag{4.16}$$

6. With the engine still removed from the vehicle, measure the noise transfer functions (NTF)

$$\text{NTF} = \frac{P}{F_i} \quad P \text{ is the sound pressure (Pa)} \tag{4.17}$$

using the force excitation from (5) and a microphone in the passenger compartment.

7. Measure the dynamic stiffness of each engine mount along with its damping. These may be combined to give a complex stiffness (Cremer and Heckl, 1988)

$$\bar{K} = K(1+i\eta) \qquad (4.18)$$

where K is the dynamic stiffness $(\mathrm{N\,m^{-1}})$ and η is the loss factor (dimensionless). A typical example of the results from a measurement of engine mount stiffness is shown in Figure 4.3.

8. Double integrate the engine side accelerations from (3) and the body side accelerations from (4) and estimate the differential displacements of the mounts. The force applied to the body at each mount is therefore

$$F_i(f) = \bar{K}(f) \cdot (x_s(f) - x_r(f)) \qquad (4.19)$$

where x_s is the displacement on the engine side of the mount (the source side) and x_r is the displacement on the body side (the receiver side).

9. Use the forces obtained from (8) along with the NTF from (6) to estimate the partial sound pressure in the passenger compartment due solely to vibration power transmitted across each engine mount.

10. Sum all of the partial sound pressure contributions from all of the identified noise paths on a polar plot (or Nyquist diagram) and compare the result with the measured total sound pressure for validation of the method.

The use of a polar plot for displaying the results is important as it shows both magnitude and phase information. Therefore, interference effects between the contributions from different paths may be investigated.

For noise paths (transfer paths) comprising only rigid (or fairly rigid) connections, the complex stiffness method is not suitable as there will be negligibly small relative displacement across the connections.

In such cases, the force imposed by the source at each input to the system is determined from the inverse of the full accelerance matrix multiplied by the accelerations measured

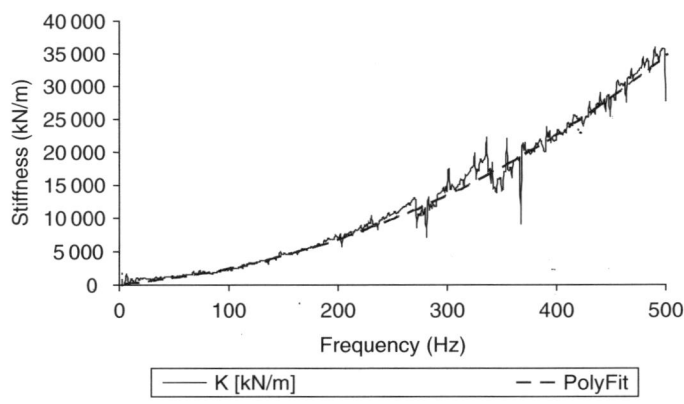

Figure 4.3 Measured engine mount stiffness (Verstraeten, 2003).

on the receiver side of the first connection under operational conditions. This is the matrix inversion method (full matrix method).

$$[F] = [A]^{-1} [\ddot{x}] \qquad (4.20)$$

where

$$[A] \text{ is the accelerance matrix } [A] = \begin{bmatrix} \dfrac{\ddot{x}_{11}}{F_1} & \dfrac{\ddot{x}_{12}}{F_2} & \cdots & \dfrac{\ddot{x}_{1n}}{F_n} \\ \\ \dfrac{\ddot{x}_{21}}{F_1} & \dfrac{\ddot{x}_{22}}{F_2} & \cdots & \cdots \\ \\ \vdots & \vdots & \vdots & \vdots \\ \\ \dfrac{\ddot{x}_{m1}}{F_1} & \cdots & \cdots & \dfrac{\ddot{x}_{mn}}{F_n} \end{bmatrix}$$

$$[\ddot{x}] = \begin{bmatrix} \ddot{x}_1 \\ \ddot{x}_2 \\ \vdots \\ \ddot{x}_m \end{bmatrix} \qquad [F] = \begin{bmatrix} F_1 \\ F_2 \\ \vdots \\ F_n \end{bmatrix}$$

For a unique solution, the number of measured responses m must be at least equal to the number of identified forces n. If $m > n$ then the equation is overdetermined and a least square estimate is found.

The matrix inversion method may be applied to the engine mounting problem thus:

1. Identify all probable vibration paths between the source (the engine) and the receiver (a microphone positioned in the passenger compartment).
2. Organise a phase reference signal for use in subsequent measurements. This might be an electrical signal taken from the tachometer or the amplified output from an accelerometer attached to the engine block.
3. On the test track, or the rolling dynamometer, measure acceleration levels in the three axes (x, y, z) on the body side of the engine mounts. Analyse these levels either as power spectral densities or as 'order-levels'.
4. Remove the engine from the vehicle and measure the input accelerances in the three axes (x, y, z) on the body side of the mounts due to force excitation applied in each of the three axes in turn. The force excitation is applied with either a shaker or using a hammer fitted with a force transducer at its tip.
5. With the engine still removed from the vehicle, measure the NTF using the force excitation from (4) and a microphone in the passenger compartment.
6. Estimate the forces applied at each engine mount from the inverse of the full accelerance matrix obtained from (4) multiplied by the body side accelerations measured in (3).

7. Use the forces obtained from (6) along with the NTF from (5) to estimate the partial sound pressure in the passenger compartment due solely to vibration power transmitted across each engine mount.
8. Sum all of the partial sound pressure contributions from all of the identified noise paths on a polar plot (or Nyquist diagram) and compare the result with the measured total sound pressure for validation of the method.

4.2.4 Non invasive methods for noise path analysis

The practical difficulty with both the complex stiffness and the matrix inversion method is that the source must at some point be disconnected from the receiver. There are at least two further methods in the literature that do not require this disconnection.

The first, the TopExpress MPSD system (Harper et al., 1993) simplifies the equations of motion by assuming that the various transmission paths can only interact via their attachments to the source and/or the structure as illustrated in Figure 4.4. The simplified matrix equation takes account of forces exerted on the paths by the source and forces exerted on the source by the paths. The matrix equation may be solved for the forces at the support ends of the structure, and the measured NTF may then be used to calculate radiated sound.

Another non-invasive technique originates from TNO (Netherlands) and ISVR (Southampton) – the equivalent forces method (Janssens et al., 1999). Here, transfer functions

$$A_{ij} = \frac{a_i}{F_j} \qquad (4.21)$$

are measured at convenient positions on the machine along with NTF

$$H_{kj} = \frac{P_k}{F_j} = \frac{a_j}{Q_R} \qquad (4.22)$$

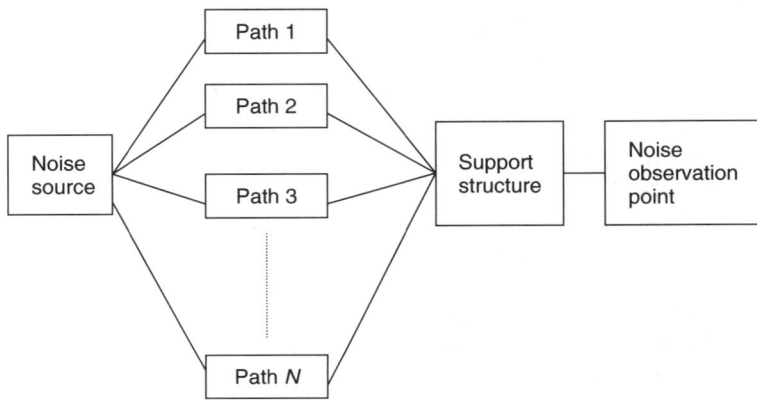

Figure 4.4 Model of the TopExpress MPSD system: after Harper et al. (1993).

where P_k is the acoustic pressure, F_j is the force and Q_R is the volume acceleration and a_j is the acceleration.

The inventive part of the method is an analytical step of determining the amplitude and phase of a set of equivalent forces $\{F_j\}_{EQ}$ which when applied at the positions j would reconstruct the vibration field due to the machine $\{a_i\}_{mach}$ as closely as possible

$$\{F_j\}_{EQ} = [A_{ij}]^{-1} \{a_i\}_{mach} \tag{4.23}$$

$$\{P_k\}_{path} = [H_{kj}] \{F_j\}_{EQ} \tag{4.24}$$

Because $\{F_j\}_{EQ}$ may be placed anywhere convenient on the system, they have no particular significance and are of little interest in themselves. For the standard (complex stiffness or matrix inversion) noise path analysis methods, the NTF are measured at the connection between the source and the receiver and therefore $\{F_j\}$ has physical significance.

4.3 Measuring the sound power of IC engines and other vehicle noise sources

4.3.1 Near and far acoustic fields

Before moving on to discussing methods by which sound power may be measured, certain restrictions concerning the acoustic near field of the sound source must be clarified.

The sound field produced by a sound source radiating in a free field may be divided into three regions (Bies and Hansen, 1996):

1. the hydrodynamic near field;
2. the geometric (or Fresnel) near field;
3. the far field.

The hydrodynamic near field is that region immediately adjacent to the vibrating surface of the source. The thickness of the hydrodynamic near field is much less than one wavelength. In this region, the fluid motion is not directly associated with sound propagation. Tangential fluid motion may occur. The local particle velocities will be out of phase with the acoustic pressures. As the propagation of sound to the far field is associated with the in-phase components of pressure and particle velocity, it follows that measurements of the acoustic pressure amplitude in the hydrodynamic near field give no indication of the sound power radiated to the far field by the source.

The sound field adjacent to the hydrodynamic near field is known as the geometric near field. In this field the particle velocities are in phase with the pressure levels and so sound propagates through the geometric near field. However, the sound pressure levels do not decrease monotonically at the rate of 6 dB per doubling of distance from the source. In fact, variations in the surface velocity of the source produce interference patterns in the geometric near field which result in a distribution of sound pressure maxima and minima.

This effect is more noticeable in tonal noise than in broadband noise. Radiated sound power can be estimated from a sufficient number of pressure measurements made in the geometric near field. However, interpretation of the measurements for directivity information should be made only with caution.

The region of the sound field extending beyond the geometric near field is known as the far field. In open free-field space, sound pressure levels decrease monotonically at a rate of 6 dB per doubling of distance in the far field. Directivity in the far field is well defined. According to Bies (1976) and reported in Bies and Hansen (1996) the far field is characterised by the satisfaction of three criteria

$$r \gg \lambda/2\pi \tag{4.25}$$

$$r \gg l \tag{4.26}$$

$$r \gg \frac{\pi l^2}{2\lambda} \tag{4.27}$$

where

r = distance between source and measurement position (m)
l = maximum source dimension (m)
λ = wavelength (m)

In this case the criteria $\gg$ is taken to mean a factor greater than 3.

It should be noted that the use of these criteria will ensure that a position is in the far field, but occasionally the far field will also occur nearer to the source. For instance, a large pulsating sphere only has a far field.

The adoption of these criteria can be difficult in anechoic chambers. For instance, a large source in a small anechoic chamber may only in theory produce far field conditions near the walls of the chamber. The absorption characteristics of the walls may disturb the field adjacent to the walls, and so the usable far field region may be quite small.

When the sound source is radiating in a reverberant or semi-reverberant space, the monotonic decrease in sound pressure level may be reduced in the far field due to the influence of the reverberant field.

4.3.2 Different methods for measuring sound power

In circumstances where the control of noise from existing machinery is to be considered, it is often advantageous to measure the sound power output of the source.

Sound power can be calculated or estimated from either:

1. pressure measurements in the far field;
2. pressure measurements in the geometric near field;
3. intensity measurements;
4. surface vibration velocity measurements.

The most precise laboratory methods involve far field pressure measurements made in anechoic or reverberant chambers. The next most precise method uses far field pressure measurements made in quiet semi-reverberant spaces in the field. Where background noise levels cause difficulties, a less precise method using near field pressure measurements may be used.

As an alternative to pressure measurements, realistic measures of sound power may be made in the field or laboratory with careful use of intensity techniques. In cases where there are several noise sources, or the use of pressure or intensity techniques is not practical, estimates of sound power can be made from measurements of surface vibration velocity. The reader is directed to Section 2.5 where a full discussion of the acoustic quantities of pressure, intensity and power is offered.

4.3.3 Measurement of sound power in a free field using sound pressure techniques

The sound power (W, Watts) of a source may be obtained from the integration of the intensity $I(\mathrm{W\,m^{-2}})$ over a notional spherical surface of area $S\,(\mathrm{m^2})$ surrounding it.

$$W = IS = 4\pi r^2 I \tag{4.28}$$

The integration of the acoustic intensity over the spherical surface is achieved by determining the time averaged, squared acoustic pressures at a number of measurement points arranged to uniformly sample the integration surface.

In an environment in which there are no reflecting surfaces, the maximum intensity is given by Bies and Hansen (1996) as

$$I_{\max} = \frac{1}{T}\int_0^T p \cdot u\ \mathrm{d}t \tag{4.29}$$

where p is the fluctuating (acoustic) pressure (Pa) and u is the particle velocity $(\mathrm{m\,s^{-1}})$ and

$$I_{\max} \simeq \frac{\overline{p^2}\ \mathrm{rms}}{\rho_0 c} \tag{4.30}$$

ρ_0 is the undisturbed density of fluid (air) $(\mathrm{kg\,m^{-3}})$
c is the speed of sound in the fluid (air) $(\mathrm{m\,s^{-1}})$

It can be shown (see Section 2.5.1) that for free-field conditions at room temperature

$$L_\mathrm{p} \simeq L_\mathrm{I} \tag{4.31}$$

where the reference sound pressure level is $20 \times 10^{-6}\,\mathrm{Pa}$ and the reference intensity is $10^{-12}\,\mathrm{W\,m^{-2}}$.
Therefore, the spatial average intensity is

$$\bar{L}_\mathrm{p} \simeq \bar{L}_\mathrm{I} = 10\ \log_{10}\left[\frac{1}{N}\sum_1^N 10^{L_{\mathrm{p}i}/10}\right]\mathrm{dB} \tag{4.32}$$

where $L_{\mathrm{p}i}$ is the sound pressure level (dB) recorded within one portion of the equipartitioned surface of a sphere that completely encloses the sound source in the far field.

So, for a free-field

$$L_w = \bar{L}_p + 10 \ \log_{10}(4\pi r^2) \ (dB)$$

$$L_w = \bar{L}_p + 20 \ \log_{10} r + 10 \ \log_{10}(4\pi) \ (dB)$$

$$L_w \simeq \bar{L}_p + 20 \ \log_{10} r + 11 \ (dB) \tag{4.33}$$

For the case of a sound source in a free field but placed next to a flat surface such as a hard floor

$$L_w = \bar{L}_p + 10 \ \log_{10}(2\pi r^2) \ (dB)$$

$$L_w \simeq \bar{L}_p + 20 \ \log_{10} r + 8 \ (dB) \tag{4.34}$$

For the case of a sound source placed near the junction of two flat planes

$$L_w \simeq \bar{L}_p + 20 \ \log_{10} r + 5 \ (dB) \tag{4.35}$$

For the case of a sound source placed near the junction of three flat planes

$$L_w \simeq \bar{L}_p + 20 \ \log_{10} r + 2 \ (dB) \tag{4.36}$$

If the measurement surface is in the far field then the directivity index corresponding to location i is

$$DI = L_{pi} - \bar{L}_p \ (dB) \tag{4.37}$$

The requirements for a space in which such measurements of sound power may be made are as follows:

1. Any reverberant sound is negligible.
2. A semi-anechoic space may be used if convenient providing that a suitable correction is made.

The procedure for the measurement of sound power using pressure measurements in a free field is as follows:

1. Install the sound source in a sufficiently large anechoic or semi-anechoic space so that free-field conditions are met at a suitable distance from the absorbent surfaces of the space.
2. Calculate the minimum propagation distance from the source required to reliably expect free-field conditions.
3. Construct a notional spherical surface in the free-field with the acoustic (or geometric) centre of the source located at the centre of the sphere (a piece of string can be used for this!).
4. Divide the surface of the sphere into a reasonable number of equal area sections (perhaps 20).
5. Make time-averaged measurements of L_{pi}.
6. If any area (S_i) is of a different size to the usual (S_n) then apply this correction

$$L_{pi, corr} = L_{pi} + 10 \ \log_{10} \left[\frac{S_i}{S_n} \right] (dB) \tag{4.38}$$

7. Calculate L_w as described earlier.

4.3.4 Measurement of sound power in a diffuse acoustic field

In a diffuse acoustic field, the net sound intensity at a point is zero. Therefore, the intensity-based methods used to determine sound power in a free field cannot be used. Instead, sound power is determined by using a number of pressure measurements to estimate the spatial energy average in the room combined with knowledge of the absorption characteristics of the space.

To provide a diffuse field a test room should fulfil the following characteristics:

- The room should be of adequate volume and suitable shape.
- The boundaries over the frequency range of interest can be considered acoustically hard.

The volume of the room should be large enough so that the number of normal modes of vibration in the octave of third octave frequency band of interest is enough to provide a satisfactory state of sound diffusion. It is common to expect at least 20 modes in the lowest frequency band where Morse and Bolt (1944) suggest that

$$N = \frac{4\pi f^3 V}{3c^3} + \frac{\pi f^2 S}{4c^2} + \frac{fL}{8c} \tag{4.39}$$

where

$c =$ speed of sound $(\mathrm{m\,s^{-1}})$
$V =$ the room volume $(\mathrm{m^3})$
$S =$ total room surface area $(\mathrm{m^2})$
$L =$ total perimeter of the room (m) (in a rectangular room this is the sum of lengths of all the edges).

In order to estimate the average number of modes in a narrow frequency band, the derivatives of this equation can be used to yield a modal density.

$$\frac{\mathrm{d}N}{\mathrm{d}f} = \frac{4\pi f^2 V}{c^3} + \frac{\pi f S}{2c^2} + \frac{L}{8c} \tag{4.40}$$

Modal density increases with the square of frequency and so a large number of modes are expected at high frequencies with only a small number of modes at low frequencies. Therefore, one can expect spatial variations in sound pressure levels at low frequencies, but at high frequencies the fluctuations become small and the field becomes more diffuse.

The shape of the test room should be such that the ratio of any two dimensions is not close to an integer. A common ratio is 2:3:5. The sound power of a source in a reverberant room is determined from the spatially averaged sound pressure level. A minimum number of six measurements would be expected for inclusion in the average. These measurements should be at locations:

- more than a quarter wavelength from any reflecting surface;
- more than one half wavelength apart;
- away from the source's direct field.

The microphone may be moved manually, or traversed linearly across the room or mounted on a continuously rotating circular boom. The number of measurement locations can be reduced using a rotating diffuser in the room particularly if the source emits narrow band or tonal noise. An accuracy of measurement of ± 0.5 dB is reasonable.

The simplest method of determining sound power in a reverberant room is to use the substitution method whereby the spatially averaged sound pressure level $\bar{L}_p$ resulting from the source of unknown power output level L_w is compared to the spatially averaged sound pressure level L_p' resulting from a source of known power output L_w'.
Therefore,

$$L_w = L_w' + (\bar{L}_p - L_p') \tag{4.41}$$

The absolute method may be used as an alternative to this, whereby the sound power is obtained from the spatially averaged sound pressure level $\bar{L}_p$ and the determination of the absorption characteristic of the room (Ver and Holmer, 1971):

$$L_w = \bar{L}_p + 10 \ \log_{10} V - 10 \ \log_{10} T_{60} + 10 \ \log_{10} \left(1 + \frac{S\lambda}{8V} \right) - 13.9 \, \text{dB}, \text{re} 10^{-12} \, \text{W} \tag{4.42}$$

where

$V \ \ = $ Room volume (m^3)
$T_{60} = $ Time taken for the sound pressure level to drop by 60 dB (s)
$S \ \ = $ Surface area of the room (m^2)
$\lambda \ \ = $ Wavelength corresponding to the centre frequency in the analysis band (m)

4.3.5 Measurement of sound power in a semi-reverberant far field

Most rooms containing noise sources are neither anechoic nor truly reverberant. When determining the sound power of a source in such a space no assumptions are made about the nature of the field (unlike before, where the field was assumed to be either all direct or all diffuse). The only requirements for the room are:

1. The sound source should be in its normal position.
2. The room should be large enough to allow measurements to be made in the far field.
3. The microphone should be kept at least one quarter wavelength away from any reflecting surface of the room.

If the source is located on a hard floor, and can be moved, and is at least a half wavelength from any other reflective surfaces, a version of the substitution method described earlier may be used to determine sound power. First the spatially averaged sound pressure level $\bar{L}_{p_1}$ is obtained over a hemispherical surface around the source. Then the source is moved away and replaced with a reference source of power output L_{W2}, and the exact procedure is repeated to obtain $\bar{L}_{p_2}$. Therefore, the sound power output L_{W1} of the source is

$$L_{W1} = L_{W2} + (\bar{L}_{p_1} - \bar{L}_{p_2}) \ (\text{dB}) \tag{4.43}$$

The movement of the source is vital as it allows both sets of pressure measurements to be made at the exactly same locations and therefore with the same mix of direct and reverberant fields.

If the source cannot be moved, then the substitution method is excluded and hence a knowledge of the spatial distribution of direct and reverberant fields must be determined before sound power estimates are made.

The sound pressure level $\bar{L}_{p_1}$ averaged over a portion of a sphere around a sound source of acoustic power output L_{W1} is

$$\bar{L}_{p_1} = L_{W1} + 10 \ \log_{10} \left[\frac{D}{4\pi r^2} + \frac{4}{R} \right] \ (\text{dB}) \tag{4.44}$$

where the directivity constant is

$D = 1$ for a sphere
$D = 2$ for a hemisphere
$D = 4$ for a quarter of a sphere etc.

and R is the room constant (Kutruff, 1979).

$$R = \frac{S\bar{\alpha}}{1 - \bar{\alpha}} \tag{4.45}$$

$S = $ total room area (m^2)
$\bar{\alpha} = $ average Sabine absorption in the room (see Section 2.5.1)

The room constant may be taken to be the total absorption of the room measured in units of area (m^2).

If $\bar{L}_{p_1}$ is obtained over a hemisphere around a reference source of known sound power output L_{W1}, then the contribution to sound pressure level over the test surface made by the direct field is

$$\bar{L}_{p_d} = L_{W1} + 10 \ \log_{10} \left[\frac{2}{4\pi r^2} \right] \ (\text{dB})$$

$$\bar{L}_{p_d} = L_{W1} + 10 \ \log_{10} \left[\frac{1}{r^2} \right] + 10 \ \log_{10} \left[\frac{2}{4\pi} \right] \ \text{dB}$$

$$\bar{L}_{p_d} = L_{W1} - 20 \ \log_{10} r - 8 \ \text{dB} \tag{4.46}$$

or

$$\bar{L}_{p_d} = L_{W1} + 10 \ \log_{10} \left[\frac{1}{S_H} \right] \ \text{dB} \tag{4.47}$$

where S_H is the surface area of a hemisphere (m^2).

Now for a hemisphere

$$\bar{L}_{p_1} = L_{W1} + 10 \ \log_{10} \left[\frac{1}{S_H} + \frac{4}{R} \right] \tag{4.48}$$

$$10 \ \log_{10} \left[\frac{1}{S_H} + \frac{4}{R} \right] = \bar{L}_{p_1} - L_{W1} \tag{4.49}$$

$$\frac{1}{S_H} + \frac{4}{R} = 10^{[L_{p_1} - L_{W1}]/10} \tag{4.50}$$

Now, it is known from equation (4.47) that:

$$\bar{L}_{p_d} = L_{W1} + 10 \ \log_{10} \left[\frac{1}{S_H} \right] \tag{4.47}$$

therefore,

$$L_{W1} = \bar{L}_{p_d} - 10 \ \log_{10} \left[\frac{1}{S_H} \right] \tag{4.51}$$

so,

$$\frac{1}{S_H} + \frac{4}{R} = 10^{[L_{p_1} - (L_{p_d} + \log_{10}(1/S_H))]/10}$$

$$\frac{1}{S_H} + \frac{4}{R} = \frac{10^{[L_{p_1} - L_{pd}]/10}}{S_H}$$

$$\frac{4}{R} = \frac{1}{S_H}[10^{[L_{p_1} - L_{pd}]/10} - 1] \tag{4.52}$$

Now that the room constant is known, $\bar{L}_p$ can be determined over a hemisphere around the sound source, and the sound power determined directly according to:

$$L_w = \bar{L}_p - 10 \ \log_{10} \left[\frac{1}{2\pi r^2} + \frac{4}{R} \right] \tag{4.53}$$

A third alternative way of determining sound power levels involves obtaining the spatially averaged sound pressure levels $\bar{L}_{p_1}$ and $\bar{L}_{p_2}$ over two different hemispherical surfaces of areas S_1 and S_2 both of which have the acoustic source at their centres.

$$\bar{L}_{p1} = L_w + 10 \ \log_{10} \left[\frac{1}{S_1} + \frac{4}{R} \right] \tag{4.54}$$

$$\bar{L}_{p2} = L_w + 10 \ \log_{10} \left[\frac{1}{S_2} + \frac{4}{R} \right] \tag{4.55}$$

L_w is to be determined. $\bar{L}_{p1}$ and $\bar{L}_{p2}$ are known, measured quantities. The term $4/R$ is of no inherent interest here, so it will be removed by algebraic manipulation.

$$\frac{4}{R} = 10^{[L_{p_1} - L_w]/10} - \frac{1}{S_1} \tag{4.56}$$

$$\frac{4}{R} = 10^{[L_{p_2} - L_w]/10} - \frac{1}{S_2} \tag{4.57}$$

Therefore,

$$10^{[L_{P1} - L_w]/10} - \frac{1}{S_1} = 10^{[L_{P2} - L_w]/10} - \frac{1}{S_2} \tag{4.58}$$

$$10^{[L_{P1} - L_w]/10} - 10^{[L_{P2} - L_w]/10} = \frac{1}{S_1} - \frac{1}{S_2} \tag{4.59}$$

Multiply both sides by $10^{L_w/10}$,

$$10^{L_{P1}/10} - 10^{L_{P2}/10} = 10^{L_w/10}\left[\frac{1}{S_1} - \frac{1}{S_2}\right] \tag{4.60}$$

Take logarithms on both sides,

$$\log 10 \left[10^{L_{P1}/10} - 10^{L_{P2}/10}\right] = \frac{L_w}{10} + \log_{10}\left[\frac{1}{S_1} - \frac{1}{S_2}\right] \tag{4.61}$$

Multiply both sides by 10

$$10 \log_{10}\left(10^{L_{P1}/10} - 10^{L_{P2}/10}\right) = L_w + 10\log_{10}\left[\frac{1}{S_1} - \frac{1}{S_2}\right] \tag{4.62}$$

Now,

$$10 \log_{10}\left(10^{L_{P1}/10} - 10^{L_{P2}/10}\right) = 10 \log_{10}\left[10^{L_{P2}/10}\left(10^{(L_{P1} - L_{P2})/10} - 1\right)\right]$$

$$\bar{L}_{P2} + 10 \log_{10}\left[10^{(L_{P1} - L_{P2})/10} - 1\right] = L_w + 10 \log_{10}\left[\frac{1}{S_1} - \frac{1}{S_2}\right] \tag{4.63}$$

and finally

$$L_w = \bar{L}_{P2} - 10 \log_{10}\left[\frac{1}{S_1} - \frac{1}{S_2}\right] + 10 \log_{10}\left(10^{(L_{P1} - L_{P2})/10} - 1\right) \tag{4.64}$$

This equation can be used directly to determine the sound power from a source when

$$\bar{L}_{P1,2} = 10 \log_{10}\left[\frac{1}{N}\sum_{i=1}^{N} 10^{(L_{Pi}/10)}\right] \text{dB} \tag{4.65}$$

4.3.6 Measurement of sound power in the near field

All the earlier methods of determining sound power have required that

(i) the room is large enough for pressure measurements to be made in the far field;
(ii) background noise is negligible (i.e. the sound source produces an increase of at least 3 dB over background noise levels at all microphone positions).

Near field techniques can be used in cases that do not satisfy the criteria above. Near field techniques consist of making sound pressure level measurements in the near field of the source (typically 1 m away or less if reduced accuracy is acceptable) across a notional test surface (which is often parallel piped).

An average sound pressure level is obtained from N measurement locations, thus

$$\bar{L}_p = 10 \ \log_{10} \left[\frac{1}{N} \sum_{i=1}^{N} 10^{(L_{pi}/10)} \right] \qquad (4.66)$$

and the following equation is used to determine an approximate value for sound power level ([Pobol, 1976; Jonasson and Elson 1981] and reported in Bies and Hansen [1996]).

$$L_w = \bar{L}_p + 10 \ \log_{10} S - \Delta_1 - \Delta_2 \qquad (4.67)$$

where

S = area of test surface (m²)
Δ_1 = correction factor to account for the absorption characteristics of the room
Δ_2 = correction factor to account for possible tangential wave propagation
Δ_1 can be obtained from

$$\Delta_1 = 10 \ \log_{10} \left[1 + \frac{4S_1}{A\bar{\alpha}} \right] \qquad (4.68)$$

where

S_1 = test measurement surface area (m²)
A = total area of room surfaces (m²)
$\bar{\alpha}$ = mean acoustic Sabine absorption coefficient

Alternatively, L_{p_1} and L_{p_2} can be obtained for two concentric surfaces around the machine and

$$\Delta_1 = L_{p_1} - L_{p_2} - 10 \ \log_{10} \left[10^{(L_{p_1} - L_{p_2})/10} - 1 \right] + 10 \ \log_{10} [1 - S_1/S_2] \qquad (4.69)$$

Pobol (1976), reported in Bies and Hansen (1996), suggests the following values for Δ_1 in terms of the volume of the room over the area of the test surface shown in Table 4.2.

Table 4.2 Values for Δ_1 suggested by Pobol (1976)

	V/S(m)			
Usual room	20–50	50–90	90–3000	>3000
Highly reflective room	50–100	100–200	200–600	>600
Δ_1 dB	3	2	1	0

Table 4.3 Values for Δ_2 suggested by Johansson and Eston (1981)

S/Sm	Δ_2 dB
1–1.1	3
1.1–1.4	2
1.4–2.5	1
>2.5	0

Johansson and Eston (1981), reported in Bies and Hansen (1996), suggest values for Δ_2 in terms of the ratio between the test surface areas S and to the area of the smallest parallelepiped surface Sm which just encloses the source. These are shown in Table 4.3.

Care should be taken to avoid errors due to the directional sensitivity of the microphone at higher frequencies. Note should be made as to whether a microphone has been calibrated for an expected angle or incidence (free-field calibration) which is a function of angle of incidence or whether a single random incidence calibration has been achieved (see Section 2.2.3.3). In either case, it is prudent to set an upper limit on frequency below which the effects of angle of incidence on microphone response are small.

4.3.7 Determination of sound power using surface vibration velocity measurements

The sound power being radiated by a vibrating structure can be estimated from a mean square vibration velocity averaged over the surface.

The radiated sound power is equal to:

$$W = \rho c S \sigma \langle v^2 \rangle \tag{4.70}$$

where

ρ = density of the air ($1.2\,\mathrm{kg\,m^{-3}}$ is typical)
c = speed of sound ($343\,\mathrm{m\,s^{-1}}$ is typical)
S = surface area ($\mathrm{m^2}$)
σ = radiation ratio (efficiency = 100%, ratio = unity)
$\langle v \rangle$ = surface-averaged vibration velocity ($\mathrm{m\,s^{-1}}$)

The sound power level (re 10^{-12} W) is thus:

$$L_\mathrm{w} = 10 \, \log_{10} \left[\frac{\rho c S \sigma \langle v^2 \rangle}{10^{-12}} \right] \tag{4.71}$$

$$L_\mathrm{w} = 10 \, \log_{10} \left[\langle v^2 \rangle \right] + 10 \, \log_{10} S + 10 \, \log_{10} \sigma + 10 \, \log_{10} \left[\frac{\rho c}{10^{-12}} \right]$$

$$L_\mathrm{w} \simeq 10 \, \log_{10} \left[\langle v^2 \rangle \right] + 10 \, \log_{10} S + 10 \, \log_{10} \sigma + 146 \text{ dB re } 10^{-12} \text{ W} \tag{4.72}$$

The determination of sound power from surface vibration measurements is an approximate method due to uncertainties in:

(i) the surface-averaged vibration velocity;
(ii) the radiation efficiency.

Ideally, the surface-averaged vibration velocity should be obtained using an accelerometer and an integrating circuit, making a large number of narrow band measurements at points away from the edges of the structure, and averaging each band result as follows:

$$\bar{L}_{v_i} = 10 \ \log_{10} \left[\frac{1}{N} \sum_{i=1}^{N} 10^{Lv_i/10} \right] \tag{4.73}$$

However, care should be taken when using integrating circuits that the signal-to-noise ratio for the measurement chain is not too severely reduced. If this is the case, or if a reasonable quality integrating circuit is not available, then the vibration velocity can be estimated from the vibration acceleration as follows:

$$|v| \, e^{i\omega t} = \int_0^\infty |a| \, e^{i\omega t} \, dt \tag{4.74}$$

$$|v| = \frac{|a|}{i\omega} \tag{4.75}$$

$$|v^2| \approx \frac{|a^2|}{(2\pi f)^2} \tag{4.76}$$

f is the band centre frequency (Hz). If this approximate integration technique is used on third octave bands, then significant errors can occur at higher frequencies where the bandwidths are greater.

The radiation efficiency of a vibrating structure is notoriously difficult to determine. The key parameter is the critical frequency, which for uniform flat plates is

$$f_c = c^2/1.8 \, hc_L \ \text{(Hz)} \qquad \text{(Fahy, 1985)} \tag{4.77}$$

where

h = plate thickness (m)
c_L = phase speed of longitudinal waves in plate (m s^{-1})
c = speed of sound in fluid (air) (m s^{-1})

The radiation efficiency is generally greater than 100% at frequencies around the critical frequency. At frequencies greatly above the critical frequency, the radiation efficiency is usually 100% whereas at frequencies greatly below the critical frequency the radiation efficiency can be as low as a fraction of 1%.

Radiation efficiencies can be predicted using boundary element techniques (SYSNOISE), or approximations may be obtained from generalised curves such as those presented by Vér and Holmer (1971), reported in Fahy (1985), reproduced in Figure 4.5.

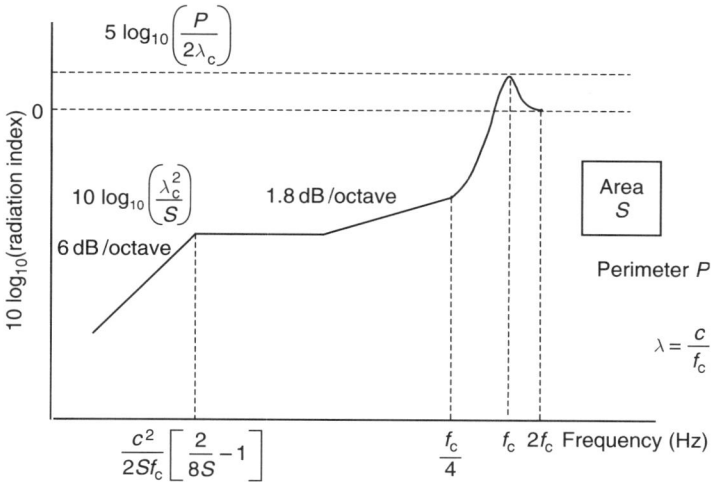

Figure 4.5 Theoretical modal average radiation efficiency of a baffled rectangular panel: after Vér and Holmer (1971).

4.3.8 Determination of sound power using an intensity meter

It has already been stated that sound power is directly related to sound intensity by the bounding area, thus

$$W = I \cdot S$$

The fact that for free-field conditions at room temperature

$$L_p \simeq L_i$$

was also employed when determining the sound power of a source using sound pressure measurements. This relationship breaks down for non-freefield conditions, but the relationship between sound power and sound intensity still holds. Therefore, if one can measure sound intensity directly then sound power may be determined even in non-free-field conditions.

Sound intensity is the long-term average integral of acoustic pressure and acoustic particle velocity

$$I = \lim_{T \to \infty} \frac{1}{T} \int_0^T p \cdot u \, dt \qquad (4.29)$$

An intensity meter has been developed by Fahy and others (Fahy, 1989) which measures intensity directly by using two phase-matched microphones at a known spacing apart to infer acoustic particle velocity from the pressure gradient. It should be appreciated that sound intensity is a vector product having both magnitude and direction.

Sound power is determined from the intensity averaged over a notional surface placed around the source. The same restrictions on accuracy regarding the number of measurement positions apply equally to intensity techniques as they do to sound pressure techniques.

The determination of sound power using an intensity meter has the following advantages over sound pressure techniques to the same aim:

1. Sound pressure techniques must assume free-field conditions where the sound is only travelling away from the source. The sound intensity meter determines the direction of propagation and therefore sound power may be determined even in the geometric near field.
2. Some areas of vibrating surfaces may act as radiators of sound while other areas may act as absorbers of sound. A sound intensity meter scanning a surface will detect such phenomena.
3. The output of an intensity meter gives magnitude and direction which can usefully be used for source location. The output from pressure measurements over a surface is less informative.

Although the intensity meter is a very flexible tool, there are conditions under which it will not perform well. These are mostly where pressures are large and the pressure gradient between the two microphones is small resulting in a poor estimate of particle velocity. Such a condition will occur near to highly reflective surfaces where the incident power is approximately equal to the reflected power. Therefore, the use of an intensity meter near to highly reflective surfaces should be avoided.

The other, unexpected, difficulty with sound intensity measurements is that due to the high detail of the output, a large number of measurements must be made at different locations to ensure that a realistic spatial average of sound intensity has been achieved. This problem gets worse the nearer to the source one gets. As long as background noise levels are not a problem, there are occasions where a more meaningful estimate of sound power can be obtained from a few quick far field sound pressure measurements rather than a detailed and laborious survey in the nearer field with an intensity meter.

4.3.9 Standard methods for measuring sound power under different circumstances

There are UK national (BS) and internationally (ISO) recognised standard methods for measuring sound power. They broadly follow the methods already discussed so far, and therefore will not be discussed further. The list of standards includes, but is not necessarily limited to (source www.bsi.org.uk),

BS 4196-0:1981 (ISO 3740:1980)
Sound power levels of noise sources. Guide for the use of basic standards and for the preparation of noise test codes.

BS 4196-1:1991 (EN 23741:1991 ISO 3741:1988)
Sound power levels of noise sources. Precision methods for determination of sound power levels for broad-band sources in reverberation rooms.

BS 4196-2:1991 (EN 23742:1991 ISO 3742:1988)
Sound power levels of noise sources. Precision methods for determination of sound power levels for discrete-frequency and narrow-band sources in reverberation rooms.

BS 4196-5:1981 (ISO 3745:1977)
Sound power levels of noise sources. Precision methods for determination of sound power levels for sources in anechoic and semi-anechoic rooms.

BS 4196-7:1988 (ISO 3747:1987)
Sound power levels of noise sources. Survey method for determination of sound power levels of noise sources using a reference sound source.

BS 4196-8:1991 (ISO 6926:1990)
Sound power levels of noise sources. Specification for the performance and calibration of reference sound sources.

BS EN ISO 3743-1:1995
Acoustics. Determination of sound power levels of noise sources. Engineering methods for small, movable sources in reverberant fields. Comparison for hard-walled test rooms.

BS EN ISO 3743-2:1997
Acoustics. Determination of sound power levels of noise sources. Engineering methods for small, movable sources in reverberant fields. Methods for special reverberation test rooms.

BS EN ISO 3744:1995
Acoustics. Determination of sound power levels of noise sources using sound pressure. Engineering method in an essentially free field over a reflecting plane.

BS EN ISO 3746:1996
Acoustics. Determination of sound power levels of noise sources using sound pressure. Survey method using an enveloping measurement surface over a reflecting plane.

BS EN ISO 9614-1:1995
Acoustics. Determination of sound power levels of noise sources using sound intensity. Measurement at discrete points.

BS EN ISO 9614-2:1997
Acoustics. Determination of sound power levels of noise sources using sound intensity. Measurement by scanning.

4.4 Engine noise

4.4.1 Introduction to engine noise

In this section, the term engine noise will be taken as the noise produced by a combination of the gas loads in the cylinders and the mechanical motions in the base engine. Intake and exhaust noise shall be considered as separate problems as will be the noise caused by engine ancillaries (alternators, fans, pumps, motors, etc.).

Engine noise is the sum of two elements:

1. combustion noise
2. mechanical noise.

The relative mix of the two will vary between engines but as a general rule:

- mechanical noise dominates the engine noise produced by spark ignition (gasoline) engines;
- combustion noise is a more significant contributor to the engine noise produced by compression ignition (diesel) engines.

Engine noise is dependent on engine speed and may also depend on engine load for some types of engine (the normally aspirated direct injection diesel and the gasoline engine in particular).

4.4.2 Combustion noise

Combustion noise results from gas forces in the cylinders applied to the structure of the engine, causing vibration to occur which is then radiated as noise. It is produced therefore by an indirect noise-generating mechanism (see Section 2.2.2).

The gas forces in each cylinder vary during the working cycle of the engine (two or four stroke). They are highest during the combustion period where the cylinder pressure is rising quickly.

The vibration response of the engine is greatest when the forcing caused by the rate of pressure rise is greatest. This is intuitively obvious: if the rate of pressure rise is zero, then the forces due to cylinder pressure will be in equilibrium with the restraining forces in the engine structure, and hence with no net force there will be no net acceleration of the structure. However, with a rapidly changing cylinder pressure, the response of the structure lags behind the causal force, equilibrium is never reached, and a net force results producing vibration. The more rapid the rate of change of pressure, the greater the net force and hence the greater the vibration and the noise. A more rapid rise in pressure also increases the high-frequency content of the force, and hence of both the vibration and the noise.

The tendency to produce combustion noise of different engine types can be reasonably ranked according to their typical rates of cylinder pressure rise during combustion.

Starting with the noisiest for combustion noise:

- NA, DI diesel   (4+ bar/degree crank)
- NA, indirect injection (IDI) diesel (3–4 bar/degree crank)
- Turbocharged DI diesel   (2–3 bar/degree crank)
- Gasoline engine   (<2–3 bar/degree crank)

The spectrum of the cylinder pressure is a more useful/reliable indicator of combustion noise. Typical spectra for the NA-DI diesel engine at full load are shown in Figure 4.6 (data taken from Nelson (1987), originally published in Russell (1979)).

The effect of increasing speed can be seen in Figure 4.6 as:

- a shift in the spectrum towards the higher frequencies;
- an increase in spectral levels in each third octave band as a result of the shift to the higher frequencies.

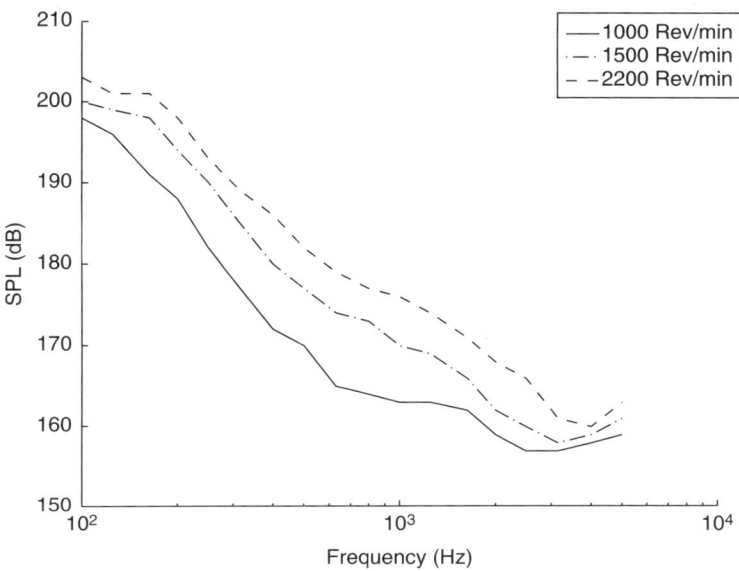

Figure 4.6 Effect of engine speed on cylinder pressure spectra. NA-DI diesel engine at full load: data obtained from Nelson (1987) and Russell (1979).

It should be appreciated that the slope of the cylinder pressure spectrum provides an indication of the speed dependence of combustion noise. Typical slopes per decade (tenfold change in frequency) are given as (Lilly, 1984):

- NA-DI diesel   25–30 dB/decade
- NA-IDl diesel   40–50 dB/decade
- Turbo DI diesel   40–50 dB/decade
- Gasoline   50–60 dB/decade

The higher the slope, the greater the speed dependence. It can be seen that the noisiest engines for combustion noise have the lowest speed dependency. This explains why heavy trucks remain noisy even when used at low speeds, and why this effect is not noticed in gasoline-powered vehicles.

The portion of each spectrum in Figure 4.6 in the 1–4 KHz range is responsible for the diesel knock commonly associated with diesel engines. The rate of pressure rise is greatest in diesel engines with the greatest ignition delay. This ignition delay is extended when the injection timing is advanced. With advanced injection, the fuel has more time to pre-mix with air before combustion occurs, yielding a larger pre-mixed charge whicl will burn quickly producing a rapid pressure rise.

So, advancing injection timing increases combustion noise. This effect is commonly used to separate combustion noise from mechanical noise. The injection timing can be slowly advanced until the change in exterior noise spectrum matches the change in cylinder pressure spectrum. At that point the exterior noise is dominated by combustion noise. This technique may not be practical with certain fuel injection equipment (it also may yield very high peak cylinder pressures) and so variants have been developed that require smaller swings in timing (Atkins and Challen (1979), reported in Nelson (1987)).

Of course the opposite applies and a retarding of injection produces a reduction in combustion noise. This noise control technique may be used only to a small extent due to the impact on engine performance and emissions (retarding the injection increases bsfc and the formation of smoke/soot/particulates). A compromise between smoke and noise emissions may be found by retarding the injection timing but increasing the injection rate (Glikin, 1985).

The modelling of the wave dynamics in the fuel injection system that strongly influences the rate of fuel injection is reported in Russell and Lee (1994) and the effect that the rate of delivery has on combustion is reported in Russell (1997). The recent advent of common rail diesel fuel injection equipment with solenoid controlled injectors allows more than one pulse of diesel fuel per cylinder per cycle and therefore fuel flow rate modulation as a means of noise control (as well as emissions control) can be explored readily.

4.4.3 Mechanical noise

The crank mechanism (pistons, conrods, crankshaft, bearings) experiences externally applied forces due to gas forces and internally generated forces due to its own inertia (see Section 6.4 for a complete discussion of these). The reaction of the engine structure to the sum of these forces produces mechanical noise by an indirect noise-generating mechanism (see Section 2.2.2).

Around TDC there is a rapid reversal in side force produced by the slider-crank mechanism. This produces piston slap as the piston impacts on the cylinder-liner. Piston slap is normally the dominant source of mechanical noise in the diesel engine (Lalor et al., 1980). There is side force throughout the cycle, along with other force reversals but the one at TDC yields the highest rate of change of side force. Piston slap noise increases with engine speed. It also increases with turbocharging. It is mostly controlled by reducing clearance between the piston and the cylinder-liner.

In gasoline engines, piston/liner clearances are relatively small, and mechanical noise tends to be dominated by impacts in the crankshaft bearings made through the oil film (Lalor et al., 1980). At low engine speeds these are magnified by increasing engine load. At high engine speeds, the inertia effects of the crank mechanism dominate so there is little load dependency.

Other sources of mechanical noise include:

- timing drive;
- valve train;
- fuel injection equipment.

4.4.4 The effects of engine speed and load on noise

The total noise emission (combustion and mechanical) for the DI diesel changes only slightly over the normal operating speed range. For the NA engine the slope is around 30 dB per decade and for the turbocharged engine around 20 dB per decade (Priede, 1975). There is modest load dependency for the NA-DI diesel engine (4–5 dB) and little for the turbo DI diesel unless excessive mechanical noise occurs due to the boost pressure.

Small high-speed NA-IDI diesel engines with smooth combustion show little load dependence and a greater speed dependence (around 40 dB per decade) than the DI diesel engines.

Gasoline engines have two sets of noise characteristics. At low speeds (up to say 2500 rev min^{-1}) they have modest load dependence (around 5 dB increase in noise level due to increasing load) and slight speed dependence (20 dB per decade). At higher speeds there is little load dependence and greater speed dependence (50 dB per decade) due to the effects of inertial forces on the mechanical noise. This explains the sudden onset of roaring engine noise commonly experienced as gasoline engines are revved hard.

4.4.5 Measuring engine noise

The most universal parameter for quantifying the noise emission from any source (including the engine) is sound power (see Section 2.5). However, it is difficult to measure (see Section 4.3).

An alternative scheme is to measure sound pressure level at specified locations around the engine and use this for rating engine noise. The most commonly used standard method of this kind is detailed in SAE J1074. The important information in J1074 is as follows:

- The engine is tested either outdoors in a flat, open space or in an acoustically treated test cell that replicates the outdoor environment (commonly a semi-anechoic cell with large sound absorbing wedges on the walls and ceiling and a flat concrete floor).
- The engine is either tested in its bare state (with just enough equipment to run – pumps and manifolds are fitted but the intake/exhaust noise is ducted away) or in its fully equipped state (everything fitted including ancillaries and sometimes full intake and exhaust systems).
- The engine is tested at the maximum power point, at the maximum torque point, at the point of maximum speed but minimum load and also at idle.
- Sound pressure levels (slow response, both 'A'- and 'C'-weightings) are measured at three positions for each engine operating condition. These are at 1.0 m from the longitudinal centres of the vertical planes forming the smallest rectangular box which completely encloses the bare engine. The measuring points are on both sides and in front of the engine at the height of the exhaust manifold and at least 1 m off the ground.
- The noise levels at the three specified locations are reported. Octave band results are also reported for the location with the highest 'A'-weighted level.
- A survey is made of 'A'-weighted sound pressure level at the same height and distance from the box as the specified locations. If the survey reveals readings more than 3 dB above the highest reading at the specified locations, then the survey readings are also reported.
- The reported results should be the averaged results of two or more test results within 2 dB of each other.

If a semi-anechoic cell is used it must be large enough to undertake the measurements and each microphone should be at least one-quarter wavelength from the walls and from the ceiling to avoid the near-acoustic fields of these absorbing surfaces (see Section 4.3.1).

4.4.6 Engine noise source ranking

Various noise source ranking techniques are discussed in Section 3.2. All are in common use for engine noise source ranking.

At higher frequencies (<300 Hz say) the shielding technique generally gives reliable results and is easy to use. However, it is far less reliable at low frequencies and commonly results in the under estimation of the contribution made by engine components at these frequencies (Crocker et al., 1980).

Therefore, at low frequencies, sound intensity (Crocker et al., 1980) or noise from vibration techniques might be preferred (Dixon and Phillips, 1998). The noise source rankings for contemporary powertrains are as follows, with the noisiest item at the top of the list (March and Croker, 1998).

	Diesel engines	*Gasoline engines*
	Oil Pan	Transmission
	Other Sources	Other Sources
	Fuel System	Intake System
Similar effect	Cylinder Block	Ancillary Drive
	Ancillary Drive	Oil Pan
	Transmission	Exhaust Manifold
	Intake System	Cam Cover
	Exhaust Manifold	Front Cover
	Front Cover	Fuel System
	Cam Cover	Cylinder Block

See also Section 1.6.2.1 for an alternative noise source ranking.

4.4.7 Engine noise control

The options for controlling engine noise are the usual ones available to the noise control engineer, namely:

- stiffen structures to push resonant frequencies above the highest forcing frequency;
- isolate components from sources of excitation;
- encapsulate noise sources with massive panels;
- add damping where resonances occur.

Engine-specific noise control measures include the following:

- Oil pan – the use of an isolating gasket between the oil pan and the crankcase. The adoption of structural aluminium oil pans to replace the traditional pressed steel components has made oil pan noise more significant in spite of improvements to the crankcase to reduce its noise radiation.
- Rocker cover – the use of rubber isolating gaskets (Querengasser et al., 1995).
- Fuel injection equipment – the adoption of common rail systems and unit injector systems which are more compact and quieter have brought about significant improvements (March and Croker, 1998).

- Cylinder block – engine blocks with separate crankshaft bearing endcaps at the bulk-heads between cylinders exhibit the lowest frequency for the first bending mode of the engine. For larger engines this may be as low as 200 Hz. The use of a ladderframe or bedplate bottom end to join the bearing caps together with a locally stiff structure can push the first bending mode above 300 Hz (Querengasser et al., 1995), reduce the axial excursion of the endcaps at resonance (commonly around 1000 Hz) and generally reduce the low-frequency modal density (March and Croker, 1998). Careful design of the crankcase and the block to reduce the effect of panel modes is also beneficial for frequencies around 800+ Hz (Russell, 1972).
- Intake system – the avoidance of large planar surfaces on intake components can reduce noise emissions along with general stiffening of the structures.
- Noise shields – well-damped, isolated engine covers can reduce noise radiated by the engine structure (Russell, 1972).
- Engine bay enclosures – engines may be effectively enclosed within their engine bay in the vehicle thus encapsulating the noise sources. Problems with ventilation and cooling are common (Thien et al., 1984).

4.5 Road noise

4.5.1 Introduction to road noise

The term road noise might be replaced by the more complete description 'road and tyre noise' as it is taken here to include:

- Interior noise resulting from the contact between the tyres and the road, being trans-mitted to the interior by both airborne and structure-borne paths. This is often labelled as road noise and is the subject here.
- Exterior noise resulting from the contact between the tyres and the road. This is often labelled as tyre noise and was the subject of an earlier section (Section 3.4).

4.5.2 Interior road noise

Vehicle interior road noise is mainly a low-frequency noise problem (<1000 Hz). Contributions are made by

- structure-borne noise paths through the vehicle suspension (<500 Hz);
- direct airborne noise paths from the tyre through the vehicle structure (>500 Hz); often confused with wind noise.

The structure-borne components tend to dominate overall noise levels except on the smoothest of road surfaces. The structure-borne element becomes clear when a vehicle is operated on a rolling road with the body jacked on air jacks and the suspension links disconnected (the so-called disconnect test). Interior noise levels below 500 Hz drop significantly during such tests whilst the levels of airborne tyre noise above 500 Hz remain fairly constant.

Interior road noise levels might peak at around 60–65 dBA at 250 Hz and fall off with increasing frequency. Levels at frequencies greater than 1000 Hz are typically 15 dBA lower than those at frequencies less than 300 Hz.

Interior road noise is typically measured with the vehicle operating at fixed speeds over different road surfaces. A standard measurement scheme is often employed such as BS 6086 (see Section 4.1.3). It is common to make measurements at the inner ear positions for all seats. Note that the 'A'-weighting scale is commonly employed. This has the benefit of avoiding the overload of the measurement chain when the vehicle passes over bumps in the road (the so-called bump thump). However, when measuring bump thump deliberately, the 'A'-weighting scale might be replaced by the 'C'-weighting scale, or no weighting may be applied at all in order to preserve signal-to-noise ratio at low frequencies.

Low frequency peaks in interior road noise spectra are commonly attributable to:

- Acoustic modes of the cockpit/interior space (see Section 2.5).
- Modes of vibration of the tyre structure, typically <200 Hz. These are often called breaker modes as they are modified by the breaker: a composite layer directly below the tread constructed from steel cords embedded in rubber. The breaker provides structural stiffness to the tyre.
- Acoustic modes of the tyre cavity (typically around 250 Hz)

4.5.3 Analysing structure-borne road noise

A disconnect test can separate structure-borne and airborne road noise in the laboratory. However, on the test track a noise path analysis (see Section 4.2) technique is required. The complex stiffness method is usually used. The noise path analysis for road noise is far more complicated than for powertrain noise as it is difficult to find a single suitable reference signal to relate the phase information to. As a result, measured force data is grouped into several sets of coherent signals (Vandenbroeck and Hendricx, 1994; Storer et al., 1998) using a technique known as principal component analysis.

Because of the complexities of noise path analysis for road noise, disconnect tests are commonly used. Shakers are attached to the wheel hubs and the noise levels (or vibration velocity amplitude at certain structural elements) produced inside the vehicle are measured. The effect of disconnecting elements of the suspension and hence severing potential noise paths can then be ascertained readily.

There is common use of measured NTF – the ratio of noise (Pa) to force input (N). With a shaker (and force gauge) attached at every suspension point in turn, the NTF to a microphone in the vehicle interior can be found using a two-channel FFT analyser. A typical NTF target for road noise is a maximum of 0.01 Pa/N.

4.5.4 Controlling interior road noise

Some design guidance has emerged for controlling interior road noise:

- Choose quieter tyres (!).
- Achieve an NTF of 0.01 Pa/N at all suspension attachment points. In practice this generally requires a body stiffness of more than 10 kN mm^{-1} (longitudinal and lateral bracing may be required at strut tops).

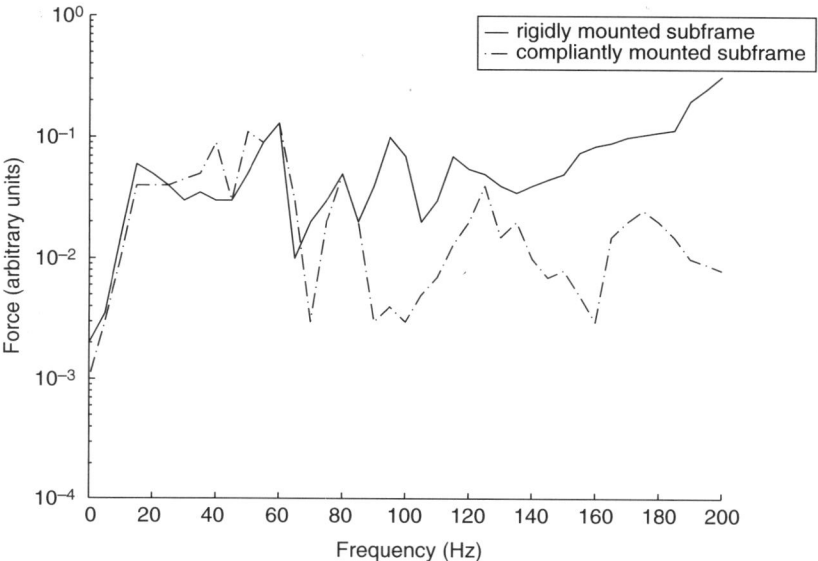

Figure 4.7 The beneficial effects of the compliant subframe: data taken from Hardy (1997).

- Use suspension bushes with a stiffness that is between one-fifth and one-tenth of the body stiffness.
- Use a compliantly mounted subframe (Hardy, 1997) where the suspension is mounted to a modified subframe, which is resiliently mounted to the body (most resilience in the longitudinal direction). Too much lateral compliance affects vehicle handling. The beneficial effects of the compliant subframe are shown in Figure 4.7, the data taken from Hardy (1997).
- Use of palliative treatments to reduce interior noise levels (see Sections 4.2.1, 4.9, 4.10) such as isolated barrier materials.
- Reduce the noise caused by shock absorbers (Cheng and Akin, 1995).

4.6   A note on aerodynamic (wind) noise

Aerodynamic (wind) noise is a significant source of interior noise for many vehicles travelling at higher speeds. It is easily confused with road noise, being generally broadband in nature but with a strong low-frequency bias.

Aerodynamic noise is caused through a variety of mechanisms:

- Aerodynamic excitation of the so-called 'greenhouse surfaces' on the car (the glass-work and the roof panel) causing structure-radiated noise in the interior.
- Airflow over the underside of the vehicle causing transmission of airborne sound to the interior, particularly in the wheel-arch areas.
- Noise transmission through door and glass seals due to aspiration (leakage) or due to aerodynamic excitation of doors and glass caused by disturbed airflow over the seals.

- Vortex shedding over protrusions in the bodywork (such as aerials, roof bars, etc.) causing tonal airborne noise.
- Cavity flows through partially open windows and sunroofs causing intense low-frequency (below 25 Hz commonly) noise and buffeting.

Aerodynamic noise is best investigated at full scale in a windtunnel although some promising results have been obtained at model scale (Kim, 2003). The use of a windtunnel allows the car to be stationary thus removing noise contributions from the powertrain and from the tyres. Frequently, aerodynamic noise development is undertaken on a subjective basis. Windows and doors are taped up, lead sheeting is added to wheel arches and to the underfloor, bodywork protrusions are removed and, at every stage, binaural recordings are made in the car using a dummy head system. The resulting recordings can be played to juries, or Zwicker loudness or Articulation Index are frequently used as a measure of improvement (see Section 1.6.3.2).

Publications by Anderson (2002) and by Coney et al. (1999) detail objective methods of assessment. These include the use of an intensity probe inside the vehicle to check for aspiration noise. An inexpensive and traditional alternative is to use a length of plastic tubing, to place one end in your ear and to pass the other end around the glasswork and door seals. The intensity probe can also be used outside the vehicle to map the contours of aerodynamic noise from bodywork protrusions and from the greenhouse surfaces. A form of noise path analysis is also possible by making measurements of structural vibration on the greenhouse surfaces and then combining these with noise transfer functions measured between positions inside the interior and the various surfaces (see Section 4.2.3 for a discussion of the NTF). The character of the aerodynamic excitation can be determined using a combination of static pressure tappings and flush mounted microphones embedded into thin rubber mats that are simply laid on the greenhouse surfaces.

4.7 A note on brake noise

Brake noise has been an issue of concern to vehicle manufacturers for decades. There are several distinct categories of brake noise (Betella et al., 2002):

- Brake squeal (occurring at higher speeds and having a tonal character with components well above 1000 Hz).
- Brake moan (occurring at moderate speeds and characterised by frequency components around 100 Hz).
- Brake creep-groan (occurring at speeds less than walking pace and characterised by frequency components around 100 Hz).
- Brake judder (occurring at speeds less than walking pace and characterised by frequency components around 10 Hz).

The source for all four classes of brake noise is the pairing of friction surfaces at the pad and rotor disk. Brake squeal noise is radiated by the brake components themselves (the disk, the pads and the caliper assembly) as is clearly seen using laser vibrometry techniques (see Section 5.2). However, because the resonant frequencies of the brake components are typically so high (seldom less than 1000–2000 Hz) it is the vehicle suspension that acts as the resonant system in the case of brake moan, creep-groan and judder.

There are few options for the control of brake noise. Careful choice of friction pairs is the most common (it is the difference between static and dynamic coefficients of friction that causes creep-groan, judder and moan [Bettella et al., 2002]) whereas the use of layers of metal plates riveted together on the back of the brake pads can act as efficient dampers for brake squeal.

4.8 A note on squeak, rattle and tizz noises

The control of squeak, rattle and tizz noises is extremely important to the passenger car industry. For some car manufacturers these faint but annoying noises are the number one cause of dealer returns (vehicles that have been returned to the manufacturer because the retail customer has not been satisfied by the remedial works undertaken by the dealer and has effectively given up on the vehicle). The main reason for their resulting in high levels of dealer returns is that they are easy to detect but their source is usually hard to find. To further complicate the matter, once the source is found a cure is often elusive.

Tizz noises are caused by high-frequency tactile vibration (such as that occurs on the gear lever knob) and are relatively easy to find and cure. By comparison, squeaks (which can be individual tick noises or sequences of ticks that sound like a squeak) are caused by relative motion between material pairs and these pairs can be buried deep within the fabric of the car interior (such as somewhere inside the dashboard assembly). The noise source results from a stick-slip process caused by a difference in the values of static and dynamic coefficient of friction.

Squeaks are known to be influenced by (Juneja et al., 1999):

- material choice;
- surface finish;
- frequency of excitation;
- amplitude of excitation;
- interference levels between the two materials;
- normal loads;
- temperature;
- humidity.

The worst combination of these seems to be high amplitude and high-frequency excitation applied to a pair of materials with high degree of interference at low humidities. Typically material pairs are tested on a rig, where one material is kept still and the other is shaken, and the rig is routinely situated in a climatic chamber. Squeak levels are frequently measured in terms of loudness (see Section 1.6.3.2). This type of pragmatic testing is now commonplace with the suppliers of vehicle interior trim systems, and much of the contractual risk caused by dealer returns is placed on these organisations.

Rattle noises can be caused by loose fitting (and hard) interior trim items but also in the vehicle transmission line (see Section 6.4.5). Due to the rather chaotic nature of the rattle phenomena, statistical correlations between subjective response to rattle sounds and objective measures (Croker et al., 1990) such as sound pressure level, loudness or speech interference level tend to be inconclusive (see Sections 1.6.3.1 and 1.6.3.2 for details on these).

4.9 Control of sound through absorption within porous materials

4.9.1 Practical approach

Interior noise levels (particularly those related to body booms and those at high frequencies) may be controlled by the addition of sound absorbing material to the passenger compartment.

From Section 2.5, for a diffuse space

$$T_{60} = \frac{0.161V}{S\bar{a}} \tag{4.78}$$

where

T_{60} = reverberation time at a particular frequency
V = volume in the space (m³)
S = surface area in the space (m²)
$\bar{a}$ = average Sabine absorptivity.

The reverberation time in the vehicle interior is not constant throughout the space (see Section 2.5) and therefore the acoustic field is not diffuse, but accepting this lack of realism and to make the calculation simple, for the purpose of illustration only, the average T_{60} @ 1000 Hz for a small European wagon style car can be estimated, thus:

Estimated volume = 3.2 m³
Estimated surface area of the interior space (without passengers)= 21.7 m²

From Bies and Hansen (1996), maximum absorption coefficients as measured in an impedance tube are:

- 0.89 @ 1000 Hz for an unoccupied, well-upholstered seat;
- 0.61 @ 1000 Hz for an unoccupied, leather-upholstered seat;
- 0.03 @ 1000 Hz for glass.

From Haines (1987), maximum absorption coefficients (method of measurement unknown) for a molded glass fibre headliner are:

- 0.60 @ 1000 Hz for 13-mm thick liner.

From Saha and Baker (1987), maximum absorption coefficients (measured in a reverberation room) are

- 0.18 @ 1000 Hz for a 6-mm thick carpet; and
- 0.42 @ 1000 Hz for a 12-mm thick carpet.

For cloth seats and leather seats see Tables 4.4 and 4.5.

The surface average absorption coefficient is in this case for cloth seats is (without passengers) $\bar{a} = 0.41$. Therefore, the average T_{60} @ 1000 Hz would be 58 ms for upholstered seats (without passengers). The surface average absorption coefficient in this

Table 4.4 Sound absorption in a typical European wagon style passenger car with cloth seats (estimates for illustration only)

Surface	Surface area (m²)	A	Sa
Headliner	3.5	0.6	2.1
Carpet	1.47	0.42	0.618
Footwell	1.05	0.42	0.441
Trunk	1.12	0.42	0.4704
Front screen	1.05	0.03	0.0315
Rear screen	1.05	0.03	0.0315
Seat squabs	1.96	0.89	1.7444
Seat backs	2.8	0.89	2.492
Side glass	3.0	0.03	0.09
Side trim	3.0	0.18*	0.54
Dash	0.84	0.18*	0.1512
Rear trim	0.84	0.18*	0.1512
Total	21.68		8.8612

* estimated.

Table 4.5 Sound absorption in a typical European wagon style passenger car with leather seats (estimates for illustration only)

Surface	Surface area (m²)	A	Sa
Headliner	3.5	0.6	2.1
Carpet	1.47	0.42	0.618
Footwell	1.05	0.42	0.441
Trunk	1.12	0.42	0.4704
Front screen	1.05	0.03	0.0315
Rear screen	1.05	0.03	0.0315
Seat squabs	1.96	0.61	1.1956
Seat backs	2.8	0.61	1.708
Side glass	3.0	0.03	0.09
Side trim	3.0	0.18*	0.54
Dash	0.84	0.18*	0.1512
Rear trim	0.84	0.18*	0.1512
Total	21.68		7.5284

* estimated.

case for leather seats is (without passengers), $\bar{a} = 0.35$. Therefore, the average T_{60} @ 1000 Hz would be 68 ms for leather seats (without passengers).

The two reverberation times estimated above broadly agree with the typical, measured reverberation times shown (Quian and Vanbuskirk, 1995), these being in the range of 70–90 ms.

From Section 4.3.5 it is known that the contribution to interior sound pressure levels from the reverberant field in a semi-diffuse environment is given by

$$\Delta Lp_{rev} = 10 \ \log_{10}\left(\frac{4}{R}\right) \tag{4.79}$$

where the room constant R is

$$R = \frac{S\bar{a}}{1 - \bar{a}} \tag{4.80}$$

Using the data from Tables 4.4 and 4.5:

$$R_{\text{cloth}} = 14.98$$
$$R_{\text{leather}} = 11.53$$

So the likely effect of fitting cloth seats rather than leather seats would be (without passengers)

$$10 \log_{10} \left(\frac{R_{\text{cloth}}}{R_{\text{leather}}} \right) = 1.1\,\text{dB reduction in the sound due to the reverberant field.}$$

This seems a rather modest reduction in reverberant sound level (remember, the level due to direct sound is not considered in the above analysis), although Quian and Vanbuskirk (1995) suggest that making the change in seat covering *resulted in measurable and noticeable differences in interior sound levels during some modes of actual vehicle operation... Speech intelligibility was improved.*

One may decide whether the driver is likely to be influenced most by direct of reverberant sound using the relationship (Section 2.5)

$$\frac{I_r}{I_d} = \frac{16\pi r^2}{S\bar{a}} \tag{4.81}$$

where I_r and I_d are the reverberant and direct sound intensities respectively. It can be seen that for the earlier case with cloth seats, the reverberant and direct intensities are equal (according to this simple model, assuming a diffuse space) at a distance of 0.4 m. So, when the driver's ears are less than 0.4 m from a noise source (like the side windows and the roof) the direct field will dominate and noise levels will be little affected by changes in the total absorption of the space. However, at longer distances from sources (the windscreen, the rear screen, the footwell, the rear floor) an increase of absorption will reduce the interior noise levels to at least some degree.

Sound absorbing materials make a contribution to controlling the levels of higher frequency noise in vehicle interiors. Their usefulness in controlling lower-frequency sound (say below 500 Hz) is limited as the thicknesses of absorbing material required would be bulky and add too much weight to the vehicle.

Note: With porous materials, the absorption coefficient α tends to increase rapidly at low frequencies and approach unity at moderate frequencies as shown in Figure 4.8.

Thickness of material increases α up to a point. As a general rule, the material thickness should be at least one quater wavelength of the lowest frequency of interest. Placing an air gap between the rear of the material and its backing changes the apparent thickness. The low-frequency absorption increases with the increasing air layer thickness but the high-frequency absorption deteriorates. The maximum absorption occurs around the one quarter wavelength resonant frequency of the air layer.

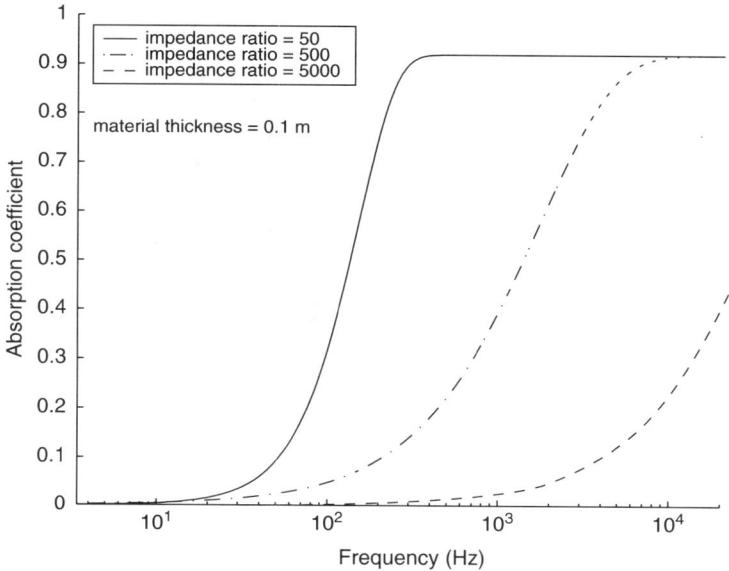

Figure 4.8 Typical absorption characteristics of a porous material: adapted from data provided by Bies and Hansen (1996). Impedance ratio is the specific acoustic impedance of the material divided by the characteristic impedance of the air.

The effectiveness of sound absorbing materials can be assessed in two ways:

1. Calculating the effect by combining the known absorption performance of individual absorbing items.
2. Measuring the effectiveness of the whole sound absorption package fitted to the vehicle.

The effectiveness of the whole vehicle sound package may be measured by:

1. Determining insertion loss, where the effect that the addition of each sound absorbing component has on interior noise levels is measured during certain vehicle operating conditions.
2. Determining the reverberation time (see Section 2.5 for details). Reverberation time (frequency variable) tends to be in the range $\ll 100$ ms for typical sedan cars. There is significant spatial variation in reverberation times so it is commonly measured at many locations and an average is then sought.

The seats and the headliner contribute most to the absorption of sound (as shown in Figure 4.9). This is due to their appropriate flow resistance (neither too high so as to appear reflective, nor too low so as to appear acoustically transparent) and their large surface areas.

Individual sound absorbing components may be characterised by their characteristic specific acoustic impedance

$$z_c = \frac{p}{u'} = \sqrt{\kappa \rho'} \qquad (4.82)$$

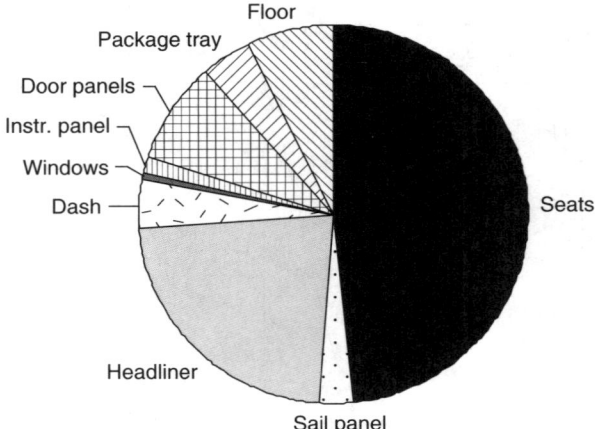

Figure 4.9 The absorption of sound by vehicle interior trim: data taken from [Qian and Vanbuskirk, 1995].

measured in either:

1. an impedance tube (Bies and Hansen, 1996; Chung and Blaser, 1980); or
2. a free field (Allard and Sieben, 1985).

A commonly quoted index is that of absorption coefficient. This varies with both frequency and angle of incidence of the impinging sound (as does the characteristic specific impedance of the material). It may be measured in many ways:

1. Random incidence α: Measured in the reverberation chamber room (see Section 4.3.4 for an introduction to such spaces).
2. Specific angle of incidence α: Measured in anechoic chamber (Ingard and Bolt, 1951).
3. Normal incidence α: Measured in the impedance tube (Bies and Hansen, 1996; Chung and Blaser, 1980)

The random impedance α seems to agree better with observed effects in road vehicles than the normal impedance α (Qian and Vanbuskirk, 1995).

4.9.2 Physical processes for sound absorption within porous materials

The physical mechanisms by which the sound is absorbed are widely acknowledged to be rather complex (Qian and Vanbuskirk, 1995). There is more than one mechanism and these are generally classified as follows (Fahy and Walker, 1998):

1. viscous losses due to oscillating flow within the internal spaces of the material;
2. heat conduction within the material;
3. vibration of the material (particularly in closed cell plastic foams).

Porous materials have rather complex geometric structures that defy mathematical description using deterministic models. Rather, bulk properties of the material are used to characterise its sound absorption (Fahy and Walker, 1998). Neglecting the vibration of the material itself, these are:

- flow resistivity (r);
- porosity (h);
- structure factor (s).

One may gain useful insight into the behaviour of sound within porous materials by inspecting a modified form of the linear plane wave acoustic equation (see Section 2.1.2 for a derivation).

$$\frac{\partial^2 p}{\partial t^2} = c^2 \frac{\partial^2 p}{\partial x^2} \qquad (4.83)$$

This equation relates the spatial variation of acoustic pressure p with the temporal distribution via the speed of sound c (m s^{-1}). It only applies to a free field. In the restricted internal spaces of a porous material, additional relationships will be sought for the spatial distribution of acoustic pressure.

4.9.3 Flow resistivity

Consider the steady flow of air through a porous material. The air is flowing with a volume velocity per unit cross section area of u' (m s^{-1}) (Fahy and Walker, 1998)

$$u' = \frac{m^3/s}{m^2/1^2} = \frac{m^3}{s} \cdot \frac{1^2}{m^2} = \frac{m}{s} \qquad (4.84)$$

The flow resistivity is defined as:

$$\frac{\partial p}{\partial x} = -ru' \qquad (4.85)$$

Bies and Hansen (1996) propose a method of measuring the flow resistivity (MKS rayls per metre) in accordance with ISO 9053-1991, by measuring the pressure drop across a sample of material in a tube through which there is a steady flow of air.

$$r = \frac{\rho \Delta P A}{\dot{m}l} \qquad (4.86)$$

ρ = density of gas (kg m^{-3})
ΔP = differential pressure (Nm^{-2})
A = gross sectional area of material (m^2)
$\dot{m}$ = mass flow rate of air (kg s^{-1})
l = specimen thickness (m)

The limitation of using flow resistivities measured in this way is that one must assume that the DC flow resistivity is the same as the oscillatory acoustic flow resistivity. For low speed internal flows this assumption generally holds (Fahy and Walker, 1998).

An alternative method of measuring flow resistivity does not suffer from such limitations as it uses acoustic field in an impedance tube to produce oscillatory flow in the

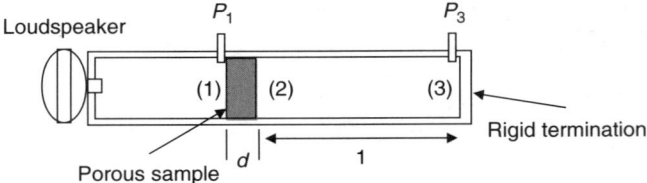

Figure 4.10 Impedance tube method for measuring flow resistivity: after Ingard and Dear (1985).

material (Ingard and Dear, 1985). In this technique, the base of the material sample is placed at a point in the tube that is an odd number of quarter wavelengths from the closed end of the tube. Two pressures are recorded, p_1 and p_3 as shown in Figure 4.10. Under these conditions,

$$\frac{r}{\rho_0 c_0} \approx \left| \frac{p_1}{p_3} \right|$$ (4.87)

at low frequencies (Ingard and Dear, 1985).

The frequency of the single tone (best below 100 Hz) driving the loudspeaker is adjusted until this is the case. The reason for the $\frac{n\lambda}{4}$ $(n = 1, 3, 5, \ldots)$ separation is explained in Appendix 4E (see also Kinsler et al. (1982)).

For fibrous materials Bies proposes (reported in Bies and Hansen [1996]):

$$\frac{rl}{\rho c} = 27.3 \left(\frac{\rho_m}{\rho_f} \right)^{1.53} \left(\frac{\mu}{d\rho c} \right) \left(\frac{l}{d} \right)$$ (4.88)

ρ = density of gas (kg m^{-3})
ρ_m = porous material bulk density (kg m^{-3})
ρ_f = fibre material density (kg m^{-3})
μ = gas viscosity (Ns m^{-1})
d = fibre diameter (m)
c = speed of sound in gas (m s^{-1})
l = specimen thickness (m)

Delaney and Bazely (reported in Bies and Hansen (1996)) offer a relationship between flow resistivity in the range of 10^3–5×10^4 MKS rayls per metre and the characteristic acoustic impedance of a porous material Z_M

$$Z_M = \rho c \left[1 + 0.0571 X^{-0.754} - i 0.087 X^{-0.732} \right]$$ (4.89)

$$X = \frac{\rho f}{r}$$ (4.90)

where

ρ = gas density (kg m^{-3})
f = frequency (Hz)
r = flow resistivity (MKS rayls per metre)

4.9.4 Porosity

Porosity is defined as (Fahy and Walker, 1998):

$$h = \frac{\text{volume of voids in the material}}{\text{total volume of the material including voids}} \tag{4.91}$$

For acoustic absorbing materials $h \to 0.90$ or 0.95.

The linearised equation of mass conservation in free air is given as (Kinsler et al., 1982):

$$\frac{\partial u}{\partial x} = -\frac{1}{\kappa_0}\frac{\partial p}{\partial t} \tag{4.92}$$

$$\kappa_0 = \text{bulk modulus} = \rho_0 c_0^2 \tag{4.93}$$

$$\kappa_0 = \rho_0 \left(\frac{\partial p}{\partial \rho}\right)_{\rho_0} \tag{4.94}$$

$$p = \kappa s, \qquad s = \frac{\rho - \rho_0}{\rho_0} \tag{4.95}$$

The effect of the porosity is to introduce a modified bulk modulus

$$\kappa = \frac{\kappa_0}{h} \tag{4.96}$$

so that

$$\frac{\partial u'}{\partial x} = -\frac{1}{\kappa}\frac{\partial p}{\partial t} \tag{4.97}$$

The derivation of the linearised mass conservation equation is given in Appendix 4F.

4.9.5 The structure factor

The structure factor (s) expresses the influence of the geometric form of the structure on the effective density of the fluid (Fahy and Walker, 1998).

There are four mechanisms for this:

1. Side pockets within the material (outside the flow stream) reduce the effective bulk modulus of the fluid. This produces a smaller acoustic pressure for a given strain gradient in the pore (see above equation) which leaves the impression that the effective density of the fluid is higher in the material than in free space.
2. Non-uniform pores causing sudden expansions that leave the impression of added mass.
3. Non-axial pore orientation. The inviscid momentum equation normal to the orientation of the pore is

$$\frac{\partial p}{\partial h} = -\rho_0 \frac{\partial u_n}{\partial t} \tag{4.98}$$

or for orientation angle θ to the surface of the material:

$$\frac{\partial p}{\partial x}\cos\theta = -\rho_0 \frac{\partial u'}{\partial t}\frac{1}{h\cos\theta}$$

$$\frac{\partial p}{\partial x} = \frac{-\rho_0}{h\cos^2\theta}\frac{\partial u'}{\partial t}$$

$$\frac{\partial p}{\partial x} = -\frac{s\rho_0}{h}\frac{\partial u'}{\partial t} \tag{4.99}$$

$$s \propto \alpha \frac{1}{\cos^2\theta} \tag{4.100}$$

The effective density is increased by factor (s). For a random orientation of pores $s = 3$. Generally s is in the range of 1.2–2.0 (Fahy and Walker, 1998).
4. The velocity profile across each pore acts to reduce volume acceleration produced by a given gradient and hence contributes to s.

4.9.6 The modified 1-D linear plane wave equation

The modified equation of motion becomes:

$$\frac{\partial p}{\partial x} = -\left(\frac{s\rho_0}{h}\right)\frac{\partial u'}{\partial t} - ru' \tag{4.101}$$

Reminder: u' is the volume flow rate/unit cross section area *not* the flow velocity in the pores.

For simple harmonic motion of frequency ω

$$\frac{\partial p}{\partial x} = -\left[\frac{s\rho_0}{h} + \frac{r}{i\omega}\right]\frac{\partial u'}{\partial t} \tag{4.102}$$

This is a modified Euler equation with the complex terms in the brackets being the effective density (Fahy and Walker, 1998). The derivation of the 1-D Euler equation is given in Appendix 4G.

Now, if this equation is combined with the linearised mass conservation equation (4.92)

$$\frac{\partial u'}{\partial x} = -\frac{1}{\kappa_0}\frac{\partial p}{\partial t} \tag{4.92}$$

a modified wave equation is obtained in this manner:

• Differentiate both sides of the modified Euler equation with respect to x

$$\frac{\partial^2 p}{\partial x^2} = -\left[\frac{s\rho_0}{h} + \frac{r}{i\omega}\right]\frac{\partial}{\partial x}\left(\frac{\partial u'}{\partial t}\right)$$

- Now take the time derivative of the linearised mass conservation equation

$$-\frac{1}{\kappa_0}\frac{\partial^2 p}{\partial t^2} = \frac{\partial}{\partial t}\left(\frac{\partial u'}{\partial x}\right) = \frac{\partial}{\partial x}\left(\frac{\partial u'}{\partial t}\right)$$

- So, combination gives:

$$\frac{\partial^2 p}{\partial x^2} = -\left[\frac{s\rho_0}{h}+\frac{r}{i\omega}\right]\times -\left[\frac{1}{\kappa_0}\frac{\partial^2 p}{\partial t^2}\right]$$

$$\kappa_0\frac{\partial^2 p}{\partial x^2} = \left[\frac{s\rho_0}{h}+\frac{r}{i\omega}\right]\frac{\partial^2 p}{\partial t^2}$$

$$\rho_0 c_0^2\frac{\partial^2 p}{\partial x^2} = \left[\frac{s\rho_0}{h}+\frac{r}{i\omega}\right]\frac{\partial^2 p}{\partial t^2}$$

$$\frac{c_0^2}{s}\frac{\partial^2 p}{\partial x^2} = \left[\frac{1}{h}+\frac{r}{i\rho_0 s\omega}\right]\frac{\partial^2 p}{\partial t^2}$$

- And finally the modified wave equation is

$$c_1^2\frac{\partial^2 p}{\partial x^2} = \frac{1}{h}\frac{\partial^2 p}{\partial t^2}+\left(\frac{r}{\rho_0 s}\right)\frac{\partial p}{\partial t} \qquad c_1^2 = \frac{c_0^2}{s} \qquad (4.103)$$

The effect of parameters s, r, h is to lower the speed of sound within the material (compared with the speed of sound in free space) and to attenuate the acoustic wave as it propagates.

$$p(x, t) = A e^{i\omega t}e^{-\gamma x} \qquad (4.104)$$

γ = propagation constant = $\alpha + i\beta$

$$p(x, t) = A e^{i(\omega t - \beta x)}e^{-\alpha x} \qquad (4.105)$$

α = attenuation constant

$$z_c = \frac{p}{u'} = \sqrt{\kappa\rho'} \qquad (4.106)$$

where z_c is the characteristic specific acoustic impedance of the gas in porous material.

$$\gamma = i\omega\sqrt{\frac{\rho'}{\kappa}} \qquad (4.107)$$

$$\kappa = \frac{\rho_0 c_0^2}{h} \qquad (4.108)$$

$$\rho' = \frac{s\rho_0}{h}+\frac{r}{i\omega} \qquad (4.109)$$

4.10   Control of sound by minimising transmission through panels

4.10.1   Introduction

The encapsulation of a noise source using panels with high TL is a valuable tool for the refinement engineer. When used appropriately it can produce significant reductions in interior noise level (more than 10 dB). It is used during the shielding technique for noise source ranking (see Section 3.2).

The principles of encapsulation are also used when designing noise barrier panels to fit under carpet in order to isolate the passenger compartment from the noise in the engine bay. Although these are a powerful way of controlling interior noise levels, a note of caution is given that they are a heavy solution. One published benchmarking exercise (Wentzel and VanBuskirk, 1999) identified nearly 39 kg of such noise barrier materials in a sedan and nearly 48 kg in a mini-van.

Encapsulation techniques can, along with noise control at source, be used as part of a general noise control strategy. A noise control problem can be split into three components – a noise source, noise propagation and the reception of noise as illustrated in Figure 4.11.

Encapsulation allows for the interruption of the airborne noise path. It should be noted that encapsulation will only remain effective whilst any structure-borne paths (or other flanking transmission) remains insignificant. Therefore, in many cases the structure-borne paths must also be interrupted or controlled for encapsulation to be fully effective. The interruption of structure-borne paths through the use of vibration isolation and the control of vibration through the use of damping is discussed elsewhere (see Sections 6.3 and 6.2 respectively).

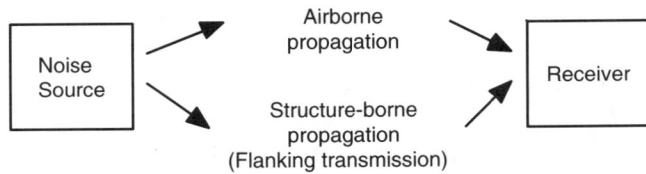

Figure 4.11 A typical noise control problem.

4.10.2   The measurement of the acoustic performance of enclosures

The effectiveness of an acoustic enclosure may be assessed according to a number of different parameters. The first is termed noise reduction and is simply the arithmetic difference between the sound pressure level at a point (or the average over a number of points) within an enclosure and the sound pressure level at a prescribed point outside the enclosure as shown in Figure 4.12:

$$NR = SPL_1 - SPL_2 \text{ (dB)} \tag{4.110}$$

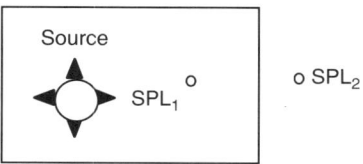

Figure 4.12 Measuring noise reduction.

This method is experimentally convenient, and an adequate means of comparing the acoustic performance of two geometrically similar enclosures. However, it is of limited use as an absolute indicator of acoustic performance as the value for noise reduction obtained is valid only for the precise microphone locations chosen.

A more generally applicable method of assessment uses TL as a parameter. TL in decibels is obtained from the ratio of incident and transmitted acoustic intensities across the boundary of the enclosure (Figure 4.13).

$$TL = 10 \ \log_{10} \left[\frac{I_i}{I_t} \right] \text{dB} \tag{4.111}$$

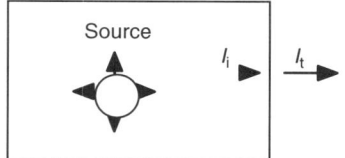

Figure 4.13 Measuring TL.

The TL across sections of enclosures may be obtained from measurements made in the field using an intensity meter, or the TL of single panels can be obtained from measurements made in a pair of special interconnected laboratory rooms known as transmission suite.

A third and intuitive method of assessment uses insertion loss as a parameter. Insertion loss is obtained by comparing sound pressure levels at a point in space with and without the enclosure in place (see Figure: 4.14).

$$IL = SPL_1 - SPL_2 \ (\text{dB}) \tag{4.112}$$

Only TL is a general and reliable performance indicator for comparing the acoustic transmission through individual panels with the performance of an entire enclosure. The main reason for this is that both the alternative noise reduction and insertion loss methods can produce results with negative values in the cases where the sound pressure levels outside the enclosure actually increase at certain frequencies once the enclosure is installed. This seemingly unexpected situation will occur when resonant acoustic modes within the enclosure result in standing wave ratios that are greater than the TL of the enclosure boundary. This phenomenon can catch the novice refinement engineer out.

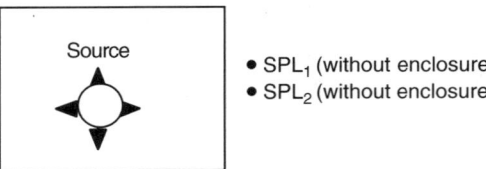

Figure 4.14 Measuring insertion loss.

4.10.3 Interpreting the acoustic performance data supplied by enclosure and panel manufacturers

Manufacturers of acoustic enclosures and high TL panels will often publish details of the acoustic performance of individual panels in terms of

- TL;
- absorption coefficient;
- sound transmission class.

It is important to ascertain the conditions under which these data were obtained. They will generally be representative yet conservative values for results obtained from a number of laboratory tests made according to relevant material or international standards.

TL data may be obtained from tests performed in a transmission suite in accordance with ISO 140/1: 1990, ISO 140/3: 1990. The sound transmission loss or sound reduction index (SRI) for a panel is described in terms of the ratio of sound power incident on the panel to the sound power transmitted through the panel.

$$TL = 10 \ \log_{10} \frac{\Pi_i}{\Pi_t} \ \text{dB} \tag{4.113}$$

Sound transmission loss is a quantity that depends only on the frequency of the sound and the properties of the panel. Consider a panel, with fixed sound transmission loss mounted in a transmission suite (Figure 4.15).

A given spatially averaged sound pressure level in the source room (L_i) gives rise to a spatially averaged velocity on the surface of the panel (u) and hence a spatially averaged sound pressure level in the receiving room (L_t) according to the relationship

$$L_t \propto \sigma S u^2 \tag{4.114}$$

where

S = panel area (m)
σ = acoustic radiation efficiency

Therefore, it can be seen that L_t is dependent on the area of the panel.

Also, L_t is found to be dependent on the absorption (A) of the receiving room for a given intensity (I) in the source room. A receiving room with little absorption will have a higher sound pressure level than one with a large absorption.

The sound transmission loss for the panel and the noise reduction in the suite can be thus related.

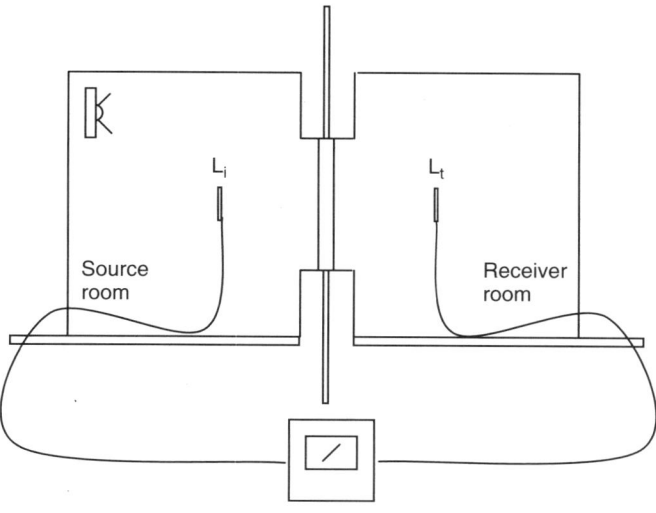

Ideal case
- Walls have infinite transmission loss.
- Rooms are perfectly isolated from each other.
- Test panel is simply supported and edges sealed to be airtight.

Figure 4.15 The transmission suite.

Power incident on the panel

$$\Pi_i = I_i S \tag{4.115}$$

Power absorbed by the receiving room

$$\Pi_t = I_t A \tag{4.116}$$

$$TL = 10 \; \log_{10} \frac{\Pi_i}{\Pi_t}$$

$$TL = 10 \; \log_{10} \left(\frac{I_i S}{I_t A} \right)$$

$$TL = L_i - L_t + 10 \; \log_{10} \left(\frac{S}{A} \right) \tag{4.117}$$

The difference in levels between the two rooms is known as the Noise Reduction (NR)

$$NR = 10 \; \log_{10} \left(\frac{I_i}{I_t} \right) = L_i - L_t \tag{4.118}$$

$$TL = NR + 10 \; \log_{10} \left(\frac{S}{A} \right) \tag{4.119}$$

According to the relations developed by Sabine, the absorption of the receiving room is related to its reverberation time (T) – the time in seconds required for the level of the sound to drop by 60 dB (Section 2.5).

$$A = \frac{0.161 \, V}{T} \qquad (4.120)$$

where (V) is the volume of the receiving room (m^3). The reverberation time for each 1/3 octave band can readily be measured in the receiving room using a real time analyser.

The transmission suite is constructed in such a way as to:

- restrict the transmission of sound to paths passing directly through the test panel;
- provide a source field which impinges with random angles of incidence on the test panel.

For such tests, a panel could be mounted and sealed in an aperture within the brick dividing wall between two rectangular reverberant rooms, both of which may be constructed from 215-mm brick with reinforced concrete floors and ceilings. In one particular case (by way of example only), the adjoining brick wall is 4.8-m wide and 3.1-m high with 550-mm nominal thickness, and forms the whole of the common area between the two rooms. One of the rooms is larger and is termed the receiving room. It has a depth of 20.2 m and a volume of 200 m^3. The adjoining room has a depth of 4 m and a volume of 60 m^3 and is known as the source room. The receiving room is isolated from the surrounding structure and adjoining room by the use of resilient mountings and seals.

Broadband white noise is produced in the source room using a minimum of two loudspeakers. The average sound pressure levels in each room are determined using distributed arrays of microphones, connected to a suitable third octave analyser. The difference in the averaged sound pressure levels forms the NR (shown as the upper curve in the left hand plot of Figure 4.16). From this, and the knowledge of the 1/3 octave reverberation times (also shown in Figure 4.16), the measured SRI or TL can be found (shown as the lower curve in the left-hand plot of Figure 4.16).

The absorption coefficient of the test panel may be measured according to standard procedures in an impedance tube or in a reverberation chamber (Section 4.9.1). The impedance tube method is restricted to small samples of material and sound of normal incidence and therefore when testing the acoustic absorption of enclosure panels, a test based on measurements is made in a reverberation chamber.

It should be noted that both TL and absorption data are obtained from tests performed in reverberation chambers with diffuse fields where the sound impinges on the material samples with an essentially random angle of incidence. These ideal conditions may not be found in the case of a practical enclosure. In addition to this, the laboratory tests may have been performed on a single panel in isolation. The TL, and, to a lesser extent, absorption characteristics of a panel depend on the panel dimensions and on its mounting arrangement. Therefore, the published laboratory specifications may not be fully realistic in the case of an enclosure made up of a number of panels mounted in a variety of ways.

When examining the acoustic performance of enclosure panels it is important to consider absorption characteristics as fully as the TL qualities, if the detrimental effects of acoustic resonances within the enclosure are to be avoided. In addition to TL data, manufacturers of acoustic enclosures might also quote sound transmission class (STC) values for panels. The STC parameter was developed based on studies made with noise

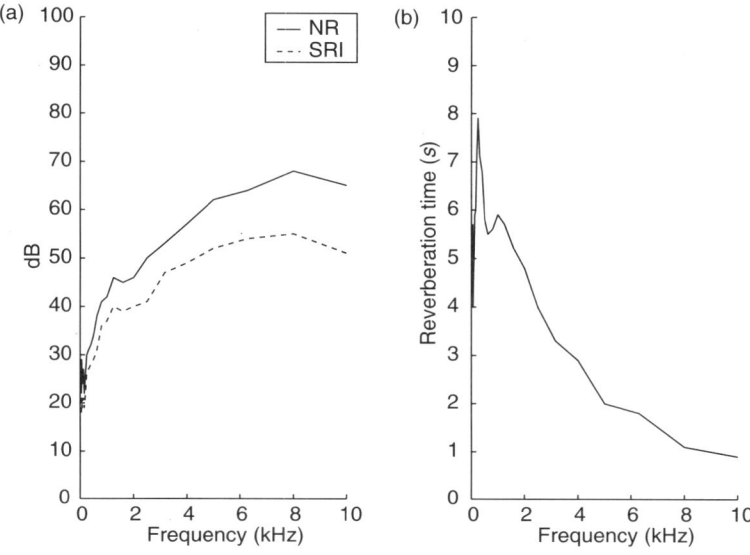

Figure 4.16 Measured NR and SRI (TL) curves along with reverberation times.

sources typical of multifamily dwellings, and provides a convenient and fairly successful single index specification of the transmission characteristics of a panel. To determine the STC of a panel, the TL is measured in the sixteen contiguous 1/3 octave bands between 125 Hz and 4000 Hz inclusive. These measured values are then compared to a family of reference curves as shown in Figure 4.17.

To determine the STC of a panel, the reference contour is chosen so that the maximum deficiency (deviation of the data below the contour) at any one frequency does not exceed 8 dB and the total deficiency at all frequencies does not exceed 32 dB. The STC of the panel is then the value of TL corresponding to the value of the chosen reference curve in the 500 Hz band. STC values are published widely in the literature, for example in Kinsler et al. (1982).

4.10.4   The significance of acoustic seals and controlling flanking transmission

The need for tight acoustic seals between adjoining panels cannot be overstated. Even the smallest air gap will limit the maximum TL that can be achieved, particularly at higher frequencies. A simple yet effective way of detecting sound leaking around a seal between panels is to use a length of narrow bore (5 mm) plastic tube, place one end in one ear and pass the other end over the edges of the panel. Any acoustic leaks become immediately apparent. The fitting of quality acoustic seals is one of the reasons for the relatively high cost of acoustic encapsulation. One of the major challenges in enclosure design is the provision of adequate ventilation and cooling without causing leakage of sound.

Flanking transmission must be controlled if an enclosure is to be fully effective. A common flanking path is due to vibration from the noise source being transmitted into

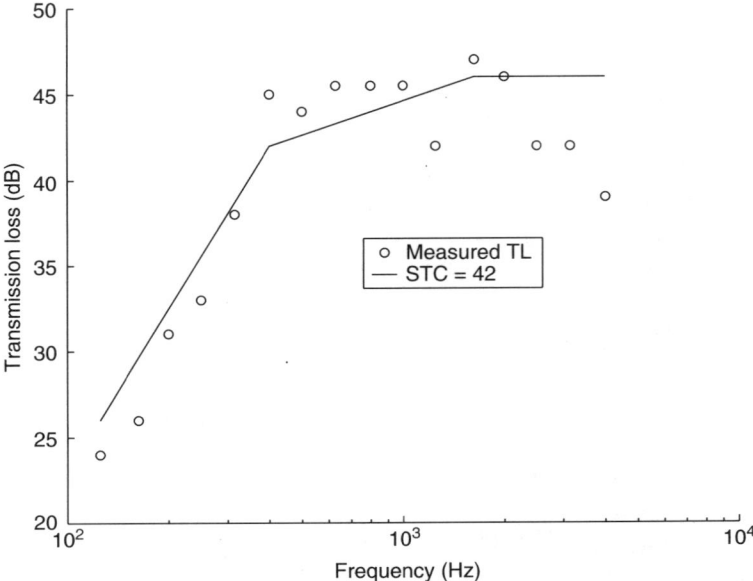

Figure 4.17 STC curves: data after Kinsler et al. (1982).

the floor of the enclosure and either producing re-radiated sound outside the enclosure or forcing the walls of the enclosure into vibration and this seriously limits their TL characteristics. An obvious example is the interior noise due to an IC engine where structure-borne noise is in effect a flanking path.

Care should be taken that pipes and service ducts crossing the boundary of the enclosure are well isolated from the structure of the enclosure. Particular care should be taken if forced ventilation is being provided as in the case of an enclosure around an IC engine.

4.10.5 The transmission of sound through panels

Vibration may propagate through a structure in the form of compressional, shear or torsional waves (see Section 5.3.1). In structures constructed with thick members all three types of propagation may be significant. However in thin panels pure compressional propagation is not likely, and in the audio frequency range such panels are usually excited by sound to form bending waves which are a combination of compression and shear motions. Bending waves result in a deflection of the panel in a direction that is normal to its surface. The spatial distribution of this deformation is a function of the bending wave speed.

The speed of propagation of sound in air c (m s^{-1}) is a function of the composition of the fluid and of temperature (Section 2.1.1)

$$c = \sqrt{\gamma RT}$$

(4.121)

where

$\gamma = c_p/c_v$ ratio of specific heats
$R =$ gas constant
$T =$ temperature (K)

Interestingly, the speed of propagation of bending waves is dependent not only on the mechanical characteristics of the material (Poison's Ratio, Young's Modulus), but also on the shape and particularly the thickness of the panel. Thus (Section 5.3.1): for large elements, the bending stiffness B is

$$B = EI \qquad (4.122)$$

where E is the Young's Modulus (N m^{-2}) and I is the moment of inertia (m^4 or beware, commonly mm^4).

For plates of thickness h the moment of inertia per unit width I' is used

$$I' = \frac{h^3}{12} \qquad (4.123)$$

and the bending stiffness becomes

$$B' = \frac{I'E}{1 - \mu^2} \qquad (4.124)$$

The velocity at which one must travel to remain always at the same phase of one infinite sinusoidal wave is the phase velocity of bending waves

$$c_B = \sqrt[4]{\frac{B}{m'}} \sqrt[2]{\omega} \ (\text{m s}^{-1}) \qquad (4.125)$$

where m' is the mass per unit length (ρS) and ω is the radial frequency (rad s^{-1}).

$$\omega = 2\pi f \quad f \text{ is the frequency (Hz)} \qquad (4.126)$$

For plates of thickness h, this becomes (providing that the wavelength $\lambda \gg 6h$) (Section 5.3.1.3)

$$c_B \approx \sqrt{1.8 c_{L_1} h f} = c_{L_1} \sqrt{\frac{1.8h}{\lambda_{L_1}}} \ (\text{m s}^{-1}) \qquad (4.127)$$

$$c_{L_1} = \sqrt{\frac{E}{\rho_m (1 - \mu^2)}} \qquad (4.128)$$

where ρ_m is the material density and μ is Poisson's ratio.

Note that the phase velocity is frequency-dependent and therefore the wavefront distorts (dispersion) as higher-frequency components propagate with higher phase velocity than lower-frequency components.

One may demonstrate that energy propagates at the group velocity C_B (see p. 106 of Cremer and Heckl (1988) for details)

$$C_B = 2c_B \text{ (m s}^{-1}) \tag{4.129}$$

It should be noted that practical panels are unlikely to be isotropic in construction and therefore bending wave speed could vary with direction making the panel orthotropic to some extent. However, the isotropic assumption is convenient and therefore will be pursued further here, particularly as the assumption tends to hold true as frequency increases and the flexural wavelength tends towards the characteristic dimension (usually thickness) of the panel.

In the case of isotropic panels there exists a frequency, named the critical frequency, at which the flexural wavelength in the panel matches the acoustic wavelength in the air. Orthotropic panels will have more than one critical frequency. The critical frequency or frequencies in either case are given by (Bies and Hansen, 1996):

$$f_c = \frac{c^2}{2\pi}\sqrt{\frac{m}{B'}} \text{ Hz} \tag{4.130}$$

where c is the speed of sound in air (m s^{-1}) and m is the surface density (kg m^{-2}).

At frequencies around the critical frequency an effect known as coincidence is noted. At coincidence frequencies the panel is strongly coupled to the fluid so that sound impinging on the panel from any angle of incidence will produce a strong flexural response in the panel (Figure 4.18). Applying reciprocity indicates that the converse is true with a panel being a strong radiator of sound of any angle of emission at coincidence frequencies. It therefore follows that the coincidence effect greatly reduces the TL of a panel at frequencies near the critical frequencies. The response of the panel at the critical frequencies is a resonant phenomena and such a response is strongly dependent on the damping in the system.

At frequencies above the critical frequency an angle of incidence may be found so that the trace of the sound wave matches that of a flexural wave and good coupling

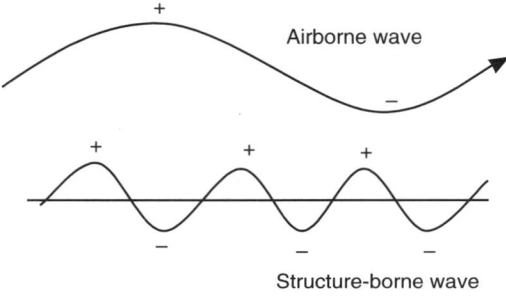

Figure 4.18 Strong coupling between the airborne and structure-borne waves occurs around the coincidence frequency.

results. At frequencies below the critical frequency, the wavelength of sound is longer than that of any flexural waves and poor coupling results due to local cancellation effects. Panels are therefore poor radiators of sound at low frequencies except at discontinuities or boundaries in their surfaces (such as at edges or ribs) where the cancellation effect is not present.

At these regions of localised coupling, the panel may be driven by sound of normal incidence. Curves showing the typical TL behaviour of isotropic and orthotropic panels are shown in Figure 4.19.

At very low frequencies, the TL of an isotropic panel is controlled by the stiffness of the panel. At the frequency of the first panel resonance the TL dips and is in part controlled by the damping of the system. At moderate frequencies, TL is controlled by the mass or surface density of the panel and increases at a rate of 6 dB per octave (the so-called mass law relationship) up to the coincidence dip around the critical frequency. At frequencies above the critical frequency the TL is said to be damping-controlled and rises at a rate of 9 dB per octave.

The corresponding typical TL curve for orthotropic panels is characterised by a wide coincidence region caused by the presence of more than one critical frequency. For this reason orthotropic panels should be avoided when noise control is important. However,

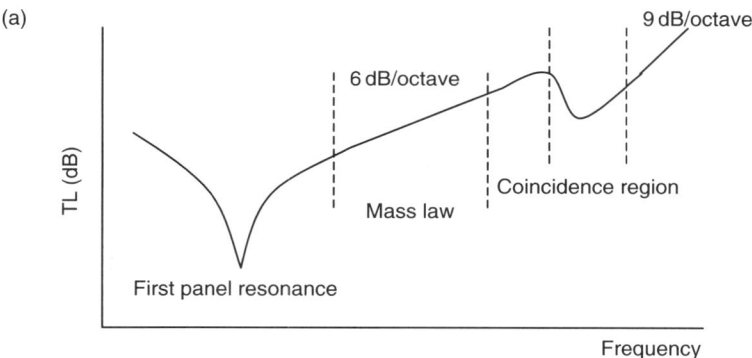

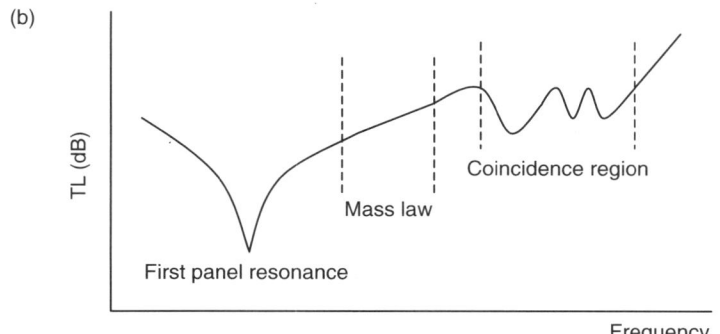

Figure 4.19 Typical panel TL (after Bies and Hansen, 1996). (a) isotropic panel (b) orthotropic panel.

the TL characteristics of a heavily damped orthotropic panel will tend towards that of an isotropic panel.

It should be noted that the range of frequencies for which the mass law operates is controlled by the panel stiffness. Stiffening a panel tends to move the first panel resonance f_0 up in frequency, and the critical frequency f_c down. For a panel of uniform thickness h (Bies and Hansen, 1996)

$$f_0 = 0.453 c_L h \, (a^{-2} + b^{-2}) \, \text{Hz} \qquad (4.131)$$

$$f_c = \frac{c^2}{1.81 c_L h} \, \text{Hz} \qquad (4.132)$$

The mass law is dependent on the angle of incidence of the impinging sound. Sharp (reported in Bies and Hansen (1996)) suggests that for an infinite panel

$$TL_\theta = 10 \, \log_{10} \left[1 + \left(\frac{\pi f m}{\rho c} \cos \theta \right)^2 \right] \, \text{dB} \qquad (4.133)$$

where ρ is the density of air (kg m^{-3}).

Pierce (reported in Fahy (1985)) shows that the random incidence (diffuse field) transmission loss (TL_D) is obtained from a weighted average of all angles of incidence:

$$TL_D = TL_N - 10 \, \log_{10}(0.23 TL_N) \, \text{dB} \qquad (4.134)$$

where

TL_N is the normal incidence transmission loss (dB).

The field TL is not necessarily equal to the TL_D. Sharp (reported in Bies and Hansen (1996)) suggests the following relationship for the field TL:

$$TL = TL_N - 5 \, \text{dB} \qquad (4.135)$$

The difference between TL and TL_D is probably due to the finite size of panels tested for the case of TL_D.

Sharp (reported in Bies and Hansen (1996)) suggests the following mass law:

$$TL = 20 \, \log_{10}[(\pi f m / \rho c)] - 5 \, \text{dB} \qquad (4.136)$$

where f is the third octave centre frequency and for the case when $\dfrac{fm}{\rho c} > 1$.

Sharp (reported in Bies and Hansen (1996)) also suggests the following relationship for the field incidence TL of an isotropic panel above the critical frequency:

$$TL = 20 \, \log_{10} \left(\frac{\pi f m}{\rho c} \right) + 10 \, \log_{10} \left[\frac{2 \eta f}{\pi f_c} \right] \, \text{dB} \qquad (4.137)$$

where η is the panel loss factor and f is the frequency (Hz) which this time is independent of bandwidth.

Higher levels of TL may be achieved if a double-skinned enclosure is used. This construction is usually more cost-effective than constructing a very massive single-walled enclosure. For best results the two skins should be mechanically and acoustically isolated from each other. Mechanical isolation can be achieved by mounting the two skins on separate beams or by using neoprene rubber between the skins and the common studs. Acoustic isolation can be achieved by filling the air gap with an absorptive material. The material should be at least $15/f$ metres thick with an impedance of $3\text{--}5\ \rho c$ (too high an impedance might result in a mechanical path through the material).

The TL characteristics of a single panel has been shown to be influenced by two frequency bands, the first centred on the lowest-order panel resonance and the second on the critical frequency. In the double leaf case, the influence of the lowest-order resonance of a single panel is replaced with the lowest-order acoustic resonance within the cavity at f_2 where (Sharp reported in Bies and Hansen (1996))

$$f_2 = \frac{c}{2L} \tag{4.138}$$

$L =$ longest cavity dimension (m)

The critical frequencies f_{c1} and f_{c2} are still important and added consideration must be given to the mass-air-mass resonance at f_0 of the two panels on the compliance of the cavity and a limiting frequency f_1 related to the width of the air gap (d) between panels. Sharp (reported in Bies and Hansen [1996]) gives the following relationships for panels which are totally mechanically and acoustically isolated from each other.

$$f_0 = \frac{1}{2\pi} \left(\frac{1.8\rho c^2 (m_1 + m_2)}{dm_1 m_2} \right)^{1/2} \text{Hz} \tag{4.139}$$

$$f_1 = \frac{c}{2\pi d} \text{Hz} \tag{4.140}$$

Below $f_{c1,2}/2$

$$TL_M = 20\ \log_{10} \left[\frac{\pi f M}{\rho c} \right] - 5\,\text{dB} \tag{4.141}$$

Above $f_{c1,2}$

$$TL_M = 20\ \log_{10} \left[\frac{\pi f M}{\rho c} \right] + 10\ \log_{10} \left[\frac{2\eta f}{\pi f_c} \right] \text{dB} \tag{4.142}$$

where

$$M = m_1 + m_2 \tag{4.143}$$

And finally

$$TL = TL_M \quad f < f_0 \tag{4.144}$$

$$TL = TL_1 + TL_2 + 20\ \log_{10} fd - 29 \quad f_0 < f < f_1 \tag{4.145}$$

where TL_1 and TL_2 are found by using m_1 or m_2 respectively in the mass law.

$$TL = TL_1 + TL_2 + 6 \quad f < f_1 \tag{4.146}$$

These equations relate to the ideal case where the two leaves are mechanically isolated from one another and absorptive material is introduced between the leaves to eliminate the effect of acoustic resonances in the cavity. Sharp presents some algorithms to predict the effect that the mounting of the leaves (line, line-point or point-point) has on the ideal TL, which will not be presented here (Sharp reported in Bies and Hansen (1996)).

Fahy presents an alternative model for the transmission loss of double-leaved panels in Fahy (1985). Here he presents a 1-D model that assumes mechanical isolation between the leaves, but takes account of the acoustic resonances in the air gap. It should be noted that Fahy's method produces predictions of normal incidence TL while Sharp's method relates to field TL. Fahy's equations are:

$$\omega_0 = \left[\frac{\rho_0 c^2}{d} \left(\frac{m_1 + m_3}{m_1 m_2} \right) \right]^{1/2} \tag{4.147}$$

(note the omission of the factor 1.8 which appears in Sharp's equation).

$$TL_{M,0} = 20 \ \log_{10} M + 20 \ \log_{10} f - 20 \ \log_{10} \left(\frac{\rho_0 c}{\pi} \right) \text{ dB} \tag{4.148}$$

where $M = m_1 + m_2$ and $TL_{1,0}$ and $TL_{2,0}$ are obtained by substitution of m_1 and m_2 respectively.

Below ω_0

$$TL_0 = TL_{M,0} \text{dB} \tag{4.149}$$

At ω_0

$$TL_0 = TL(0, M, \omega') + 20 \ \log_{10} \eta \text{ dB} \tag{4.150}$$

for the case of $m_1 = m_2$, and ω' is the natural frequency of the panels which Bies and Hansen (1996) suggest to be

$$\omega'_{i,n} = \pi^2 \sqrt{\frac{B'}{m}} \left[\frac{i^2}{a^2} + \frac{n^2}{b^2} \right] (\text{rad s}^{-1}) \ i, n = 1, 2, 3, \ldots \tag{4.151}$$

where a, b are panel width and length, B' is the bending stiffness per unit width and the natural frequency occurs when $i = n = 1$.

Between ω_0 and $kd = \pi/4$ where

$$k = \frac{\omega}{c} \tag{4.152}$$

$$TL_0 = TL_{1,0} + TL_{2,0} + 20 \ \log_{10}(2kd) \text{ dB} \tag{4.153}$$

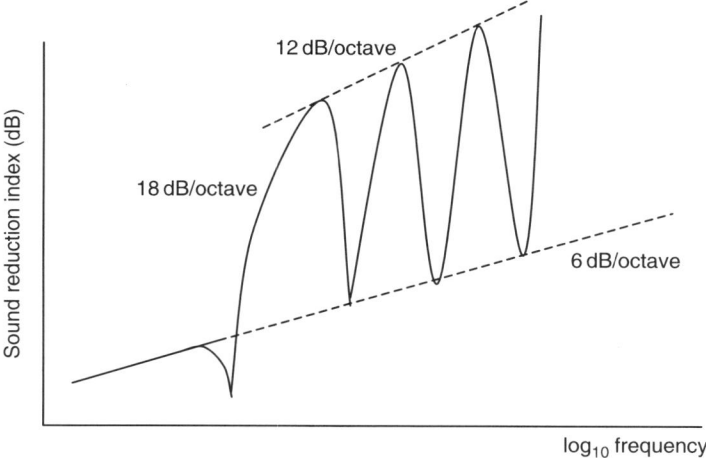

Figure 4.20 Generalised TL curves: after Fahy (1985).

At frequencies corresponding to acoustic anti-resonances of the air gap the TL is maximum at

$$kd = (2n-1)\frac{\pi}{2} \quad n = 1, 2, 3, \ldots \tag{4.154}$$

$$TL \simeq TL_{1,0} + TL_{2,0} + 6\,\text{dB} \tag{4.155}$$

At frequencies corresponding to resonances of the air gap, the TL dips down to the combined mass law for the two leaves:

$$TL_0 = 20\ \log_{10} M + 20\ \log_{10} f - 20\ \log_{10}\left(\frac{\rho_0 c}{\pi}\right)\,\text{dB} \tag{4.156}$$

when

$$kd = n\pi$$

These equations can be used to form a generalised TL curve as shown in Figure 4.20.

4.10.6 Sound inside and outside large enclosures

Enclosures are deemed to be large if they are not designed to be close fitting around a noise source. An enclosure might be considered large if a complete wavelength of sound at a low frequency separated the nearest wall of the enclosure from the noise source.

As already discussed, the installation of an enclosure around a noise source will produce a reverberant field within the enclosure. The distribution of sound pressure level within the enclosure can be ascertained, given its absorption characteristics and a noise source of known power output using the relationships discussed in Section 2.5.

The sound field outside the enclosure may be assumed to comprise a contribution from the direct field of the source reduced in amplitude by the normal TL (TL_N) of the boundary of the enclosure and a contribution from the reverberant field within the enclosure reduced in amplitude by the field TL of the enclosure.

Now

$$TL_n = -10 \log_{10} \tau_n \text{ dB} \tag{4.157}$$

$$TL = -10 \log_{10} \tau \text{ dB} \tag{4.158}$$

where τ_n and τ are transmission coefficients.

Also

$$TL_N - TL = 5 \text{ dB} \tag{4.159}$$

therefore

$$\tau_n = 0.3\tau \tag{4.160}$$

Now, the total sound power radiated by the enclosure is given by (Bies and Hansen, 1996):

$$S_E \frac{\langle p_1^2 \rangle}{\rho c} = W_{\tau_n} + W(1 - \bar{a}_i) \frac{S_E}{S_i \bar{a}_i} \tau \tag{4.161}$$

which is the fraction of sound power transmitted by the direct field added to the fraction of sound power transmitted by the reverberant field. In this equation

$\bar{a}_i$ = average internal Sabine absorptivity
S_i = internal surface area including the surface area of the noise source
S_E = external surface area of the enclosure

This equation can be re-written as

$$S_E \frac{\langle p_1^2 \rangle}{\rho c} = W \tau_E \tag{4.162}$$

where

$$\frac{\langle p_1^2 \rangle}{\rho c} = \text{sound intensity (Bies and Hansen, 1996)} \tag{4.163}$$

$$\tau_E = \tau [0.3 + S_E(1 - \bar{a}_i/S_i\bar{a}_i] \tag{4.164}$$

The sound pressure directly outside the enclosure can therefore be obtained from (Bies and Hansen, 1996)

$$L_{p_1} = L_w - TL - 10 \log_{10} S_E + C \tag{4.165}$$

where

$$C = 10 \log_{10}[0.3 + S_E(1 - \bar{a}_i)/(S_i\bar{a}_i)] \text{ dB} \tag{4.166}$$

It should be noted that this equation gives very approximate results only .

The sound pressure level at a point in free space some distance r away from the enclosure can be determined from (Bies and Hansen, 1996):

$$L_{WE} \simeq L_{p_1} + 10 \ \log_{10} S_E \ dB \tag{4.167}$$

$$L_{p_2} = L_{WE} + 10 \ \log_{10} \left[\frac{D_\theta}{4\pi r^2} \right] dB \tag{4.168}$$

When the enclosure is positioned on a hard floor:

$$D_\theta = 2 \tag{4.169}$$

The sound pressure level at a point in a reverberant space can equally be found using (see Section 4.3.5)

$$L_{p_2} = L_{WE} = 10 \ \log_{10} \left[\frac{D_\theta}{4\pi r^2} + \frac{4(1-\bar{a})}{S\bar{a}} \right] dB \tag{4.170}$$

By performing this calculation twice, once with the enclosure in place and once without, it can be shown that

$$NR = TL - C \ dB \tag{4.171}$$

A similar calculation may be performed to estimate the sound pressure field within an enclosure sited in a reverberant field. The power flow into the enclosure is equal to

$$W_i = S_E \frac{\langle p_1^2 \rangle}{4\rho c} \tau \tag{4.172}$$

so that

$$L_{Wi} = L_{p1} + 10 \ \log_{10} S_E - TL - 6 \ dB \tag{4.173}$$

so

$$L_{pi} = L_{Wi} + 10 \ \log_{10} \left[\frac{1}{S_E} + \frac{4(1-\bar{a}_i)}{S_i \bar{a}_i} \right] dB \tag{4.174}$$

It can be shown that

$$NR = TL - C \ dB \tag{4.171}$$

The problem is not so simple when the direct field is dominant at one or more walls of the enclosure. In this case, the enclosure should be treated as a barrier.

4.10.7 Sound inside and outside close fitting enclosures

Jackson produced two useful papers on the performance of close fitting enclosures (Jackson, 1962; 1966). He developed a 1-D model of such an enclosure by treating the noise source and the enclosure as a pair of concentric pulsating boxes. The potentially complex three-dimensional problem was reduced to that of a pair of flat, infinite parallel plates separated by a distance l and immersed in air as shown in Figure 4.21.

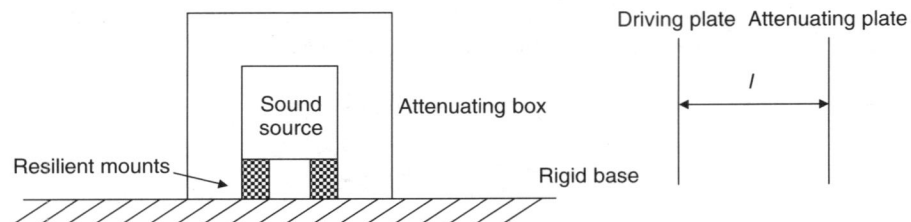

Figure 4.21 Jackson's models (Jackson, 1962).

Jackson made the following assumptions:

1. That generally large radiating areas of machinery are involved in cases using close fitting enclosures which encourage the propagation of acoustic waves normal to their surface.
2. The enclosure does not touch any part of the body it encloses.
3. The presence of the enclosure does not affect the magnitude of vibration of the enclosed surfaces.
4. Direct transmission of vibration through the support of the sound source does not occur.

Jackson developed an equation describing the attenuation produced by such an enclosure.

$$\left|\frac{Y_1}{Y_0}\right| = A = \left[1 - \frac{2\sin\theta(X\cos\theta - R\sin\theta)}{\rho c} + \frac{\sin^2\theta(X^2 + R^2)}{\rho^2 c^2}\right]^{1/2} \tag{4.175}$$

where

$$X = \omega M - \frac{S}{\omega} \tag{4.176}$$

$M =$ mass per unit area (kg m^{-2})
$R =$ mechanical resistance (damping)
$\theta = \dfrac{\omega L}{c}$
$L =$ distance between plates (m)
$\rho =$ density of air (kg m^{-3})
$\omega =$ angular frequency (rad s^{-1})
$S =$ uniform elastic restraint per unit area (stiffness)

Theoretical results are shown in Figure 4.22.

Jackson concluded that, providing the vibration source is sealed from free space, good low-frequency performance is possible if the wall stiffness is made high. Also, in a sealed system in which little mechanical damping is associated with the mass, if no stiffness is present then a magnification of sound will occur at low frequencies due to the presence of the hood, an effect not predicted by the mass law.

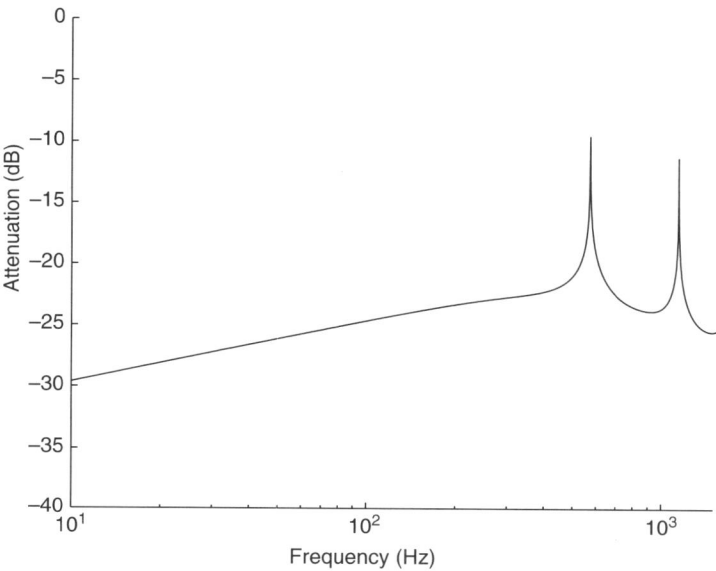

Figure 4.22 Results for a steel box with a surface density of 16 kg m^{-2} and $l = 0.3$ m and $s = 1 \times 10^8$ Nm^{-1} using Jackson's method (Jackson, 1966).

References

BS 6086: 1981 (ISO 5128 – 1980), Method of measurement of noise inside motor vehicles, British Standards Institution, 1981

J1074, Engine sound level measurement procedure, Society of Automotive Engineers, 1987

ISO 140/1, Acoustics: measurement of sound insulation in buildings and of building elements. Part 1: Requirements for laboratories, International Standards Organisation, 1990

ISO 140/3, Acoustics: measurement of sound insulation in buildings and of building elements. Part 3: Laboratory measurements of airborne sound insulation of building elements (amendment 1 to ISO 140/3 – 1978), International Standards Organisation, 1990

Allard, J.F., Sieben, B., Measurements of acoustic impedance in a free field with two microphones and a spectrum analyser, *Journal of Acoustical Society of America*, 77(4), 1617–1618, 1985

Anderson, C.G., A review of phenomena and associated experimental methods in automotive aero-acoustics, IMechE Paper No. C605/017/2002, 2002

Atkins, K.A., Challen, B.J., A practical approach to truck noise reduction, Institution of Mechanical Engineers, C131/79, 1979

Betella, M., Harrison, M.F., Sharp, R.S., Investigation of automotive creep groan noise a distributed-source excitation technique, *Journal of Sound and Vibration*, 255(3), 531–547, 2002

Bies, D.A., Uses of anechoic and reverberant rooms, *Noise Control Engineering Journal*, 7, 154–163, 1976

Bies, D.A., Hansen, C.H., Engineering noise control – Theory and practice – Second edition, E&FN Spon, 1996

Cheng, J.G., Akin, T., Air and structure borne noise reduction of automotive dampers, SAE Paper No. 951256, Proceedings of the 1995 noise and vibration conference – Vol. 1, Society of Automotive Engineers, 1995

Chung, J.Y., Blaser, D.A., Transfer Function Method of Measuring In-duct Acoustic Properties. II. Experiment, *Journal of Acoustical Society of America*, 68(3), 914–921, 1980

Coney, W.B., Her, J.Y., Tomaszewicz, K., Zhang, K.Y., Moore, J.A., Experimental evaluation of wind noise sources, SAE Paper No. 1999-01-1812, 1999

Cremer, L., Heckl, M., Structure borne sound – Structural vibrations and sound radiation at audio frequencies – Second edition, Springer Verlag, 1988

Crocker, M.J., Zockel, M., McGary, M., Reinhart, T., Noise source identification under steady and accelerating conditions on a turbocharged diesel engine, Vehicle Noise Regulation and Reduction – SP-456, Society of Automotive Engineers, Paper No. 800275, 1980

Croker, M.D., Greer, R.J., Hilbert, D., Granstrom, J., The development of transmission rattle indices, IMEchE Paper No. C420/024, Published in Quiet Revolutions – powertrain and vehicle noise refinement, international conference 9–11 October 1990, pp. 129–135, 1990

Dixon, J., Phillips, A.V., Power unit low frequency airborne noise, European Conference on Vehicle Noise and Vibration, Institution of Mechanical Engineers, C521/032/98, 1998

Fahy, F., Sound and structural vibration – radiation, transmission and response, Academic Press, 1985

Fahy, F.J., Sound intensity, Elsevier Applied Science, London, 1989

Fahy, F., Walker, J. (eds), Fundamentals of Noise and Vibration, E&FN Spon, 1998

Glikin, P. E., Fuel injection in diesel engines, Proceedings of the Institution of Mechanical Engineers, Vol. 199, D3, 161–174, 1985

Haines, J.C., Sound absorbing properties of molded fibreglass panels for use in vehicle noise control, SAE Paper No. 870987, Society of Automotive Engineers, 1987 Noise & Vibration Conference, 1987

Halvorsen, W.G., Bendat, J.S., Noise Source Identification using Coherent Output Power Spectra, *Journal of Sound and Vibration*, pp. 15–24, August 1975

Hardy, M., Whole vehicle refinement, Paper No. S477/001/97, IMechE Seminar – Automotive modeling and NVH – techniques and solutions, Institution of Mechanical Engineers, 1997

Harper, M.F.L., Dorling, C.M., Allwright, D.J., A non-intrusive method of signature analysis in systems with multiple noise paths, *International Journal of Modeling and Simulation*, 13 (4), 146–151, 1993

Ingard, K.U., Bolt, R.H., A free field method of measuring the absorption coefficient of acoustic materials, *Journal of Acoustical Society of America*, 23(5), 509–516, 1951

Ingard, K.U., Dear, T.A., Measurement of acoustic flow resistance, *Journal of Sound and Vibration*, 103(4), 567–572, 1985

Jackson, R.S., The performance of acoustic hoods at low frequencies, Acustica, 12, 139–152, 1962

Jackson, R.S., Some aspects of the performance of acoustic hoods, *Journal of Sound and Vibration*, 3(1), 82–94, 1966

Janssens, M.H.A., Verheij, J.W., Thompson, D.J., The Use of an Equivalent Forces Method for the Experimental Quantification of Structural Sound Transmission in Ships, *Journal of Sound and Vibration*, 226(2), 305–328, 1999

Johnson, C.M., Vehicle exhaust sound specification development, SAE Paper No. 951259, Society of Automotive Engineers, 1995 Noise & Vibration Conference, 1995

Jonasson, H., Elson, L., Determination of sound power levels of external sources, Report SP-RAPP, National testing institute, Acoustics laboratory, Borus, Sweden, 1981

Juneja, V., Rediers, B., Kavarana, F., Kimball, J., Squeak studies on material pairs, SAE Paper No. 1999-01-1727, 1999

Kim, J., Vehicle cavity wind noise experiments on a scale model in a windtunnel, MSc Thesis, Cranfield University, 2003

Kinsler, L.E., Frey, A.R., Coppens, A.B., Sanders, J.V., Fundamentals of acoustics – Third edition, John Wiley & Sons Inc., 1982

Kuttruff, H., Room Acoustics – Second edition, E&FN Spon, 1979

Lalor, N., Grover, E.C., Priede, T., Engine noise due to mechanical impacts at pistons and bearings, Society of Automotive Engineers, Paper No. 800402, 1980

Lilly, L.R.C., Diesel engine reference book, Butterworths, London, 1984 (Second edition now available – Challen, Baranescu 1999)

LMS International – Application Note Transfer path analysis – the qualification and quantification of vibro-acoustic transfer paths, LMS International, 1998

March, J.P., Croker, M.D., Present and future perspectives of powertrain refinement, European Conference on Vehicle Noise and Vibration, Institution of Mechanical Engineers, C521/023/98, 1998

Maunder, R.M.S., Practical applications of sound quality analysis techniques, Paper No. C498/13/067/95, Institution of Mechanical Engineers, 1996

Morse, P.M., Bolt, R.H., Sound waves in rooms, *Reviews of Modern Physics*, 16, 65–150, 1944

Morse, P.M., Ingard, K.U., Theoretical acoustics, McGraw-Hill Book Co, New York, 1968

MSX International, I-DEAS noise path analysis, MTS Systems Corporation, 1998

Naylor, S., Willats, R., The development of a sports tailpipe noise with predictions of its effect on interior vehicle sound quality, IMEchE Paper No. C577/002/2000, 2000

Nelson, P.M., (ed.) Transportation noise reference book, Butterworths, London, 1987

Otto, N., Amman, S., Eaton, C., Lake, S., Guidelines for jury evaluations of automotive sounds, SAE Paper No. 1999-01-1822, 1999

Piersol, A.G., Use of coherence and phase data between two receivers in evaluation of noise environments, *Journal of Sound and Vibration*, 56(2), 215–228, 1978

Pobol, O., Method of measuring noise characteristics of textile machines, Measurement techniques, USSR, 19, 1736–1739, 1976

Priede, T., The effect of operating parameters on sources of vehicle noise, *Journal of Sound and Vibration*, 43(2), 239–252, 1975

Qian, Y., Vanbuskirk, J.A., Sound absorption composites and their use in automotive interior sound control, Proceedings of the 1995 Noise and Vibration Conference, Traverse City Michigan, SAE P-291 (Vol. 1), 1995

Querengasser, J., Meyer, J., Wolschendorf, J., Nehl, J., NVH optimisation of an in-line 4-cylinder powertrain, Proceedings of the 1995 noise and vibration conference – Vol. 1, (p-291) Society of Automotive Engineers, Paper No. 951294, 1995

Russell, M.F., Reduction of noise emissions from diesel engine surfaces. Society of Automotive Engineers, Paper No. 720135 1972

Russell, M.F., Automotive diesel engine noise analysis, diagnosis and control, Lucas Engineering Review – 7(4), 1979

Russell, M.F., Worley, S.A., Young, C.D., An analyser to estimate subjective reaction to diesel engine noise, IMEchE Paper No. C30/88, 1988

Russell, M.F., Sekowski, M., Nikokiroulis, N., Subjective assessment of diesel vehicle noise Paper No. C389/044, Institution of Mechanical Engineers, 1992

Russell, M.F., Lee, H.K., Modeling diesel injection rate to control combustion noise, Institution of Mechanical Engineers, C-487/027, 1994

Russell, M.F., The dependence of diesel combustion on injection rate, Institution of Mechanical Engineers, S-490/005/97, 1997

Saha, P., Baker, R.N., Sound absorption study for automotive carpet materials, SAE Paper No. 870988, Society of Automotive Engineers, 1987 Noise & Vibration Conference, 1987

Sinha, N.K., Linear Systems, John Wiley & Sons 1991

Steel, J.A., Fraser, G., Sendall, P., A study of exhaust noise in a motor vehicle using statistical energy analysis, Proceedings of the Institution of Mechanical Engineers, Vol. 214, Part D, 75-83, 2000

Storer, D., Gatti, S., Pisino, E., Characterizing the transfer of road-surface excited vibrations through vehicle suspension systems, Paper No. C521/015/98, Institution of Mechanical Engineers, 1998

Thien, G.E., Brandl, F.K., Kirchweger, K., Winklhofer, E., Cars with closed engine compartment – effect upon exterior noise and passenger comfort, Vehicle noise and vibration – IMechE conference publications, Institution of Mechanical Engineers, C133/84, 1984

Van der Auweraer, H., Wyckaert, K., Hendricx, W., From sound quality to the engineering of solutions for NVH problems: case studies, Acustica: acta acustica, 83, 796–804, 1997

Vandenbroeck, D., Hendricx, W.S.F., Interior road noise optimization in a multiple input environment, Proceedings of the International Conference on vehicle NVH and refinement, Birmingham, 3–5 May 1994, Paper No. C487/002/94, Institution of Mechanical Engineers, 1994

Vér, I.L., Holmer, C.I., In Noise and Vibration Control (L.L. Beranek (ed.)), McGraw-Hill, New York, 1971

Verhulst, K., Verheij, J.W., Coherence Measurements in multi-delay systems, *Journal of Sound and Vibration*, 62(3), 460–463, 1979

Verstraeten, S., Dynamic bush properties and their effects on compliant subframe performance, MSc Thesis, Cranfield University, 2003

Weltner, K., Grosjean, J., Shuster, P., Weber, W.J., Mathematics for Engineers and Scientists, Stanley Thornes (Publishers) Limited, 1986

Wentzel, R.E., Saha, P., Empirically predicting the sound transmission loss of double-wall sound barrier assemblies, SAE Paper No. 951268, Society of Automotive Engineers, 1995 Noise & Vibration Conference, 1995

Wentzel, R.E., VanBuskirk, J., A dissipative approach to vehicle sound abatement, SAE Paper No. 1999-01-1668, 1999

Appendix 4A: Some background information on systems

In the simplest form a system may have only one input and one output. Such systems are often called single input, single output systems or two-pole systems (Sinha, 1991). In the more general case, systems may have several inputs and several outputs. These are called multivariable or multipole systems (as illustrated in Figure A4.1).

A zero state system is one where all the initial conditions are zero.

Linear and non-linear systems

A system is said to be linear if, and only if, it satisfies the superposition theorem. Note that superposition consists of two basic but quite distinct concepts:

1. The property of additivity – the response of a zero state system to the sum of two inputs is equal to the sum of the responses to each of the inputs acting alone.
2. The property of homogeneity – requires that the effect of multiplying the input by a constant would be to multiply the output by the same constant.

A system is said to be non-linear if it does not satisfy the properties of additivity and homogeneity.

Differential equations

The input–output relations for networks that store energy (dynamic systems) are given by differential equations.

Consider a linear, continuous-time system where :

- The output is represented by $y(t)$;
- The input is represented by $x(t)$.

The two variables will be related by a differential equation of the form:

$$a_n \frac{d^n y}{dt^n} + a_{n-1} \frac{d^{n-1} y}{dt^{n-1}} + \ldots + a_1 \frac{dy}{dt} + a_0 y = b_m \frac{d^m x}{dt^m} + \ldots + b_0 x \qquad (A4.1)$$

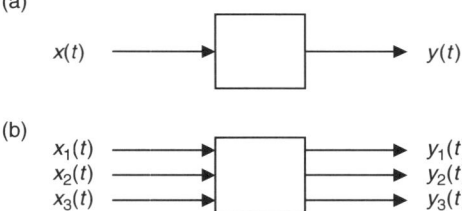

Figure A4.1 (a) Two-pole system; (b) multipole system.

This equation is called a 'linear differential equation of order n' and for most practical cases $n \geq m$.

It is convenient to replace $\dfrac{d}{dt}$ by the operator 'p' resulting in the equation

$$(a_n p^n + a_{n-1} p^{n-1} + \ldots + a_1 p + a_0) y(t) = (b_m p^m + \ldots + b_1 p + b_0) x(t) \qquad (A4.2)$$

which may be written compactly as

$$D(p)y(t) = N(p)x(t) \qquad (A4.3)$$

where $D(p)$ and $N(p)$ are polynomials in the operator 'p'

$$D(p) = a_n p^n + a_{n-1} p^{n-1} + \ldots + a_1 p + a_0 \qquad (A4.4)$$

$$N(p) = b_m p^m + b_{m-1} p^{m-1} + \ldots + b_1 p + b_0 \qquad (A4.5)$$

The operator 'p' does not satisfy the commutative property

$$py(t) \neq y(t)p \qquad (A4.6)$$

The system operator $L(p)$ or transfer function is the ratio of the two polynomials $D(p)$ and $N(p)$

$$L(p) = \frac{N(p)}{D(p)} \qquad (A4.7)$$

Any dynamic system described by a differential equation of order n can be solved uniquely only if at least n initial or boundary conditions are known.

As an example, consider that the unique solution to the differential equation characterising the input–output relationship of an electrical circuit can be obtained only if the initial values of the voltages across each capacitor and the current through each inductance are known.

The same rules apply to the well-known, second-order differential equation characterising the motion of a mass on a spring with a viscous damper

$$m\ddot{x} + c\dot{x} + kx = F(t) \qquad (A4.8)$$

If two initial conditions are known, namely $x(0)$ and $\dot{x}(0)$, the value of x at some instant t later may be found.

Appendix 4B: The convolution integral

Some aperiodic signals have unique properties and are known as singularity functions because they are either discontinuous or have discontinuous derivatives (Sinha, 1991). The simplest of these is the unit step function (as illustrated in Figure B4.1), given the symbol $\gamma(t)$

The unit impulse function or delta function $\delta(t)$ is defined as the function which after integration yields the unity step function, so (Sinha, 1991)

$$\gamma(t) = \int_{-\infty}^{t} \delta(\tau) \, d\tau \tag{B4.1}$$

Alternatively,

$$\delta(t) = \frac{\gamma(t)}{dt} \tag{B4.2}$$

The impulse function must satisfy:

$$\delta(t) = 0, \text{ for } t \text{ not equal to zero} \tag{B4.3}$$

and

$$\int_{-\infty}^{\infty} \delta(t) \, dt = 1 \tag{B4.4}$$

Therefore, the area under the impulse function is unity and it occurs over an infinitesimal interval around $t = 0$. So, as the period dt tends towards zero, the height of the impulse function approaches infinity.
Also,

$$\frac{d\delta(t)}{dt} = \infty \quad \text{at} \quad t = 0 \text{ and is zero elsewhere.}$$

Consider a unit impulse that occurs at time $t = \tau$ as illustrated in Figure B4.2
Remember that the definition of the unit impulse requires it to occur at time $t = 0$. Therefore, shift the time axis in the Figure B4.2 by the amount required for $t = \tau = 0$. Therefore, the unit impulse occurring at time $t = \tau$ is assigned with the symbol $\delta(t - \tau)$.

Define the impulse–response function $h(t - \tau)$ of a system as the response $y(t)$ of the system at time t to a unit impulse $\delta(t - \tau)$ of duration $\rightarrow 0$ input sometime earlier at

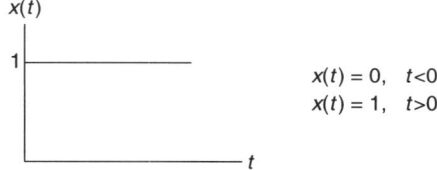

$$
\begin{aligned}
&x(t) = 0, \quad t < 0\\
&x(t) = 1, \quad t > 0
\end{aligned}
$$

Figure B4.1 The unit step function.

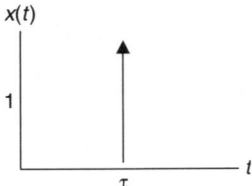

Figure B4.2 The unit impulse at time $t = \tau$.

time $t = \tau$ and remember that the definition of the impulse function dictates that input time to be time $t = 0$.

Now, remember that the area under the unit impulse function is unity so it follows that the response $y(t)$ to a non-unitary impulse (i.e. a practical pulse, one with a finite duration Δ somewhat larger than zero) at time $t = \tau$ is given approximately by the product of the area of the non-unitary pulse and the impulse–response function:

$$y(t) \approx x(\tau)\Delta \cdot h\,(t - \tau) \tag{B4.5}$$

In the limit as $\Delta \to 0$, and applying the superposition theorem for linear systems where the signal represented by a continuum of impulses is given by the sum of the individual responses to earlier impulses the convolution integral is obtained

$$y(t) = \int_{-\infty}^{\infty} x(\tau)\mathrm{h}\,(t - \tau)\,\mathrm{d}\tau \tag{B4.6}$$

An important application of the impulse function is the possibility of representing some arbitrary, continuous time signal of time $x(t)$ as a continuum of impulses as illustrated in Figure B4.3.

One approximation to the smooth function above can be obtained by representing it as a sequence of rectangular pulses where the height of each pulse is made equal to the value of $x(t)$ at the centre of each pulse. The width of the pulse is Δ.

It follows that the approximation improves as the pulse width Δ tends to zero, i.e. as the pulse tends towards the unit impulse and at this point one can write:

$$x(t) = \int_{-\infty}^{\infty} x(\tau)\delta\,(t - \tau)\,\mathrm{d}\tau \tag{B4.7}$$

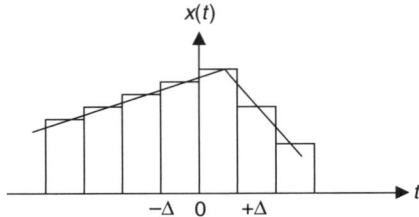

Figure B4.3 $x(t)$ represented as a continuum of impulses: after (Sinha, 1991).

This equation is the result of an interesting property of the unit impulse function known as the sifting property, whereby a time-varying signal is described as the sum of a train of impulses, each one with a strength that is equal to the value of the signal at the time of the impulse.

Another important application of the impulse–response function is that it is directly related to the transfer function of a linear, time-invariant, continuous time system. There are two possible formal definitions of the transfer function (Sinha, 1991).

Definition 1   The transfer function of a linear, time-invariant, continuous time system is the Laplace transform of its impulse response.

Definition 2   The transfer function of a linear, time-invariant, continuous time system is the ratio of the Laplace transforms of the output and input under zero initial conditions.

Appendix 4C: The covariance function, correlation and coherence

Consider some probability attributes of a random variable X (Fahy and Walker, 1998). The distribution function $F(x)$ of a random variable X is given by

$$F(x) = \int_{-\infty}^{x} p(u)du \tag{C4.1}$$

where p is the probability density function having the following attributes for a continuous distribution:

$$p(x) \geq 0$$

$$\int_{-\infty}^{\infty} p(x)dx = 1$$

$$P[a\langle x\rangle b] = \int_{a}^{b} p(x)dx$$

so

$$p(x) = \frac{dF(x)}{dx} \tag{C4.2}$$

$F(x)$ is the probability of X taking a value up to and including x.

The expected value of X is defined as

$$E[X] = \int_{-\infty}^{\infty} x \cdot p(x)dx \tag{C4.3}$$

which is also known as the mean value μ_x or the first moment of X.

If Y is a function of X, i.e. $Y = g(X)$

$$E[Y] = E[g(X)] = \int_{-\infty}^{\infty} g(x)p(x)dx \tag{C4.4}$$

Where W is a function of two variables, i.e. $W = g(X, \ Y)$

$$E[W] = \int_{-\infty}^{\infty} \int g(x, y)p(x, y)dxdy \tag{C4.5}$$

The second moment is given by

$$E[X^2] = \int_{-\infty}^{\infty} x^2 p(x)dx \tag{C4.6}$$

This is a measure of the spread relative to the origin.

The spread relative to the mean is called the variance and is given by:

$$V(x) = E[(x - \mu_x)^2] = \int_{-\infty}^{\infty} (x - \mu_x)^2 p(x)dx \tag{C4.7}$$

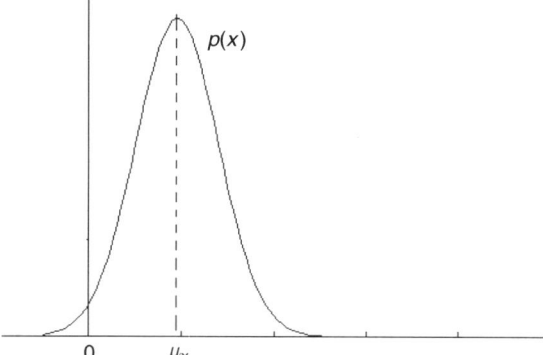

Figure C4.1 The Gaussian distribution.

The standard deviation is given by

$$\sigma_x = \sqrt{V(x)} \tag{C4.8}$$

A random variable has a Gaussian distribution as illustrated in Figure C4.1 if (Weltner et al. (1986) for example)

$$p(x) = \frac{1}{\sigma\sqrt{2\pi}} e^{-\frac{1}{2}\left(\frac{x-\mu}{\sigma}\right)^2} \tag{C4.9}$$

The second joint moment of two randomly distributed variables is

$$E[(x-\mu_x)(y-\mu_y)] = \int_{-\infty}^{\infty} \int (x-\mu_x)(y-\mu_y)p(x,y)\,dx\,dy \tag{C4.10}$$

This is called the covariance function relating x and y.
   Some useful definitions are (Fahy and Walker, 1998):

x and y are *uncorrelated* if $E(X, Y) = E(W) = E(X) \cdot E(Y)$
x and y are orthogonal if $E(W) = 0$ (i.e. X and Y do not coexist)
x and y are independent if $p(x, y) = p(x)p(y)$

   The degree of correlation between two statistical data sets might be established using the three categories above or using the correlation coefficient.

$$r = \frac{\sum xy - \frac{\sum x \sum y}{n}}{\sqrt{\left(\sum x^2 - \frac{(\sum x)^2}{n}\right)\left(\sum y^2 - \frac{(\sum y)^2}{n}\right)}} \tag{C4.11}$$

$$-1 < r > 1$$

(see http://max.econ.hku.hk/stat/hyperstat/A56626.html for example)

x, y are the measured values. All sums are formed from $i = 1$ to $i = n$, where n is the number of measurements.

However, beware, there are many potential pitfalls when using correlation coefficients. A high correlation does not imply causation. Reasons for this include:

- x and y may seem well correlated (a value near -1 or $+1$) but this may be due to the effect both of them being related to the same third variable.
- x and y may seem to be poorly correlated but there might be a causal relationship between them – it might be that the relationship is not linear or is being confounded by the effect of another variable, or that the data range of x is rather small.

(see for example http://www.math.virginia.edu/~der/usem170/Chapter05/sld040.htm)

An alternative to the use of the correlation coefficient is the use of the autocovariance function with a random process (Fahy and Walker, 1998).

$$R_{xx}(t_1, t_2) = E\left[(x(t_1) - \mu_x(t_1))(x(t_2)\mu_x(t_2))\right] \tag{C4.12}$$

This is a measure of the degree of association of the signal at time t_1 and the same signal at time t_2. Perhaps one could see it as a measure of how predictable future signal levels are based on a historic knowledge of that signal.

If the mean values are not subtracted the autocorrelation function is obtained

$$E\left[x(t_1)x(t_2)\right] \tag{C4.13}$$

With a stationary random process μ_x remains constant with time in the period $(t_1 - t_2)$

$$R_{xx}(t_2 - t_1) = E\left[(x(t_1) - \mu_x)(x(t_2) - \mu_x)\right] \tag{C4.14}$$

Commonly:

$$t_2 = t_1 + \tau \quad \text{where } \tau \text{ is the time lag}$$
$$t_1 = t$$

so

$$R_{xx}(\tau) = E\left[(x(t) - \mu_x)(x(t + \tau) - \mu_x)\right] \tag{C4.15}$$

when

$$\tau = 0, \qquad R_{xx} = V(x)$$

When $\tau \to \infty$, $R_{xx} \to 0$ as the two random samples tend to be less associated.

A typical autocorrelation function looks like that shown in Figure C4.2.

When two random variables are involved the cross covariance function is obtained.

$$R_{xy}(\tau) = E\left[(x(t) - \mu_x)(y(t + \tau) - \mu_y)\right] \tag{C4.16}$$

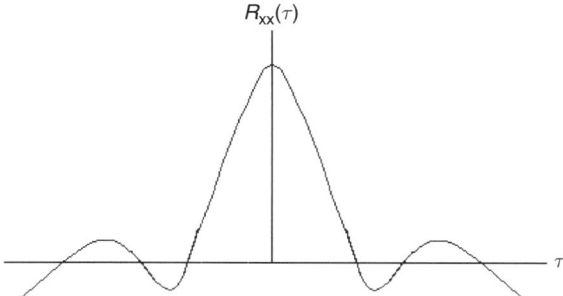

Figure C4.2 A typical autocorrelation function.

Note that as $\tau \to \infty$, the mean values $\to 0$ for random signals. So

$$R_{xx}(\tau) = E[x(t)x(t+\tau)] \tag{C4.17}$$

$$R_{xy}(\tau) = E[x(t)y(t+\tau)] \tag{C4.18}$$

When $\tau \to 0$, $R_{xx}(\tau)$ is the variance of x (i.e. x^2 when the mean is zero).
The average power in the signal over period $T \to \infty$ is (Sinha, 1991)

$$P_{\lim T \to \infty} = \frac{1}{T} \int_{-T/2}^{T/2} x^2(t)\mathrm{d}t \tag{C4.19}$$

Now (Fahy and Walker, 1998) Parseval's Theorem states that

$$\frac{1}{T} \int_{-T/2}^{T/2} x^2(t)\mathrm{d}t = \frac{1}{T} \int_{-\infty}^{\infty} x_T^2(t)\mathrm{d}t = \frac{1}{T} \int_{-\infty}^{\infty} |X_T(f)|^2 \, \mathrm{d}f \tag{C4.20}$$

where x_T is the truncated data set for $x(t)$ between times $-T/2$ and $T/2$.
Now, as $T \to \infty$

$$P_{\lim T \to \infty} = \frac{1}{T} \int_{-T/2}^{T/2} x^2(t)\mathrm{d}t = \int_{-\infty}^{\infty} \left| \frac{X_T(f)}{T} \right|^2 \mathrm{d}f \tag{C4.21}$$

So,

$$E[P_{\lim T \to \infty}] = E\left[\frac{1}{T} \int_{-T/2}^{T/2} x^2(t)\mathrm{d}t \right] = E\left[\int_{-\infty}^{\infty} \left| \frac{X_T(f)}{T} \right|^2 \mathrm{d}f \right] \tag{C4.22}$$

Now the power spectral density function S_{xx} is given by (Fahy and Walker, 1998)

$$S_{xx}(f)_{\lim T \to \infty} = \frac{E[X_T(f)]^2}{T} \tag{C4.23}$$

So, from

$$R_{xx}(\tau) = E[x(t)x(t+\tau)] \tag{C4.24}$$

$$R_{xy}(\tau) = E[x(t)y(t+\tau)] \tag{C4.25}$$

one may write

$$S_{xx}(f) = \int_{-\infty}^{\infty} R_{xx}(\tau) e^{-i(2\pi f\tau)} d\tau \qquad \text{(C4.26)}$$

$$S_{xy}(f) = \int_{-\infty}^{\infty} R_{xy}(\tau) e^{-i(2\pi f\tau)} d\tau \qquad \text{(C4.27)}$$

$S_{xy}(f)$ is the cross spectral density function.

The auto and cross correlation functions may be readily obtained from the power and cross spectral densities, which are quantities commonly measured.

The normalised cross spectral density is called the coherence between signals $x(t)$ and $y(t)$.

$$\gamma_{xy}^2(f) = \frac{\left|S_{xy}(f)\right|^2}{S_{xx}(f)S_{yy}(f)} \qquad \text{(C4.28)}$$

When one-sided power and cross spectral densities are used,

$$\gamma_{xy}^2(f) = \frac{\left|G_{xy}(f)\right|^2}{G_{xx}(f)G_{yy}(f)} \qquad \text{(C4.29)}$$

If $x(t)$ and $y(t)$ are linearly related and the only signals in a two-port system, then the coherence will be unity. The coherence will be zero if those two signals are completely unrelated by linear relationships.

Appendix 4D: The frequency response function

Take a linear, time-invariant system with input $x(t)$ and output $y(t)$.

The response due to an input starting at time t_0 is given by (Fahy and Walker, 1998) as

$$y(t) = \int_{t_0}^{t} h(t - t_1) x(t_1) dt_1 \tag{D4.1}$$

where $h(t)$ is the impulse response function of the system. This equation is the causal version of the convolution integral.

If $y(t)$ is stationary then

$$y(t) = \int_{-\infty}^{t} h(t - t_1) x(t_1) dt_1 \tag{D4.2}$$

or

$$y(t) = \int_{0}^{\infty} h(\tau) x(t - \tau) d\tau \tag{D4.3}$$

Now, for random signals as $\tau \to \infty$, the autocorrelation function R_{yy} is

$$R_{yy}(\tau) = E[y(t)y(t + \tau)] \tag{D4.4}$$

one can substitute for $y(t)$ and for $y(t + \tau)$ to give

$$R_{yy}(\tau) = E\left[\left(\int_{0}^{\infty} h(\tau_1) x(t - \tau_1) d\tau_1\right)\left(\int_{0}^{\infty} h(\tau_2) x(t + \tau - \tau_2) d\tau_2\right)\right] \tag{D4.5}$$

Re-arranging:

$$R_{yy}(\tau) = E\left[\int_{0}^{\infty}\int_{0}^{\infty} h(\tau_1) h(\tau_2) x(t - \tau_1) x(t + \tau - \tau_2) d\tau_1 d\tau_2\right] \tag{D4.6}$$

which is after substitution of

$$R_{xx}(\tau) = E[x(t)x(t + \tau)] \tag{D4.7}$$

$$R_{yy}(\tau) = \int_{0}^{\infty}\int_{0}^{\infty} h(\tau_1) h(\tau_2) R_{xx}(\tau + \tau_1 - \tau_2) d\tau_1 d\tau_2 \tag{D4.8}$$

The Fourier transform of this equation is

$$S_{yy}(f) = |H(f)|^2 S_{xx}(f) \tag{D4.9}$$

where $H(f)$ is the system frequency response function, related to the system impulse–response function thus:

$$H(f) = \int_0^\infty h(\tau)e^{-i(2\pi f\tau)}d\tau \qquad (D4.10)$$

$S_{yy}(f)$ is the power spectral density function.

The above can be repeated, this time using the cross correlation function R_{xy} for random signals as $\tau \to \infty$,

$$R_{yy}(\tau) = E\left[x(t)y(t+\tau)\right] \qquad (D4.11)$$

$$R_{xy}(\tau) = E\left[x(t)\int_0^\infty h(\tau_1)x(t+\tau-\tau_1)d\tau_1\right] \qquad (D4.12)$$

$$R_{xy}(\tau) = \int_0^\infty h(\tau_1)R_{xx}(\tau-\tau_1)d\tau_1 \qquad (D4.13)$$

$$S_{xy}(f) = H(f)S_{xx} \qquad (D4.14)$$

$S_{xy}(f)$ is the cross spectral density function.

Appendix 4E: Plane waves in a tube with a termination impedance

If the frequency of sound in a tube of length L is sufficiently low so that only plane waves propagate, then the solution to the 1-D linear acoustic wave equation (Section 2.1.4) has the form (Kinsler et al., 1982):

$$p(x) = Ae^{i[wt+k(L-x)]} + Be^{i[wt-k(L-x)]} \tag{E4.1}$$

At $x = L$ the impedance of the acoustic wave equals the mechanical impedance of the termination Z_{mL}.

Now, taking the linearised inviscid Euler equation (derived in Appendix 4G)

$$-\frac{\partial p}{\partial x} = \rho_0 \frac{\partial u}{\partial t} \tag{E4.2}$$

From this:

$$u(x, t) = -\frac{1}{\rho_0} \int \frac{\partial p}{\partial x} dt \tag{E4.3}$$

So, a relationship is found between pressure gradient and particle velocity.

For harmonic waves the integral with respect to time is given by Morse and Ingard (1968), $\frac{1}{i\omega}$.

So,

$$u(x, t) = \frac{1}{i\omega\rho_0} \frac{\partial p}{\partial x} \tag{E4.4}$$

$$\frac{\partial p}{\partial x} = ik(L-x)Ae^{i[\omega t+k(L-x)]} - ik(L-x)Be^{i[\omega t-k(L-x)]} \tag{E4.5}$$

As $k = \omega c$

$$u(x, t) = \frac{1}{\rho_0 c_0} Ae^{i(\omega t+k(L-x))} - Be^{i(\omega t-k(L-x))} \tag{E4.6}$$

Now

$$Z_{ML} = \frac{\text{Force}}{u} \tag{E4.7}$$

$$\text{Force} = P(L, t) \cdot S$$

$$Z_{ML} = \rho_0 c_0 S \frac{A+B}{A-B} \tag{E4.8}$$

The input mechanical impedance is given by

$$Z_{MO} = \rho_0 c_0 S \frac{A e^{ikL} + B e^{-ikL}}{A e^{ikL} - B e^{-ikL}} \tag{E4.9}$$

The two equations may be combined to eliminate both A and B
From equation (E4.8)

$$x = \frac{A+B}{A-B} \qquad \frac{Z_{ML}}{\rho_0 c_0 S} = x$$

$$xA - xB = A + B$$

$$xA - A = xB + B$$

$$B = \frac{(x-1)}{x+1} A \tag{E4.10}$$

Substituting equation (E4.10) into equation (E4.9)

$$\frac{Z_{MO}}{\rho_0 c_0 S} = Y = \frac{A e^{ikL} + \left(\dfrac{x-1}{x+1}\right) A e^{-ikL}}{A e^{ikL} - \left(\dfrac{x-1}{x+1}\right) A e^{-ikL}}$$

Dividing both numerator and denominator by A

$$Y = \frac{e^{ikL} + \dfrac{x-1}{x+1} e^{-ikL}}{e^{ikL} - \left(\dfrac{x-1}{x+1}\right) e^{ikL}} \tag{E4.11}$$

Multiplying both numerator and denominator of equation (E4.11) by $(x+1)$

$$Y = \frac{e^{ikL} x + e^{ikL} + e^{-ikL} x - e^{-ikL}}{e^{ikL} x + e^{ikL} - e^{-ikL} + e^{-ikL}} \tag{E4.12}$$

Now, there are these two standard relationships (see Weltner et al. [1986] for example)

$$\frac{e^{ix} - e^{-ix}}{2i} = \sin x \tag{E4.13}$$

$$\frac{e^{ix} + e^{-ix}}{2} = \cos x \tag{E4.13a}$$

Substituting equations (E4.13) and (E4.13a) into equation (E4.12)

$$Y = \frac{x \left(e^{ikL} + e^{-ikL}\right) + \left(e^{ikL} - e^{-ikL}\right)}{x \left(e^{ikL} - e^{-ikL}\right) + \left(e^{ikL} + e^{-ikL}\right)}$$

$$Y = \frac{2x \cos kL + i2 \sin kL}{2 \cos kL + i2x \sin kL} \tag{E4.14}$$

Divide both the numerator and the denominator of equation (E4.14) by $2\cos kL$ to get

$$Y = \frac{x + i\tan kL}{1 + ix\tan kL}$$

$$\frac{Z_{MO}}{\rho_0 c_0 S} = \frac{\dfrac{Z_{ML}}{\rho_0 c_0 S} + i\tan kL}{1 + i\dfrac{Z_{ML}}{\rho_0 c_0 S}\tan kL} \qquad \text{[Kinsler et al.,1982]} \qquad (E4.15)$$

Now $\dfrac{Z_{ML}}{\rho_0 c_0 S}$ is a complex term

$$\frac{Z_{ML}}{\rho_0 c_0 S} = r + ix \qquad (E4.16)$$

$r =$ acoustic (flow) resistance
$x =$ acoustic (flow) reactance

Substituting equation (E4.16) into equation (E4.15) yields

$$\frac{Z_{MO}}{\rho_0 c_0 S} = \frac{r + ix + i\tan kL}{1 + i\left[(r + ix)\tan kL\right]} \qquad (E4.17)$$

Now separate the real and the imaginary parts. Take the denominator of equation (E4.17) first

$$D = 1 + ir\tan kL - x\tan kL$$
$$D = (1 - x\tan kL) + ir\tan kL \qquad (E4.18)$$

Multiplying equation (E4.18) by its complex conjugate yields

$$D \times D^* = \left[(1 - x\tan kL) + ir\tan kL\right] \times \left[(1 - x\tan kL - ir\tan kL)\right]$$
$$D \times D^* = (1 - x\tan kL)^2 + r^2\tan^2 kL$$
$$D \times D^* = 1 - 2x\tan kL + x^2\tan^2 kL + r^2\tan kL \qquad (E4.19)$$
$$D \times D^* = \left(x^2 + r^2\right)\tan^2 kL - 2x\tan kL + 1$$

Multiplying the numerator of equation (E4.17) by the complex conjugate of equation (E4.18) gives

$$N \times D^* = \left[r + i(x + \tan kL)\right] \times \left[(1 - x\tan kL) - ir\tan kL\right]$$
$$N \times D^* = (r - rx\tan kL) - \left(ir^2\tan kL\right) + i\left(x - x^2\tan kL + \tan kL - x\tan^2 kL\right)$$
$$\qquad\qquad + \left(xr\tan kL + r\tan^2 kL\right)$$
$$N \times D^* = r + r\tan^2 kL + i\left[x - x^2\tan kL - x\tan^2 kL + \tan kL - r^2\tan kL\right]$$
$$N \times D^* = r\left(\tan^2 kL + 1\right) + i\left[x - \left(x^2 + r^2\right)\tan kL - x\tan^2 kL + \tan kL\right]$$
$$N \times D^* = r\left(\tan^2 kL + 1\right) + i\left[x - \left(x^2 + r^2 - 1\right)\tan kL - x\tan^2 kL\right] \qquad (E4.20)$$

So equation (E4.17) becomes (using equations (E4.20) and (E4.19) for numerator and denominator respectively)

$$\frac{Z_{MO}}{\rho_0 c_0 S} = \frac{r\left(\tan^2 kL + 1\right)}{\left(x^2 + r^2\right)\tan^2 kL - 2x\tan kL + 1} + \frac{i\left[x - \left(x^2 + r^2 - 1\right)\tan kL - x\tan^2 kL\right]}{\left(x^2 + r^2\right)\tan^2 kL - 2x\tan kL + 1}$$

(E4.21)

Consider what happens to equation (E4.15) when there is a truly rigid termination to the tube, i.e. $|Z_{ML}| \to \infty$

$$\frac{Z_{MO}}{\rho_0 c_0 S} = \frac{Z_R + i\tan kL}{1 + iZ_R \tan kL} \qquad \left|\frac{Z_{ML}}{\rho_0 c_0 S}\right| = Z_R$$

$$\frac{Z_{MO}}{\rho_0 c_0 S} = \frac{(Z_R + i\tan kL)(1 - iZ_R \tan kL)}{1 + Z_R^2 \tan^2 kL}$$

$$\frac{Z_{MO}}{\rho_0 c_0 S} = Z_R - iZ_R^2 \tan kL + i\tan kL + Z_R \tan^2 kL$$

$$\frac{Z_{MO}}{\rho_0 c_0 S} = \frac{Z_R\left(1 - \tan^2 kL\right) + i\left(1 - Z_R^2\right)\tan kL}{1 + Z_R^2 \tan^2 kL}$$

$$\frac{Z_{MO}}{\rho_0 c_0 S} = \frac{Z_R\left(1 - \tan^2 kL\right)}{1 + Z_R^2 \tan^2 kL} + \frac{i\left(1 - Z_R^2\right)\tan kL}{1 + Z_R^2 \tan^2 kL}$$

(E4.22)

When $Z_R \to \infty$ the reactance $\to 0$ when

$$-i\frac{1}{\tan kL} \to 0 \qquad \text{i.e.} \qquad \cot kL = 0$$

i.e.

$$\text{for } k_n L = (2n - 1) - \frac{\pi}{2} \qquad n = 1, 2, 3, 4, \ldots$$

or

$$\frac{2\pi f}{c}L = \frac{(2n - 1)\,\pi}{2} \qquad f = \frac{2n - 1}{4}\frac{c}{L}$$

(E4.23)

Under these conditions, flow reactance is zero and the input mechanical impedance is a function of flow resistivity only.

Appendix 4F: The derivation of the linearised mass conservation equation (After Fahy and Walker, 1998)

The net rate of mass inflow into a 1-D control volume of cross-sectional area S (m^2) and length dx (m) is:

$$\rho_{TOT}uS - \left[\rho_{TOT}u + \frac{\partial(\rho_{TOT}u)}{\partial x}dx\right]S = -\frac{\partial(\rho_{TOT}u)}{\partial x}dxS \qquad \text{(F4.1)}$$

This net inflow of mass must be balanced by the increase in mass in the control volume (the principle of conservation of mass) which is given by

Rate of increase in mass within control volume $= \dfrac{\partial\rho_{TOT}}{\partial t}dxS$

So

$$\frac{\partial\rho_{TOT}}{\partial t}dx\ S = -\frac{\partial(\rho_{TOT}u)}{\partial x}dx\ S \qquad \text{(F4.2)}$$

$$\frac{\partial\rho_{TOT}}{\partial t} + \frac{\partial(\rho_{TOT}u)}{\partial x} = 0 \qquad \text{(F4.3)}$$

This equation of mass conservation may be linearised thus

$$\rho_{TOT}u = (\rho_0 + \rho)u$$
$$\rho_{TOT}u = \rho_0 u + \rho u \qquad \text{(F4.4)}$$

The ρu term is the product of two small quantities and one might choose to neglect it and so

$$\rho_{TOT}u \approx \rho_0 u \qquad \text{(F4.5)}$$

As ρ_0 is not a function of x:

$$\frac{\partial(\rho_{TOT}u)}{\partial x} \approx \rho_0\frac{\partial u}{\partial x} \qquad \text{(F4.6)}$$

Now ρ_0 is not a function of t either, so

$$\frac{\partial\rho_{TOT}}{\partial t} = \frac{\partial\rho}{\partial t} \qquad \text{(F4.7)}$$

and the linearised mass conservation equation becomes

$$\frac{\partial\rho}{\partial t} + \rho_0\frac{\partial u}{\partial x} = 0 \qquad \text{(F4.8)}$$

Re-arranging

$$\frac{\partial u}{\partial x} = -\frac{1}{\rho_0}\frac{\partial\rho}{\partial t} \qquad \text{(F4.9)}$$

Now $-\dfrac{1}{\rho_0} = -\dfrac{1}{\kappa_0}\left(\dfrac{\partial p}{\partial \rho}\right)$ from $\kappa_0 = \rho_0 \left(\dfrac{\partial p}{\partial \rho}\right)_{\rho_0}$

$$\frac{\partial u}{\partial x} = -\frac{1}{\kappa_0}\left(\frac{\partial p}{\partial \rho}\right)\left(\frac{\partial \rho}{\partial t}\right) \tag{F4.10}$$

So finally the linearised mass conservation equation is:

$$\frac{\partial u}{\partial x} = -\frac{1}{\kappa_0}\frac{\partial p}{\partial t} \tag{F4.11}$$

Appendix 4G: The derivation of the non-linear (and linearised) inviscid Euler equation

Take two plane surfaces – one at x and the other at $x+dx$, each one with unit area. There is an acoustic wave causing (Section 2.1.1):

$$p = p_0 + p' \tag{G4.1}$$

$$\rho = \rho_0 + \rho' \tag{G4.2}$$

$$T = T_0 + T' \tag{G4.3}$$

The mass of air between the two plates is given by $(\rho_0 + \rho')dx$ (Morse and Ingard, 1968). The net force acting on this mass is $p(x) - p(x+dx)$. Using Newton's second law, this must be equal to the mass times the acceleration of the fluid.

$$p(x) - p(x+dx) = -\frac{\partial p}{\partial x}dx = \frac{du}{dt}(\rho_0 + \rho')dx \tag{G4.4}$$

Now the total differential $\dfrac{du}{dx}$ may be expressed in its partial differential form where $u = f(x, t)$ and the total change in u is given by the sum of the partial changes:

$$du = \frac{\partial u}{\partial x}dx + \frac{\partial u}{\partial t}dt \quad \text{(Weltner et al. (1986) for example)}$$

So, dividing both sides by dt in the limit $dt \to 0$

$$\frac{du}{dt} = \frac{\partial u}{\partial x} \cdot \frac{dx}{dt} + \frac{\partial u}{\partial t} \cdot \frac{dt}{dt}$$

$$\frac{du}{dt} = u\frac{\partial u}{\partial x} + \frac{\partial u}{\partial t}$$

So, from equation (G4.4)

$$-\frac{\partial p}{\partial x}dx = \left[u\frac{\partial u}{\partial x} + \frac{\partial u}{\partial t}\right](\rho_0 + \rho')\,dx$$

$$-\frac{\partial p}{\partial x} = (\rho_0 + \rho')\left[\frac{\partial u}{\partial t} + u\frac{\partial u}{\partial x}\right] \tag{G4.5}$$

This is the non-linear inviscid Euler equation (Kinsler et al., 1982).

Now if $u\dfrac{\partial u}{\partial x} \ll \dfrac{\partial u}{\partial t}$ and the condensation $\left(s = \dfrac{\rho'}{\rho_0}\right) \ll 1$ (Section 2.1.2), the non-linear inviscid Euler equation reduces to its linearised form, being

$$-\frac{\partial p}{\partial x} = \rho_0\frac{\partial u}{\partial t} \tag{G4.6}$$

From this

$$u(x, t) = -\frac{1}{\rho_0}\int \frac{\partial p}{\partial x}dt \tag{G4.7}$$

So, a relationship between pressure gradient and particle velocity is found.

The measurement and behaviour of vibration

5.1 Making basic vibration measurements

5.1.1 Types of vibration transducer

The two most commonly used vibration transducers are the accelerometer and the velocity transducer (sometimes referred to as a geophone).

5.1.2 The accelerometer

The accelerometer provides a direct measure of vibration acceleration (SI unit $-\mathrm{m\,s^{-2}}$, although g, where $1\,\mathrm{g} = 9.81\,\mathrm{m\,s^{-2}}$, is sometimes used). The main points regarding accelerometers are:

- They are a piezo-electric device using a piezo-electric crystal loaded with a small mass.
- They are designed to have a natural frequency well above the anticipated excitation frequency range.
- They are usually small in size.
- The largest accelerometers have lower natural frequencies, are more sensitive but cannot be used for measuring high-frequency vibration or on lightweight panels due to the mass loading imposed by them. A range of accelerometers are shown in Figure 5.1.
- The smallest accelerometers have high natural frequencies, will not mass load lightweight panels, but have a lower sensitivity.
- Multi-axis accelerometers are available.
- Accelerometers are generally expensive and fragile.
- Accelerometers may be attached using plasticine, wax, magnetic holders for low frequencies (usually below 2 kHz). Hard epoxy or cyanoacrylate adhesive or screw holders are needed for high-frequency measurements. The mass of the mounting adds to the mass loading effect.
- Charge amplification of the signals from high impedance accelerometers is preferred. Charge amplifiers allow measurement of frequencies down to 0.2 Hz and they are relatively insensitive to connecting cable length. They allow integration of acceleration to velocity or displacement. They tend to consume batteries quickly and are generally expensive to buy.

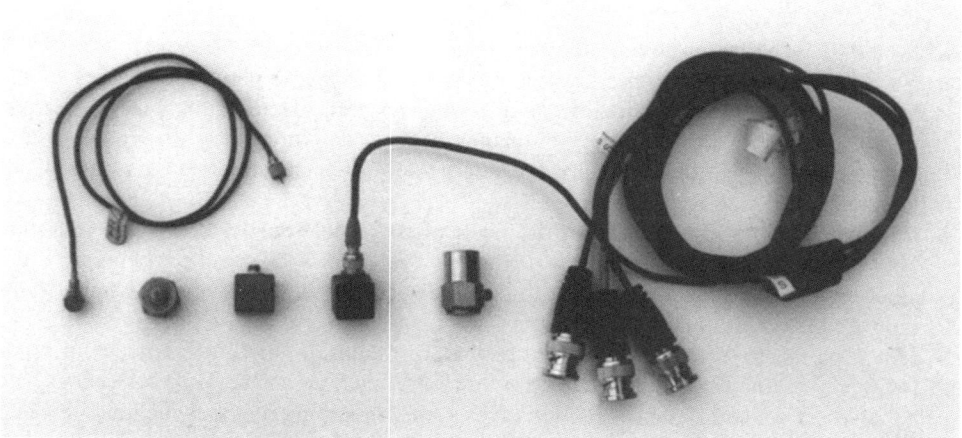

Figure 5.1 A range of accelerometers.

- Voltage amplification of the signals from low impedance accelerometers is cheaper but sensitive to the length of cable connecting the amplifier to the accelerometer (the cable impedance is a factor in the calibration – lose the calibrated cable and you are in trouble) and also sensitive to accelerometer cable movement. The lowest practical frequency (limited by how high the input impedance to the amplifier is) is usually 20 Hz. Amplifiers may be built into the accelerometer itself for ease of use. Most amplifiers work on the principle that the output signal from the accelerometer modulates the voltage on the power line of a constant-current power supply.
- Performance characteristics for appropriate instrumentation for measuring acceleration can be obtained in BS 6841 and BS 7482-Parts 1 and 3.

5.1.2.1 Selecting appropriate accelerometers

The first choice is whether to use a high-impedance accelerometer or a low-impedance one. The advantages of the high-impedance accelerometer are:

- wide dynamic range (several decades – generally difficult to 'overload');
- ability to use long connector cables between the accelerometer and the charge amplifier.

The main disadvantage of high-impedance accelerometers is the high cost.
   The advantages of the low-impedance accelerometer are:

- cheaper;
- simpler signal conditioning (often built into the transducer itself).

   The main disadvantages of low-impedance accelerometers are their size and mass (once the electronics is built in), their intolerance of being overloaded (relative to high-impedance accelerometers) and a rather high limit on the lowest frequency at which they may be used.

Subsequent choice of accelerometer will depend on:

- *Sensitivity*: The larger devices having greater sensitivity.
- *Frequency range*: The upper limit being determined by the natural frequency of the piezoelectric element. Generally, smaller devices may be used to measure higher-frequency vibrations. The lower-frequency limit is determined by the time constant of the signal conditioning. Only the charge-amplified accelerometer is capable of measuring down to (almost) 0 Hz.
- *Physical size*: As a general rule, the mass of the accelerometer should be less than 10% of the mass of the vibrating structure to which it will be attached. Otherwise, the additional mass of the attached accelerometer will alter the vibration characteristics of the structure. This is often termed the mass loading effect.
- *Mounting accessories*: The preferred method of mounting the accelerometer for high-frequency measurements is with a screw-threaded stud. However, various cyanocrylate adhesives are also available as alternatives. Other mounting methods include magnetic mounts, bees-wax, plasticine and double-sided adhesive tape. All of these, although convenient, will restrict the upper frequency limit of the measurement. Some accelerometers are made with their cable terminations on top of the casing to facilitate stud mounting.
- *Cables*: Low-impedance accelerometers with built-in amplification can be used directly with even long lengths of co-axial cable before the frequency response is affected. High-impedance devices require special low noise/high insulation resistance cables known as microdot cables which are of limited length and also fragile.

5.1.2.2 *Charge amplifiers for accelerometers*

A good-quality charge amplifier (such as the 'classic' B&K Type 2635 or the Type 2626 shown in Figure 5.2) will have most or all of the following features:

- A charge input, with a microdot cable termination on the front or rear panel.
- A facility for the user to enter the transducer sensitivity (commonly as the pC per m s^{-2}) to three significant figures.

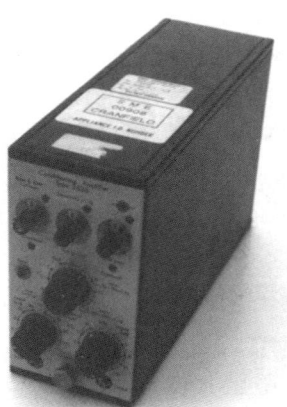

Figure 5.2 A typical charge amplifier — B&K Type 2626.

- A variable amplifier gain to allow input signals of amplitude between 0.1 mV per $m s^{-2}$ and 1 V per $m s^{-2}$.
- An indicator that warns the user when the amplifier is being overloaded.
- A low-frequency limit to suppress low-frequency noise (particularly if signal integration is to be used).
- Integration circuits to allow the direct measurement of velocity or displacement.
- A low-pass filter to remove unwanted signal components prior to amplification.
- Battery and external supply via a main transformer.

Notes: The frequency response of the amplifier will be significantly reduced at higher frequencies if the signal from the accelerometer is integrated to be proportional to velocity (velocity response is 20 dB lower than the acceleration response at 159 Hz). If the original signal is double integrated so as to be proportional to displacement, the frequency response is reduced further (displacement response is 40 dB lower than the acceleration response at 159 Hz).

5.1.2.3 Calibration of accelerometers

Precision accelerometers are supplied along with a calibration chart and details of their sensitivity. When used with a good-quality charge amplifier, this data can be used, along with the amplifier gain to calculate a calibration factor (V per $m s^{-2}$).

In the field, a calibration exciter can be used to calibrate the entire signal chain (from the accelerometer to the analyser display). Typical examples (such as the B&K Type 4294 shown in Figure 5.3) produce a reference vibration level of $10\,m s^{-2}$ RMS ($\pm 3\%$) at 159 Hz for a maximum accelerometer load of 70 g.

With high-impedance accelerometers and good-quality charge amplification, the calibration factor is independent of the (reasonable) length of microdot lead. Therefore, subsequent to initial calibration, a replacement microdot lead (perhaps of longer length) can be substituted without the need to recalibrate.

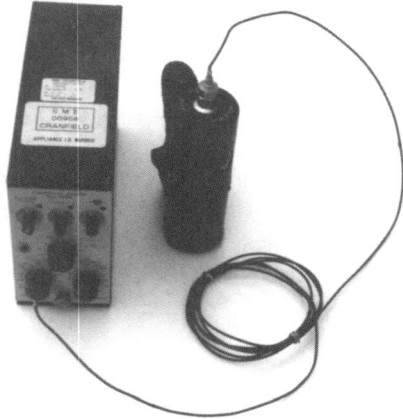

Figure 5.3 A field calibration exciter in action.

5.1.3 Velocity transducer

The velocity transducer provides a direct measure of vibration velocity (SI unit $-$ m s^{-1} although mm s^{-1} is commonly used). The main points regarding velocity transducers are:

- They are often moving coil devices – an electrical coil is suspended around a permanent magnet that is kept static.
- They are designed to have a low natural frequency (typically a few Hz), chosen to be much lower than the anticipated excitation frequency range (being typically 10–1000 Hz).
- They are usually large and fairly heavy, but robust.
- They cannot be used on lightweight structures due to the mass loading effect.
- Multi-axis transducers are available.
- They usually stand on their own base – often an adjustable tripod to cater for use on sloping or uneven surfaces.
- Simple, and low-cost voltage amplification is commonly used. Battery usage is relatively efficient and the transducer and its signal conditioning is often housed in a single rugged package.

5.2 Laser-based vibration measurements

Various laser methods for measuring vibration have been developed since the 1970s. Originally bulky and fragile and confined to the University laboratory, affordable and easy-to-use commercial systems have been adopted for use by vehicle engineering companies in more recent times. The most important laser vibration measuring methods are:

- single-point laser doppler velocimetry (LDV);
- scanning laser doppler velocimetry (SLDV);
- dual beam laser velocimetry for rotating systems;
- double-pulsed laser holographic interferometry;
- electronic speckle pattern interferometry (ESPI).

The commercial single-point LDV is a small box containing a Class II laser, optics and sensing electronics. It is a safe 'point and shoot' sensor that will measure vibration velocity directly. The frequency range is very wide (0.1 Hz–300 kHz is typical) and has good sensitivity (0.1–1000 mm s^{-1} is typical). Measurements are made without contact with the 'target' and can be made remotely over distances of several metres.

A device commonly known as a Michelson Interferometer (although there are derivatives of this with different names) is used at the heart of the LDV instrument. This splits the laser beam into two paths: an outgoing signal beam and a reference beam. The signal beam is directed at the target and the reflected light from this is directed so that it recombines with the reference beam. When the surface of the target vibrates, the path length difference between the reference and the reflected beams changes and the intensity of the recombined beam changes. In fact, the intensity modulates with target vibration. One full cycle of modulation corresponds to a target displacement amplitude equal to one half wavelength of the laser light (0.316μ m for a typical helium–neon laser source).

The frequency Fd at which the modulation occurs corresponds to the surface vibration velocity v of the target as follows (Bruel & Kjaer/Ormetron application note, 2003):

$$v = Fd\frac{\lambda}{2} \tag{5.1}$$

The recombined beam is routinely split between two independent detection channels, (one positioned half a wavelength further away from the target than the other) to allow the instantaneous direction of vibration to be determined. The outputs from these two detectors are mixed and converted to yield a single analogue voltage that is proportional to surface vibration velocity.

A single-point LDV sensor is the measurement device of choice for several common-place situations:

- when measuring surface vibration levels on lightweight structures where the mass of a contact sensor (such as an accelerometer) would influence the vibration;
- when there is a need for direct measurement of vibration velocity, and a conventional (bulky) vibration velocity transducer is inappropriate;
- when measuring vibration levels on the surfaces of foams, fabrics and other yielding surfaces;
- when measuring vibration levels in inaccessible locations;
- when measuring vibration levels on rotating components (thus avoiding the need for slip rings to carry the signals from contact sensors).

The commercial scanning LDV device includes all of the components found in the single-point LDV sensor with two added mirrors. These are digitally controlled to rotate allowing a signal beam to be directed to different locations on the target surface with a spatial resolution of fractions of 1 mm. In this way a two-dimensional surface can be scanned as a series of discrete point measurements.

Modern SLDV systems are automated. For larger two-dimensional fields, the scanning process takes several minutes and so only steady-state vibration phenomena can be assessed using the SLDV.

The SLDV system is most useful for certain types of vehicle refinement testing:

- Measuring surface variation in rms. vibration velocity for subsequent use in the calculation of structure-radiated sound levels.
- Measuring operating deflection shapes (ODS) of a vibrating target. The ODS analysis uses an SLDV system to measure sequences of vibration velocity at different locations, and uses the FFT to find their relative magnitude and phase. The phase is related back to that of a reference measurement made at a fixed point and a form of modal analysis is obtained (see Section 5.5).

Electronic speckle pattern interferometry is a laboratory-based alternative to SLDV. It uses a high-powered laser to produce a full-field measurement of spatial variation in deflection of the target surface. The main advantage of this method over the SLDV is that no scanning is required and so the method can be used to capture non-steady vibration patterns. The main disadvantages are the cost and complexity of the equipment required and the need to paint the target surface with a reflective paint (although simple white emulsion paint will often do).

Electronic speckle pattern interferometry is preferable to double-pulsed laser holography which predated it. This method produced interference fringes on a light-sensitive film plate and these could be viewed in order to distinguish the spatial pattern of vibration on the target surface. ESPI offers obvious advantages over this method as the results are near instantaneous and digital, and hence can be stored, copied and distributed easily.

The commercial laser vibrometer for rotating systems uses a beam-splitter to split the beam from a low-powered laser into two parallel beams a small distance d apart. The incident and reflected light from each beam are re-combined and a modulated intensity is found for each recombined beam. The frequency of modulation (Fa, Fb) is different for each beam. The instantaneous shaft speed of the target is related to the difference between Fb and Fa (Bruel & Kjaer application note for the Type 2523, 2003):

$$\omega = \frac{\lambda}{2d}(Fb - Fa) \tag{5.2}$$

Such a non-contact, remote sensing system must be the device of choice for measuring torsional vibrations (see Section 6.4 for a discussion of the torsional vibrations in the vehicle drivetrain).

The literature includes many examples where laser methods have been used in vehicle refinement. An early example is given by Felske et al. (1978) where holographic interferometry was used to determine the operating deflection shapes of a squealing brake disk. The use of such optical techniques (and others such as SLDV) are still of interest to more recent researchers (for example Papinniemi et al. (2002), Talbot and Fieldhouse (2002) and Burgess (2003)) who studied brake squeal phenomena.

Callow et al. (1994) demonstrated the use of an early prototype SLDV system to map the spatial distribution of vibration velocity on the body of a diesel fuel injection pump. More recently, more commercialised variants were used to map vibration levels across the block of a six-cylinder diesel engine (Cox and Fliesser, 2000) and vibration levels on the outer skin of a car door (Petniunas et al., 1999).

King demonstrated ESPI as early as 1990 showing vibration patterns on the block of a running V-8 engine, on the gearbox casing, on the differential and on mounting brackets. Krupka et al. (2002) demonstrate the use of the technique for analysing brake squeal, truck body panel vibration and the vibration on a catalytic converter.

5.3 Analysis and presentation of vibration data: quantifying vibration

5.3.1 Wave types within solids and their characteristics

Cremer and Heckl provide a very thorough review of the types of vibration waves that may propagate through solid media (Cremer and Heckl, 1988). Key points from that review are reproduced below, with supporting information and numerical examples.

5.3.1.1 Longitudinal waves

Pure longitudinal waves, in which the direction of particle displacement is in the direction of wave propagation, occur in common engineering fluids. Such waves may also occur in large solids whose dimensions in all directions are much greater than one wavelength.

The wavespeed for longitudinal waves in solids c_L is given by:

$$c_L = \sqrt{\frac{D}{\rho}} \ (\text{m s}^{-1}) \tag{5.3}$$

where the longitudinal stiffness D (N m^{-2}) of the material is given by

$$D = \frac{E\,(1-\mu)}{(1+\mu)\,(1-2\mu)} \tag{5.4}$$

where

$E = $ Young's Modulus (N m^{-2})
$\rho = $ Density (kg m^{-3})
$\mu = $ Poisson's Ratio (the ratio of lateral contracting strain to the elongation strain when a rod is stretched by a inline force at its ends).

So for mild steel, $E = 207\text{e}^9 \, \text{N m}^{-2}$, $\mu = 0.3$ and therefore $c_L = 5962 \, \text{m s}^{-1}$.

As the wavespeed is equal to the product of frequency and wavelength, this results in the wavelength of a longitudinal wave of 100 Hz in steel being 60 m. The equivalent wavelength in air is 3.4 m.

For concrete, $E = 18.5\text{e}^9 \, \text{N m}^{-2}$, $\mu = 0.2$ and therefore $c_L = 2930 \, \text{m s}^{-1}$. The wavelength of a longitudinal wave of 100 Hz in concrete is therefore 29 m. Pure longitudinal waves within solids may occur as seismic waves in the earth, but in buildings and in engineering structures they are not expected at frequencies below 10 kHz.

When a material is extended, the effect governed by Poisson's ratio determines that there must be displacement perpendicular to the line of force as well as elongation in the direction of the force. This is illustrated in Figure 5.4.

Therefore, a wave travelling along a 'slender' element cannot be a pure longitudinal wave. Such waves are called quasi-longitudinal waves. For the case of a rod or a beam, longitudinal motion occurs in the direction say x, and cross-sectional contraction is unconstrained in both y and z directions.

For rods, etc.

$$c_{L_{II}} = \sqrt{\frac{E}{\rho}} \quad \text{m s}^{-1} \text{ (the subscript } II \text{ denotes unconstrained lateral motion in 2 axes)}$$

$$\tag{5.5}$$

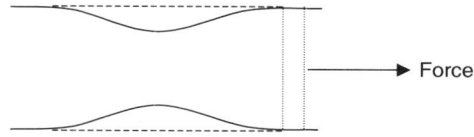

Force

Figure 5.4 Cross-sectional contraction due to Poisson's ratio not being unity results in transverse displacement.

So, for example:

Steel	$c_{LII} = 5135 \text{ m s}^{-1}$	(14% less than for the pure longitudinal case)
Concrete	$c_{LII} = 2776 \text{ m s}^{-1}$	(5% less than for the pure longitudinal case)

For plates, cross-sectional contraction can only occur in one direction and for plates

$$c_{L_I} = \sqrt{\frac{E}{\rho(1-\mu^2)}} \text{ m s}^{-1} \text{ (the subscript } I \text{ denotes unconstrained lateral motion in one axis)}$$

(5.6)

And for example

Steel	$c_{LI} = 5383 \text{ m s}^{-1}$	(10% less than for the pure longitudinal case)
Concrete	$c_{LII} = 2834 \text{ m s}^{-1}$	(3% less than for the pure longitudinal case)

5.3.1.2 Rotational waves

Transverse, shearing (or rotational) waves can occur and for circular rods only

$$c_T = \sqrt{\frac{G}{\rho}} \quad (\text{m s}^{-1})$$

(5.7)

$G =$ shear modulus (N m^{-2})
   For the case of a steel circular rod,

$$c_T = 3212 \text{ m s}^{-1} \quad (46\% \text{ less than for the pure longitudinal case})$$

For rectangular-section rods, of height h and width b see Table 5.1.
As h/b tends to infinity, c_{TI} tends to $2bc_T/h$.

5.3.1.3 Bending waves

Bending waves are also known as flexural waves.
   For large elements, the bending stiffness B is

$$B = EI$$

(5.8)

I is the moment of inertia (m^4 or commonly mm^4).
For plates of thickness h the moment of inertia per unit width I' is used

$$I' = \frac{h^3}{12}$$

(5.9)

Table 5.1 Wave speeds for rectangular rods of height h and width b

h/b	1	1.5	2	3	6	10
c_{T_I}/c_T	0.92	0.85	0.74	0.56	0.32	0.19

and the bending stiffness becomes

$$B' = \frac{I'E}{1-\mu^2} \tag{5.10}$$

The velocity at which one must travel to remain always at the same phase of one infinite sinusoidal wave is the phase velocity of bending waves

$$c_B = \sqrt[4]{\frac{B}{m'}} \sqrt[2]{\omega} \quad (\text{m s}^{-1}) \tag{5.11}$$

where m' is the mass per unit length (ρS) and ω is the radial frequency (rad s^{-1}). $\omega = 2\pi f$, f is the frequency (Hz).

For plates of thickness h, this becomes (providing that the wavelength $\lambda \gg 6h$)

$$c_B \approx \sqrt{1.8 c_{L_1} h f} = c_{L_1} \sqrt{\frac{1.8h}{\lambda_{L_I}}} \quad (\text{m s}^{-1}) \tag{5.12}$$

Note that the phase velocity is frequency-dependent and therefore the wavefront distorts (dispersion) as higher-frequency components propagate with higher phase velocity than lower-frequency components. One may demonstrate that energy propagates at the group velocity c_B (see Cremer and Heckl (1988) for details) which is equal to twice the phase velocity c_B. Consider a steel plate 10-mm thick, $c_{LI} = 5170 \, \text{m s}^{-1}$ (value quoted in Cremer and Heckl (1988))

$$\text{At } 100\,\text{Hz,} \qquad c_B = 96.5 \text{ m s}^{-1} \quad \text{and} \quad \lambda_B = 0.96\,\text{m}$$

$$\text{At } 10\,000\,\text{Hz,} \qquad c_B = 965 \text{ m s}^{-1} \quad \text{and} \quad \lambda_B = 0.096\,\text{m}$$

Consider a steel plate 1-mm thick, $c_{L_I} = 5170 \text{ m s}^{-1}$

$$\text{At } 100\,\text{Hz,} \qquad c_B = 30.5 \text{ m s}^{-1} \quad \text{and} \quad \lambda_B = 0.31\,\text{m}$$

$$\text{At } 10\,000\,\text{Hz,} \qquad c_B = 305 \text{ m s}^{-1} \quad \text{and} \quad \lambda_B = 0.031\,\text{m}$$

Compare these rather short bending wavelengths in steel plates with the rather long longitudinal wavelengths in large pieces of solid steel discussed earlier.

5.3.2 Vibration wave equations

The preceding discussion of the various forms of wave and the speed at which they travel leads to the respective wave equations.

5.3.2.1 Longitudinal wave equation

Kinsler et al. (1982) give a detailed account of the derivation of the longitudinal wave equation for the case of the longitudinal vibration of a bar. Only the result shall be presented here:

$$\frac{\partial^2 \xi}{\partial t^2} = c_{L_{II}}^2 \frac{\partial^2 \xi}{\partial x^2} \tag{5.13}$$

where ξ is the longitudinal displacement. It is clear that the above equation takes the same form as that of the linear plane-wave acoustic equation (see Section 2.1.2)

5.3.2.2 Torsional (rotational) wave equation

The equation of motion of a rod in torsional vibration is the same as that of the longitudinal vibration of a rod above (Thompson, 1993)

$$\frac{\partial^2 \theta}{\partial t^2} = c_T^2 \frac{\partial^2 \theta}{\partial x^2} \tag{5.14}$$

but for the use of θ for the twist rather than ξ for the displacement and the use of the appropriate wavespeed.

5.3.2.3 Transverse wave equation

Kinsler et al. (1982) give a detailed account of the derivation of the transverse wave equation for the case of the transverse vibration of a bar. Only the result shall be presented here:

$$\frac{\partial^2 y}{\partial t^2} = -\kappa^2 c_{L_{II}}^2 \frac{\partial^4 y}{\partial x^4} \tag{5.15}$$

where y is the transverse displacement and κ is the radius of gyration of the cross-sectional area S

$$\kappa^2 = \frac{\int r^2 \mathrm{d}S}{S} \tag{5.16}$$

where r is the distance from the neutral axis. For a circular rod of radius a, $\kappa = a/2$.

As a consequence of the fourth derivative, the transverse wave equation does not share the same form of solution as the longitudinal wave equation. In particular, dispersion occurs with transverse waves but not with longitudinal or torsional waves.

5.3.2.4 Vibration of thin plates

The derivation of the equation of motion for the vibration of a thin plate is too involved to be given here. However, for completeness, the equation for the symmetrical vibrations of a uniform circular diaphragm will be written down (Kinsler et al., 1982):

$$\frac{\partial^2 y}{\partial t^2} = -\kappa^2 c_{L_I}^2 \nabla^2 \left(\nabla^2 y \right) \tag{5.17}$$

$\nabla^2 y$ is called the Laplacian of y and is equal to

$$\nabla^2 y = \frac{\partial^2 y}{\partial x^2} + \frac{\partial^2 y}{\partial z^2} \tag{5.18}$$

Note the similarity in the form of the thin plate equation and that of the transverse vibration of the rod shown earlier – the only differences being:

- the need for a two-dimensional equation for the plate but only a one-dimensional equation for the rod;
- the use of a different form of wavespeed.

5.3.3 Quoting vibration levels

5.3.3.1 Single-value index methods

The following single-value index measures of vibration can be displayed as time histories or as rms. quantities in the manner discussed previously for sound (see Section 2.4.1):

- vibration acceleration ($m s^{-2}$ or 'g' being equal to $9.81 m s^{-2}$);
- vibration velocity ($m s^{-1}$ or commonly $mm s^{-1}$);
- vibration displacement (m or commonly mm).

In each case any frequency weighting applied to the data must be declared. Acceleration, velocity and displacement levels (dB) are often used.
Care must be taken when quoting vibration levels to:

- distinguish between vibration acceleration, velocity and displacement;
- distinguish between peak amplitude, peak-to-peak (twice the peak amplitude – not commonly used) and rms.;
- quote the key frequency (if possible) when quoting peak particle velocity (ppv);
- quote all units carefully (use SI units where possible, although the use of $mm s^{-1}$ for velocity is common).

5.3.3.2 Acceleration levels (dB)

The usual practice is to measure rms. levels in $m s^{-2}$ and the level reference is usually taken as one micrometer per second squared rms. ($10^{-6} m s^{-2}$).

$$L_a = 20 \ \log_{10} \left(\frac{a}{a_{ref}} \right) \ dB \tag{5.19}$$

5.3.3.3 Velocity levels (dB)

The usual practice is to measure rms. levels in $mm s^{-1}$ and the level reference is usually taken as one nanometer per second rms. ($10^{-9} m s^{-1}$ or $10^{-6} mm s^{-1}$).

$$L_v = 20 \ \log_{10} \left(\frac{v}{v_{ref}} \right) \ dB \tag{5.20}$$

Note that:

- the time integral of acceleration yields the velocity time history;
- the time integral of velocity yields the displacement time history.

Numerical methods may be used to post-process time histories and yield the time integral. Alternatively, electrical circuits are available to perform the integration at the point of recording the signal (these often become unreliable for double integration of acceleration, limiting signal-to-noise ratio and suffering from DC drift).

For simple harmonic motion analysed in the frequency domain the integral may be obtained by dividing the amplitude of a spectral component by $i\omega$, where i is the square root of -1 and $\omega = 2\pi f$ (f is the frequency in Hz).

5.3.3.4 *Frequency-dependent index methods*

All the methods described for the frequency analysis of sound (see Section 2.4.2) may also be used for analysing vibration.

5.3.4 Assessing vibration levels – human response to vibration

ISO 2631 Part 1 (1985) – Mechanical vibration and shock. Evaluation of human exposure to whole body vibration – Part 1 General Requirements.

This has now been superseded and therefore it should not be used as a Standard. However, Part 1 (General requirements) offered user-friendly guidance on the effects of vibration acceleration amplitude (1–80 Hz). Three boundaries were given:

1. reduced comfort boundary;
2. fatigue-decreased proficiency boundary;
3. exposure limits (for health and safety);

for various exposure periods between 1 minute and 24 hours. These boundaries often still form part of contemporary performance specifications for vehicles.

ISO 2631 Part 1 (1997) – Part 1 of ISO 2631 was revised in 1997. The guidance on safety, performance and comfort boundaries was removed. In its place:

- The main body of the text describes a means of measuring a weighted rms acceleration according to

$$a_{\mathrm{w}} = \left[\frac{1}{T} \int_0^T a_{\mathrm{w}}^2 (t) \, \mathrm{d}t \right]^{1/2} \tag{5.21}$$

where $a_{\mathrm{w}}(t)$ is the weighted acceleration time history in $\mathrm{m\,s^{-2}}$ and T is the duration in seconds.

- Different weightings are given for health, comfort, perception, the different axes and the position of the human subject (standing, seated and recumbent). W_k is used for assessing the effects of vibration on the comfort and perception of a standing person with vertical acceleration in the head-to-toe axis.
- Weighting W_k suggests that a standing person is most susceptible to vertical vibration in the frequency range of 4–8 Hz.
- Annexes to the main text give limited guidance on the health, comfort and perception and motion sickness effects of vibration. This guidance is in terms of weighted rms. acceleration.
- The Vibration Dose Value (VDV) is introduced but is not used in any guidance on the likely effects of vibration.

BS 6841:1987 – Measurement and evaluation of human exposure to whole body mechanical vibration and shock. The British Standards equivalent to ISO 2631 Part 1:

- Describes the method of obtaining weighted rms. acceleration.
- Provides very approximate indications of the likely reactions to various magnitudes of frequency-weighted rms. acceleration

$< 0.315\,\mathrm{m\,s^{-2}}$	not uncomfortable
$0.315–0.63\,\mathrm{m\,s^{-2}}$	a little uncomfortable
$0.5–1.0\,\mathrm{m\,s^{-2}}$	fairly uncomfortable
$0.8–1.6\,\mathrm{m\,s^{-2}}$	uncomfortable
$1.25–2.5\,\mathrm{m\,s^{-2}}$	very uncomfortable
$> 2.0\,\mathrm{m\,s^{-2}}$	extremely uncomfortable

ISO 2631 Part 2 (1989) – Evaluation of human exposure to whole body vibration. Continuous and shock-induced vibration in buildings (1–80 Hz). It offers guidance on the application of ISO 2631 Part 1 to human response to building vibration. It:

- does not offer guidance on complaint levels from occupants of buildings or any acceptable magnitudes or limits of building vibration;
- does contain frequency-weighting curves for human response to building vibration;
- Encourages the collection of data in a uniform format.

BS 6472:1992 – Guide to evaluation of human exposure to vibration in buildings (1–80 Hz). British Standard equivalent to ISO 2631 Part 2.

Although this Standard only refers to human exposure to vibration in buildings, because that vibration often originates from road traffic, it is discussed here for the sake of completeness. The Standard does not provide guidance on the probability of structural damage to buildings or injury to occupants of buildings subject to vibration. It makes use of the VDV where $a_w(t)$ is the weighted acceleration time history in $\mathrm{m\,s^{-2}}$ squared and T is the duration in seconds. The unit of VDV is $\mathrm{m\,s^{-1.75}}$.

$$VDV = \left[\int_0^T [a_w(t)]^4 \, dt \right]^{1/4} \tag{5.22}$$

It should be appreciated that the measurement of VDV is complex as it involves the gathering and subsequent weighting of acceleration time histories and their subsequent integration.

If the VDV is repeated n times during a period, then the total VDV period is

$$VDV_{total} = \left(\frac{t_{total}}{t_i} \right)^{1/4} \times VDV_i \tag{5.23}$$

VDV ($\mathrm{m\,s^{-1.75}}$) above which various degrees of adverse comment may be expected in residential buildings are given in Table 5.2.

Table 5.2 VDV classifications in residential buildings

Residential buildings	Low probability of adverse comment	Adverse comment possible	Adverse comment probable
16-hour day	0.2–0.4	0.4–0.8	0.8–1.6
8-hour night	0.13	0.26	0.51

5.4 Modes of vibration and resonance

5.4.1 Some preliminary concepts

The number of independent coordinates required to describe the motion of a system is called the degrees of freedom of the system. A single particle moving through space will have three degrees of freedom (x, y, z in Cartesian coordinates) and a rigid body will have six degrees of freedom (three components of translation and three of rotation). For a car, the six degrees of freedom are:

- bounce (vertical displacement)
- shunt (longitudinal displacement)
- lateral displacement
- pitch
- roll
- yaw.

A continuous elastic body will have an infinite number of degrees of freedom (three components of direction for each particle in the body) although such bodies can often be represented with a fair degree of realism by defining only a limited number of degrees of freedom.

Free vibration takes place when a system oscillates under the action of forces inherent in the system itself and external forces are absent. Such a system will vibrate at one or more of its natural frequencies, which are properties of the dynamic system as a result of the distribution of mass and stiffness.

At each natural frequency, particular types of motion occur within the body, and these are known as modes of vibration. At low frequencies and in bodies with light damping, different distinct types of motion such as bouncing, bending and twisting can be viewed at different natural frequencies.

A system under forced vibration will react with motion at the forcing frequency as well as motion at the natural frequencies. If the forcing frequency coincides with a natural frequency of the system then resonance occurs and the amplitude of vibration will tend to infinity, being controlled only by the damping in the system.

The amplitude of vibration at resonance is controlled by the damping of the system. When damping is small, it has little effect on the natural frequencies of the system and hence calculations of natural frequency are generally made on the assumption of no damping.

5.4.2 Free vibration of single degree of freedom systems

Consider the simple system where a mass m hangs on a spring of stiffness k (Figure 5.5).

In section 2.1.1 it is shown that a system with both inertia and elasticity will tend to oscillate about its usual position of rest. Neglect the effects of damping at present (these will be considered later in Sections 5.4.4 and 6.2) and consider Figure 5.6.

The system has a sdof as the motion may be wholly described in terms of the single spatial variable x. When the mass is hooked onto the spring, the spring stretches and the mass drops a distance Δx under the force of the spring to the static equilibrium position. Here the tension in the spring is equal to the weight of the mass

$$k = \frac{\text{Force}}{\text{Extension}} \, \text{N} \, \text{m}^{-1} \tag{5.24}$$

Weight $= m \, g \quad (g = 9.81 \, \text{m} \, \text{s}^{-2})$
Applying Newton's second law

$$F = ma \tag{5.25}$$

$$k\Delta x = mg \tag{5.26}$$

The variable x is taken to be positive in the downward direction. As a result, all other quantities such as Force (F), velocity ($\dot{x}$) and acceleration ($\ddot{x}$) are also positive in the downward direction.

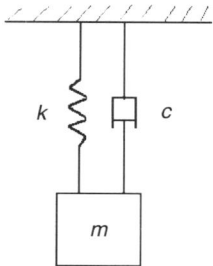

Figure 5.5 A simple, single degree of freedom (sdof) system.

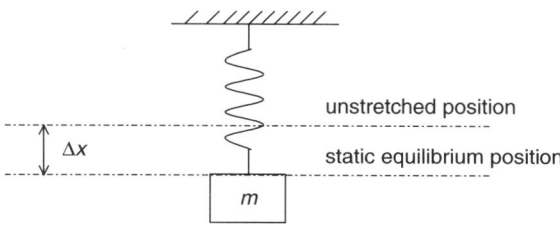

Figure 5.6 The sdof without damping and at static equilibrium.

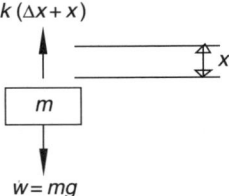

Figure 5.7 Free body diagram for the single mass-spring system.

If the mass is displaced a known distance x in the downward direction, the forces on the mass are resolved (as illustrated in Figure 5.7)

$$\Sigma F = w - k(\Delta x + x) \tag{5.27}$$

However, from above:

$$k\Delta x = mg$$

and therefore

$$\Sigma F = mg - kx - mg \tag{5.28}$$

Applying Newton's second law to the mass

$$m\ddot{x} = \Sigma F = -kx \tag{5.29}$$

the following relationship is obtained:

$$m\ddot{x} + kx = 0$$

This is the equation of motion of an undamped sdof, linear system undergoing rectilinear free vibration.

By the way, when viscous damping is added (more of this later) it becomes

$$m\ddot{x} + kx + c\dot{x} = 0 \quad \text{where c is the damping coefficient (N s m}^{-1}) \tag{5.30}$$

From observation it is clear that:

$$\omega_n^2 = 2\pi f_n = \frac{k}{m} \tag{5.31}$$

and this can be shown to be the case by assuming harmonic motion so that:

$$x = Xe^{i\omega t} \tag{5.32}$$

$$\dot{x} = i\omega Xe^{i\omega t} \tag{5.33}$$

$$\ddot{x} = -\omega^2 Xe^{i\omega t} \tag{5.34}$$

and so

$$m\ddot{x} = -m\omega_n^2 x \tag{5.35}$$

and

$$m\ddot{x} = -kx \tag{5.36}$$

so

$$-m\omega_n^2 x = -kx \tag{5.37}$$

$$\omega_n^2 = 2\pi f_n = \frac{k}{m} \tag{5.38}$$

The equation of motion is a homogenous (because the right-hand side is equal to zero), second-order, linear differential equation that has the general solution (see for instance Weltner et al. (1986)):

$$x = A \sin \omega_n t + B \cos \omega_n t \tag{5.39}$$

where A and B are constants obtained by considering the initial condition when $t = 0$ as $\sin \omega_n(0) = 0$ and $\cos \omega_n(0) = 1$

Thus

$$x(0) = B \tag{5.40}$$

$$\dot{x} = \omega_n A \cos \omega_n t - \omega_n B \sin \omega_n t \tag{5.41}$$

$$\dot{x}(0) = \omega_n A \tag{5.42}$$

$$A = \frac{\dot{x}(0)}{\omega_n} \tag{5.43}$$

and therefore

$$x = \frac{\dot{x}(0)}{\omega_n} \sin \omega_n t + x(0) \cos \omega_n t \tag{5.44}$$

The natural period of oscillation is τ:

$$\omega_n \tau = 2\pi \tag{5.45}$$

$$\tau = \frac{2\pi}{\omega_n} = 2\pi \sqrt{\frac{m}{k}} \tag{5.46}$$

and the natural frequency is:

$$f_n = \frac{1}{\tau} = \frac{1}{2\pi} \sqrt{\frac{k}{m}} \tag{5.47}$$

Remembering that $k\Delta x = mg$

$$f_n = \frac{1}{2\pi} \sqrt{\frac{g}{\Delta x}} \tag{5.48}$$

The above analysis can be repeated for a rotational sdof system. The rotational equation corresponding to Newton's second law is:

$$J\ddot{\theta} = -K\theta \tag{5.49}$$

where J is the rotational mass moment of inertia, K is the rotational stiffness and θ is the angle of rotation. The rotational equation of motion is therefore

$$\ddot{\theta} + \omega_n^2\theta = 0 \tag{5.50}$$

with

$$\omega_n = \sqrt{\frac{K}{J}} \tag{5.51}$$

It should be noted that moment of inertia in rotational motion corresponds to mass in linear motion. The moment of inertia of a body about an axis XX is defined as

$$J = \Sigma m x^2$$

where m is the mass of one of the particles and x is the distance of the particle from the axis. It is possible to define an equivalent radius known as the 'radius of gyration' k such that if all the particles were at this radius,

$$J = Mk^2 \tag{5.52}$$

where M is the total mass of the body.

There is a 'parallel axes theorem' whereby the moment of inertia of a body about axis (XX) which is at a distance x from a parallel axis (OO) is

$$J_{xx} = J_{00} + Mx^2 \tag{5.53}$$
$$k_{xx}^2 = k_{00}^2 + x^2 \tag{5.54}$$

A number of comparisons may be made between linear and rotational (angular) motion and these are shown in Table 5.3.

5.4.3   Free vibration of multiple degree of freedom systems

As an example of a multiple degree of freedom (mdof) system a two degree of freedom system will be used to predict the bounce and the pitch modes of a vehicle. A simple model is shown in Figure 5.8 that neglects:

- four of the six degrees of freedom;
- the unsprung masses;
- the effects of damping;
- non-linear stiffness characteristics.

Table 5.3 Basic relationships for rectilinear and rotational motion

Linear Motion	Angular Motion
Displacement or distance 's' in metres	Angular displacement 'θ' in radians
Velocity $v = \dfrac{ds}{dt}$	Angular velocity $\dot\theta = \dfrac{d\theta}{dt}$
Acceleration $a = \dfrac{dv}{dt} = \dfrac{d^2 s}{dt^2}$	Angular acceleration $\ddot\theta = \dfrac{d\dot\theta}{dt} = \dfrac{d^2\theta}{dt^2}$
$v = u + at$	$\dot\theta = \dot\theta_1 + \ddot\theta t$
$s = ut + 1/2at^2$	$\theta = \dot\theta_1 t + \dfrac{1}{2}\ddot\theta\, t^2$
$v^2 = u^2 + 2as$	$\dot\theta^2 = \dot\theta_1^2 + 2\ddot\theta s$
where u = initial velocity	$\dot\theta_1$ is initial velocity
$ds = \int v dt$ + constant	$d\theta = \int \dot\theta\, dt$ + constant
Momentum $= mv = \dfrac{W}{g} v$	Angular momentum $= J\dot\theta = \dfrac{w}{g} k^2 \dot\theta$
	k is the radius of gyration
Force (N) $F = \dfrac{d}{dt}(mv)$	Torque (N m) $T = \dfrac{d}{dt}(J\dot\theta)$
$\quad = v\dfrac{dm}{dt} + m\dfrac{dv}{dt}$	$\quad = \dot\theta\dfrac{dJ}{dt} + J\dfrac{d\dot\theta}{dt}$
$v_2 - v_1 = \int_{t_1}^{t_2} \dfrac{F}{m} dt$	$\dot\theta_2 - \dot\theta_1 = \int_{t_1}^{t_2} \dfrac{T}{J} dt$

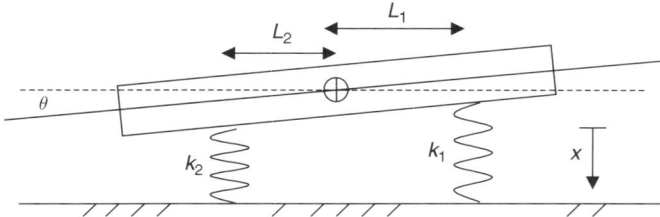

Figure 5.8 The simplified vehicle ride model.

This is a two degree of freedom system as it requires two variables x and θ to fully describe the motion of the body mass m.

First consider the rectilinear motion. Newton's second law gives the inertial force

$$F = m\ddot x \tag{5.55}$$

The centre of gravity is displaced downwards a distance x with an acceleration $\ddot x$. This motion causes (different) compressions of the two springs, Δx_1 and Δx_2. The spring compressions result in upward restoring forces that are deemed to be negative in sign as their direction opposes that of positive x.

$$\Delta x_1 = x - \theta L_1 \tag{5.56}$$

$$\Delta x_2 = x + \theta L_2 \tag{5.57}$$

$$F_1 = -k_1\Delta x_1 = -k_1(x - \theta L_1) \tag{5.58}$$

$$F_2 = -k_2\Delta x_2 = -k_2(x + \theta L_2) \tag{5.59}$$

At equilibrium, the inertial force is balanced by the combined restoring forces in the springs

$$m\ddot{x} = -k_1 (x - L_1\theta) - k_2 (x + L_2\theta) \tag{5.60}$$

The inertial force due to rotation is $J\ddot{\theta}$. At equilibrium, this is balanced by a positive torque generated by spring 1

$$T_1 = k_1 L_1 (x - L_1\theta) \tag{5.61}$$

and a negative torque generated by spring 2

$$T_2 = -k_2 L_2 (x + L_2\theta) \tag{5.62}$$

so

$$J\ddot{\theta} = k_1 L_1 (x - L_1\theta) - k_2 L_2 (x + L_2\theta) \tag{5.63}$$

Assuming harmonic motion

$$
\begin{aligned}
x &= A \sin \omega t & \theta &= B \sin \omega t \\
\dot{x} &= A\omega \cos \omega t & \dot{\theta} &= B\omega \cos \omega t \\
\ddot{x} &= -\omega^2 A \sin \omega t & \ddot{\theta} &= -\omega^2 B \sin \omega t
\end{aligned}
$$

Substituting these relationships into equations (5.60) and (5.63) gives

$$-m\omega^2 x = -k_1 (x - L_1\theta) - k_2 (x + L_2\theta) \tag{5.64}$$
$$-J\omega^2\theta = k_1 L_1 (x - L_1\theta) - k_2 L_2 (x + L_2\theta) \tag{5.65}$$

so

$$-m\omega^2 x + k_1 (x - L_1\theta) + k_2 (x + L_2\theta) = 0 \tag{5.66}$$
$$-J\omega^2\theta - k_1 L_1 (x - L_1\theta) + k_2 L_2 (x + L_2\theta) = 0 \tag{5.67}$$

This pair of simultaneous equations may be thus expressed in matrix form:

$$[K]\begin{Bmatrix} x \\ \theta \end{Bmatrix} = \begin{Bmatrix} 0 \\ 0 \end{Bmatrix} \tag{5.68}$$

Taking the rectilinear equation first

$$\begin{bmatrix} x \text{ terms} & \theta \text{ terms} \\ 0 & 0 \end{bmatrix}\begin{Bmatrix} x \\ \theta \end{Bmatrix} = \begin{Bmatrix} 0 \\ 0 \end{Bmatrix} \tag{5.69}$$

$$\begin{bmatrix} -\omega^2 m + k_1 + k_2 & -k_1 L_1 + k_2 L_2 \\ 0 & 0 \end{bmatrix}\begin{Bmatrix} x \\ \theta \end{Bmatrix} = \begin{Bmatrix} 0 \\ 0 \end{Bmatrix} \tag{5.70}$$

and adding the rotational equation gives:

$$
\begin{bmatrix} -\omega^2 m + k_1 + k_2 & -k_1 L_1 + k_2 L_2 \\ -k_1 L_1 + k_2 L_2 & -\omega^2 J + k_1 L_1^2 + k_2 L_2^2 \end{bmatrix} \begin{Bmatrix} x \\ \theta \end{Bmatrix} = \begin{Bmatrix} 0 \\ 0 \end{Bmatrix}
\tag{5.71}
$$

A non-trivial solution to this matrix equation occurs when (Weltner et al., 1986)

$$
\mathrm{Det}\,[K] = 0
\tag{5.72}
$$

where

$$
\mathrm{Det}\,[A] = ad - cb \quad A = \begin{bmatrix} a & b \\ c & d \end{bmatrix}
\tag{5.73}
$$

This is easy to compute and the use of only two degrees of freedom and the neglect of damping has been deliberate so that this may be so. Unfortunately, the proceeding arithmetic (although easy to perform) only yields the natural frequencies and mode shapes (neglecting damping) and the model cannot predict the time history of the forced response of the system.

Let $\omega^2 = \lambda$, so

$$
\begin{vmatrix} -\lambda m + k_1 + k_2 & -k_1 L_1 + k_2 L_2 \\ -k_1 L_1 + k_2 L_2 & -\lambda J + k_1 L_1^2 + k_2 L_2^2 \end{vmatrix} = 0
\tag{5.74}
$$

So,

$$
(k_1 + k_2 - \lambda m)(k_1 L_1^2 + k_2 L_2^2 - \lambda J) - (k_2 L_2 - k_1 L_1)(k_2 L_2 - k_1 L_1) = 0
\tag{5.75}
$$

Multiplying out

$$
\begin{aligned}
& \left[k_1^2 L_1^2 + k_1 k_2 L_2^2 - k_1 \lambda J + k_2 k_1 L_1^2 + k_2^2 L_2^2 - k_2 \lambda J - \lambda m k_1 L_1^2 - \lambda m k_2 L_2^2 + \lambda^2 m J \right] \\
& - \left[k_2^2 L_2^2 - k_2 L_2 k_1 L_1 - k_1 L_1 k_2 L_2 + k_1^2 L_1^2 \right] = 0
\end{aligned}
\tag{5.76}
$$

$$
\begin{aligned}
& \lambda^2 m J - \lambda \left(m k_1 L_1^2 + m k_2 L_2^2 \right) - \lambda \left(J k_1 + J k_2 \right) + k_1 k_2 L_2^2 + k_1 k_2 L_1^2 \\
& + 2 k_2 k_1 L_2 L_1 = 0
\end{aligned}
\tag{5.77}
$$

$$
\lambda^2 m J - \lambda \left(m k_1 L_1^2 + m k_2 L_2^2 + J k_1 + J k_2 \right) + k_1 k_2 \left(L_2^2 + L_1^2 + 2 L_1 L_2 \right) = 0
\tag{5.78}
$$

Equation (5.78) is the characteristic equation. This is obviously a quadratic equation, the roots of which may be found using the well-known formula:

$$
\mathrm{roots}_{1,2} = \frac{-b \pm \sqrt{b^2 - 4ac}}{2a}
\tag{5.79}
$$

$a = mJ$

$b = -\left[m k_1 L_1^2 + m k_2 L_2^2 + J k_1 + J k_2 \right]$

$c = k_1 k_2 \left(L_2^2 + L_1^2 + 2 L_1 L_2 \right)$

Consider a practical example of a vehicle where:

Mass $(m) = 1500$ kg
$J = 2160$ kg m^2
$K_1 = 35\,000$ N m^{-1}
$K_2 = 38\,000$ N m^{-1}
$L_1 = 1.4$ m
$L_2 = 1.7$ m

Thus

$a = 32\,40\,000$
$b = -42\,53\,10\,000$
$c = 1278\,13\,00\,000$

or

$a = 3.240$
$b = -425.31$
$c = 12\,781.3$

$$\text{roots}_{1,2} = \frac{425.31 \pm \sqrt{1\,80\,888.6 - 1\,65\,645.6}}{6.48}$$

$$\lambda_{1,2} = 46.582 \text{ and } 84.69$$

so,

$\omega_1 = \sqrt{46.582} = 6.83\,\text{rad}/\text{s}^{-1} = 1.09\,\text{Hz}$
$\omega_2 = \sqrt{84.69} = 9.20\,\text{rad}/\text{s}^{-1} = 1.46\,\text{Hz}$

The roots of the characteristic equation are known as the eigenvalues and they yield the natural frequency of each mode of vibration. In this case there are two, it being a two degree of freedom system.

The location of the bounce oscillation centre can be found by analysing the rectilinear equation of motion. This leads to being able to attribute mode shapes to each of the natural frequencies.

$$-m\omega^2 x + k_1 (x - L_1\theta) + k_2 (x + L_2\theta) = 0 \qquad (5.80)$$

$$x (k_1 + k_2 - m\omega^2) + \theta (k_2 L_2 - k_1 L_1) = 0 \qquad (5.81)$$

$$\frac{x}{\theta} = \frac{k_1 L_1 - k_2 L_2}{k_1 + k_2 - m\omega^2} \qquad (5.82)$$

With $\omega = 6.83\,\text{rad}/\text{s}^{-1}$,

$$\frac{x}{\theta} = \frac{49\,000 - 64\,600}{35\,000 + 38\,000 - 69\,973} = -5.15\,\text{m/rad}^{-1}$$

This result shows that $x = -5.15\theta$ and therefore the oscillation radius is 5.15 m with such a large radius, a given vertical displacement x will result in only a small angle θ. In other words, the 1.09 Hz natural frequency corresponds to a mode shape where the vehicle is mostly bouncing with only very little pitching motion.

With $\omega = 9.20 \, \text{rad/s}$,

$$\frac{x}{\theta} = \frac{-15\,600}{35\,000 + 38\,000 - 126\,960} = +0.28 \, \text{m/rad}^{-1}$$

This describes a mostly pitching mode with only very little bounce.

The bounce and pitch motions are pure or uncoupled when $k_1 L_1 = k_2 L_2$. In this case, the oscillation centre of rotation is at the centre of gravity and the bounce oscillation radius is infinitely long. The uncoupled ride characteristic is not desirable.

Simplified, uncoupled bounce and pitch equations may be used to predict approximate ride frequencies:

$$\omega_n = \sqrt{\frac{k_1 + k_2}{m}} \tag{5.83}$$

$$\omega_n = \sqrt{\frac{k_1 L_1^2 + k_2 L_2^2}{J}} \tag{5.84}$$

$\omega_1 = 6.98 \, \text{rad/s}^{-1}$, $\omega_2 = 0.09 \, \text{rad/s}^{-1}$

The uncoupled ride frequencies lie inside the coupled frequencies by 1–2%.

So far, the free vibration of sdof and mdof systems has been discussed. With transient forcing, such systems will respond at their natural frequencies displaying their respective mode shapes.

5.4.4 Forced vibration of sdof systems

Attention will now be to the case where the sdof is being forced harmonically by a force $F_0 \sin \omega t$

$$F_0 \sin \omega t = m\ddot{x} + c\dot{x} + kx \tag{5.85}$$

When F_0 is zero, the problem reduces to the complementary function which is the solution to the homogenous equation for free vibration of a damped sdof system. The solution to the non-zero F_0 case yields the particular integral.

The particular solution to the equation of motion is a steady-state oscillation at the same frequency ω as the exciting force.

A solution of the form

$$x = X \sin(\omega t) \tag{5.86}$$

can be assumed where X is the amplitude of oscillation.
Now

$$\dot{x} = \omega X \cos(\omega t) = i\omega X \sin(\omega t) \tag{5.87}$$

$$\ddot{x} = -\omega^2 X \sin(\omega t) \tag{5.88}$$

therefore

$$F_0 \sin \omega t = -m\omega^2 X \; \sin(\omega t) + i\omega c X \; \sin(\omega t) + kX \; \sin(\omega t) \qquad (5.89)$$

Now let

$$a = kX - m\omega^2 X \qquad (5.90)$$

$$b = ic\omega X \qquad (5.91)$$

so

$$F_0 = \sqrt{a^2 + b^2} = \sqrt{(kX - m\omega^2 X)^2 + (ic\omega X)^2}$$
$$F_0 = x\sqrt{(k - m\omega^2)^2 + (ic\omega)^2} \qquad (5.92)$$

and

$$X = \frac{F_0}{\sqrt{(k - m\omega^2)^2 + (ic\omega)^2}} \qquad (5.93)$$

$$\tan \varphi = \frac{b}{a} \quad \text{so} \quad \varphi = \tan^{-1} \frac{ic\omega}{k - m\omega^2} \qquad (5.94)$$

where φ is the phase of the displacement relative to the force.

Dividing the numerator and the denominator of both equations (5.93) and (5.94) by k, and making the substitutions (5.95)–(5.98)

$$\omega_n = \sqrt{\frac{k}{m}} \qquad (5.95)$$

$$c_c = 2m \, \omega_n \qquad (5.96)$$

$$\xi = \frac{c}{c_c} \qquad (5.97)$$

$$\frac{c\omega}{k} = \frac{c}{c_c} \frac{c_c \omega}{k} = 2\xi \frac{\omega}{\omega_n} \qquad (5.98)$$

the following are obtained:

$$\frac{xk}{F_0} = \frac{1}{\sqrt{\left[1 - \left(\dfrac{\omega}{\omega_n}\right)^2\right]^2 + \left[2\xi\left(\dfrac{\omega}{\omega_n}\right)\right]^2}} \qquad (5.99)$$

$$\tan \varphi = \frac{2\xi\left(\dfrac{\omega}{\omega_n}\right)}{1 - \left(\dfrac{\omega}{\omega_n}\right)^2} \qquad (5.100)$$

When $\omega/\omega_n \ll 1$ the inertia and damping forces are small which leads to a small φ. The spring force is nearly equal to the applied force.

When $\omega/\omega_n = 1$ the phase angle is 90° and the inertia force balances the spring force. The displacement amplitude at this resonant condition is given by

$$X = \frac{F_0}{c\omega_n} = \frac{F_0}{2\xi k} \qquad (5.101)$$

At large values of $\omega/\omega_n \gg 1$ the phase lag approaches 180° and the applied force is expended almost entirely in overcoming the large inertia force.

Equations (5.99) and (5.100) are used to produce Figure 5.9.

5.4.5 Forced vibration of mdof systems

Some attention will now be paid to the continuous forcing of an (mdof) system, this leading to the build-up of resonant response at the natural frequencies. Note, the damping within the system at that particular mode (the modal damping) will control:

- the rate at which the amplitude of response builds up once the forcing at that particular frequency begins;
- the peak forced response amplitude at that frequency;
- the rate at which the response decays once the forcing ceases.

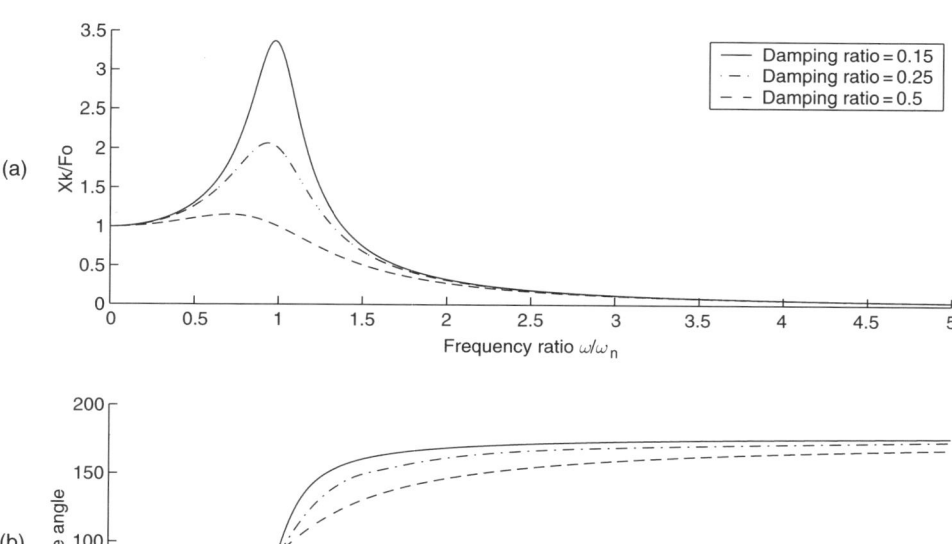

Figure 5.9 Forced vibration response of an sdof system produced using equations (5.99) and (5.100) respectively.

Consider the simplified model of a quarter vehicle, being shaken at one wheel by a force F_1 as shown in Figure 5.10.

Say $x_1 > x_2 > x_3$. A free body diagram can be drawn for each of the three masses thus (Figure 5.11):

Now an equation of motion can be written for each of the three bodies, thus

$$M_w\ddot{x}_1 = F_1 \sin \omega t - x_1 k_t - k_p (x_1 - x_2) - c_p (\dot{x}_1 - \dot{x}_2)$$

$$M_w\ddot{x}_1 + x_1 (k_t + k_p) - k_p x_2 + c_p (\dot{x}_1 - \dot{x}_2) = F_1 \sin \omega t \qquad (5.102)$$

$$M_b\ddot{x}_2 = -k_s (x_2 - x_3) + k_p (x_1 - x_2) - c_s (\dot{x}_2 - \dot{x}_3) + c_p (\dot{x}_1 - \dot{x}_2)$$

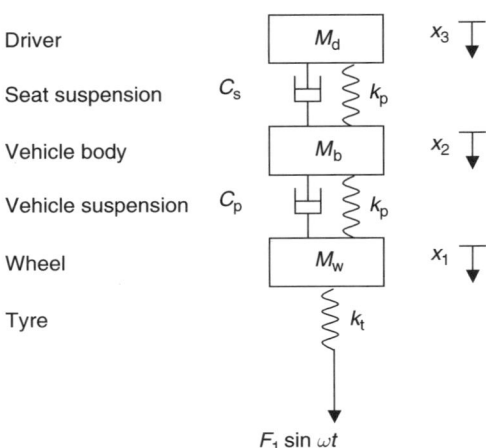

Figure 5.10 A simplified model of a quarter vehicle and its driver, being shaken at one tyre by a force F_1.

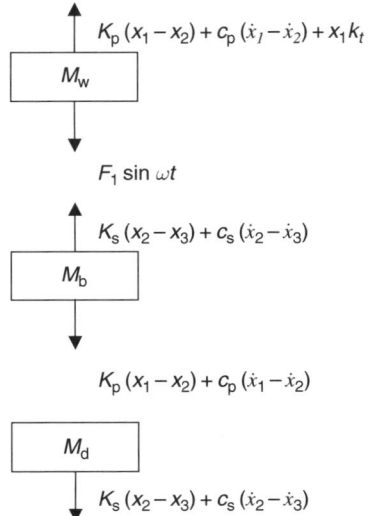

Figure 5.11 Free body diagrams for the quarter vehicle model with driver.

$$M_b\ddot{x}_2 = -x_2\left(k_s + k_p\right) + k_s x_3 + k_p x_1 - \dot{x}_2\left(c_s + c_p\right) + c_s \dot{x}_3 + c_p \dot{x}_1$$

$$M_b\ddot{x}_2 + x_2\left(k_s + k_p\right) - k_s x_3 - k_p x_1 + \dot{x}_2\left(c_s + c_p\right) - c_s \dot{x}_3 - c_p \dot{x}_1 = 0 \qquad (5.103)$$

$$M_d\ddot{x}_3 = k_s\left(x_2 - x_3\right) + c_s\left(\dot{x}_2 - \dot{x}_3\right)$$

$$M_d\ddot{x}_3 - k_s x_2 + k_s x_3 - c_s\left(\dot{x}_2 - \dot{x}_3\right) = 0 \qquad (5.104)$$

Putting equations (5.102), (5.103) and (5.104) into matrix form

$$
\begin{bmatrix} M_w & 0 & 0 \\ 0 & M_b & 0 \\ 0 & 0 & M_d \end{bmatrix}
\begin{bmatrix} \ddot{x}_1 \\ \ddot{x}_2 \\ \ddot{x}_3 \end{bmatrix}
+
\begin{bmatrix} c_p & -c_p & 0 \\ -c_p & c_s + c_p & -c_s \\ 0 & -c_s & c_s \end{bmatrix}
\begin{bmatrix} \dot{x}_1 \\ \dot{x}_2 \\ \dot{x}_3 \end{bmatrix}
$$

$$
+
\begin{bmatrix} k_p + k_t & -k_p & 0 \\ -k_p & k_s + k_p & -k_s \\ 0 & -k_s & k_s \end{bmatrix}
\begin{bmatrix} x_1 \\ x_2 \\ x_3 \end{bmatrix}
=
\begin{bmatrix} F_1 \\ 0 \\ 0 \end{bmatrix} \qquad (5.105)
$$

For harmonic excitation

$$F_1 = F_0 e^{i\omega t} \qquad (5.106)$$

$$\dot{x}_1 = i\omega a_1 e^{i\omega t} \qquad (5.107)$$

$$\ddot{x}_1 = -\omega^2 a_1 e^{i\omega t} \qquad (5.108)$$

so equation (5.105) becomes

$$\left[-\omega^2 [M] + i\omega [c] + [k]\right]\{a\} = \begin{Bmatrix} F_0 \\ 0 \\ 0 \end{Bmatrix} \qquad (5.109)$$

The displacement vector $\{a\}$ can be found from the inverse of the left-hand side of equation (5.110) which is a complex matrix. Such a calculation is very straightforward in Matlab™, requiring only one line of code:

$$a = inv(equ)*f \qquad (5.110)$$

where equ is the name given to left-hand side matrix, and a and f are displacement and force vectors respectively. The inversion of complex matrices in virtually any other computing language would involve writing many lines of code (or the use of proprietary libraries such as the LAPACK method in FORTRAN).

Typical results of the calculation for the driver's seat position are shown in Figure 5.12, where the displacement vector has been differentiated twice with respect to frequency to obtain the acceleration spectrum. The acceleration at the driver's seat exhibits three peaks:

1. one at the frequency corresponding to the vehicle bounce mode (around 1 Hz);
2. one at the frequency corresponding to a mode where the driver bounces on the seat (around 4 Hz);
3. one at the frequency corresponding to the wheel hop mode (around 12 Hz).

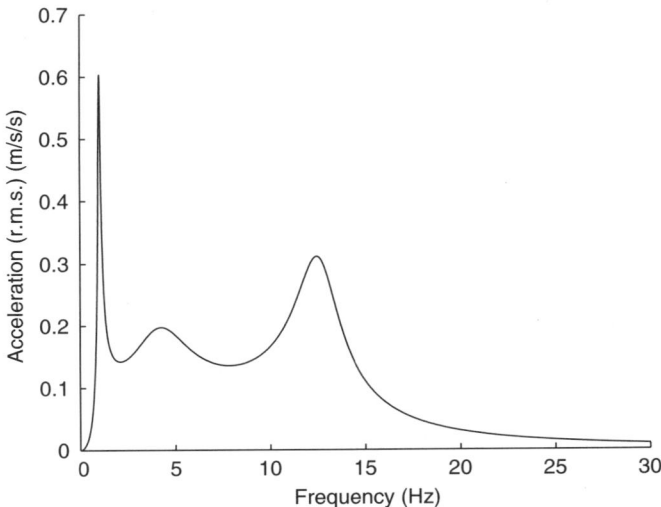

Figure 5.12 Typical results at the driver's seat obtained from the quarter vehicle model with driver.

5.4.6 Transverse vibration of single span beams

The modes of vibration of lumped element systems with discrete parcels of mass, stiffness and damping have been considered. Now attention turns to the modes of vibration of continuous systems (beams and plates, etc.) that have an infinite number of natural frequencies and mode shapes.

Blevins (1979) provides a valuable set of formulae for calculating the first few modes of most common forms of engineering element – plates, beams, etc. reproduced in Tables 5.4. and 5.5 for convenience:

Table 5.4 After (Blevins, 1979)

Description	$\lambda_i i = 1, \ 2, \ 3, \dots$	Mode shape $y_i \left(\dfrac{x}{L} \right)$	$\sigma_i i = 1, \ 2, \ 3, \dots$
Free-free beam	4.730 04 07 4	$\cosh \dfrac{\lambda_i x}{L} + \cos \dfrac{\lambda_i x}{L}$	0.982 50 22 15
	7.853 20 46 2	$-\sigma_i \left(\sinh \dfrac{\lambda_i x}{L} + \sin \dfrac{\lambda_i x}{L} \right)$	1.000 77 73 12
	10.995 60 78 0		0.999 96 64 50
Clamped-free beam	1.875 10 40 7	$\cosh \dfrac{\lambda_i x}{L} - \cos \dfrac{\lambda_i x}{L}$	0.734 09 55 14
	4.694 09 11 3	$-\sigma_i \left(\sinh \dfrac{\lambda_i x}{L} - \sin \dfrac{\lambda_i x}{L} \right)$	1.018 46 73 19
	7.854 75 74 4		0.999 22 44 97
Pinned-pinned beam	$i\pi$	$\sin \dfrac{i\pi x}{L}$	
Clamped-clamped beam	4.730 04 07 4	$\cosh \dfrac{\lambda_i x}{L} - \cos \dfrac{\lambda_i x}{L}$	0.982 50 22 15
	7.853 20 46 2	$-\sigma_i \left(\sinh \dfrac{\lambda_i x}{L} - \sin \dfrac{\lambda_i x}{L} \right)$	1.000 77 73 12
	10.995 60 79 0		0.999 96 64 50V

Table 5.5 After (Blevins, 1979)

		Mode sequence: λ_{ij}^2 and (ij)		
Description	a/b	1	3	3
Free on all edges	0.4	3.463	5.288	9.622
		(13)	(22)	(14)
	2/3	8.946	9.602	20.74
		(22)	(13)	(23)
	1.0	13.49	19.79	24.43
		(22)	(13)	(31)
Clamped on all edges	0.4	23.65	27.82	35.45
		(11)	(12)	(13)
	2/3	27.01	41.72	66.14
		(11)	(12)	(21)
	1.0	35.99	73.41	73.41
		(11)	(21)	(12)

$$\text{Natural frequency (Hz)} \ f_i = \frac{\lambda_i^2}{2\pi L^2} \left(\frac{EI}{m} \right)^{1/2} \quad \text{for } i = 1, 2, 3, \ldots \tag{5.111}$$

where

X = distance along beam (m)
m = mass per unit length of beam (kg m^{-1})
E = modulus of elasticity (N m^{-2})
I = area moment of inertia of beam about neutral axis (m^4)
L = span of beam (m)

5.4.7 Transverse vibration of rectangular plates

$$\text{Natural frequency (Hz)} \ f_{ij} = \frac{\lambda_{ij}^2}{2\pi a^2} \left(\frac{Eh^3}{12\gamma (1 - v^2)} \right)^{1/2}$$
$$\text{for } i = 1, 2, 3, \ldots \quad j = 1, 2, 3, \ldots \tag{5.112}$$

where

a = length of plate (m)
b = width of plate (m)
h = thickness of plate (m)
i = number of half-waves in mode shape along horizontal axis
j = number of half-waves in mode shape along vertical axis
γ = mass per unit area of plate (kg m^{-2})
v = Poisson's ratio

5.5 Modal analysis

Consider a mdof (see Section 5.4.5). If it were excited with an impulsive force, the subsequent motion would be determined by:

1. the natural frequencies of the system – there is one for each degree of freedom. These are also known as the modal frequencies;
2. the damping at each natural frequency (modal damping);
3. the spatial distribution of displacement vibration amplitude at each natural frequency (the mode shape).

For an mdof with N degrees of freedom it is often possible to construct a theoretical model consisting of N equations of motion (see Section 5.4.5). The simultaneous solution of these equations in matrix form can lead deliberately to a solution in two parts sometimes known as the modal properties (Ewins, 1984). These are:

1. A diagonal $N \times N$ matrix of eigenvalues $[\lambda_r^2]$ each value being the natural frequency and damping of one of the modes.
   For hysteretic damping (proportional to x^2 – also known as structural damping)

$$\lambda_r^2 = \omega_r^2 \,(1 + i\eta_r) \qquad (5.113)$$

η is the loss factor and $\omega = 2\pi f (\text{rad s}^{-1})$.
2. An $N \times N$ matrix of eigenvectors $[\phi_r]$ the rth column of which describes the mode shape that corresponds to the rth eigenvalue.

When the mdof is excited continuously with a harmonic (sinusoidal) force it is possible to solve the equations of motion to yield a single-solution matrix – the mobility matrix $[Y_{jk}(\omega)]$. This is an $N \times N$ matrix.

$\text{Mobility} = \dfrac{\dot{x}_j}{F_k}$ complex, frequency-dependent ratio of vibration velocity to force.

It is possible to derive a single equation that links the forced and free vibration characteristics of the mdof (Ewins, 1984):

$$Y_{jk}(\omega) = \frac{\dot{x}_j(\omega)}{F_k(\omega)} = i\omega \sum_{r=1}^{N} \frac{(r\phi_j)(r\phi_k)}{\lambda_r^2 - \omega^2} \qquad (5.114)$$

$r\phi_j$ is the jth element of the rth column eigenvector (ϕ_r).
This equation requires some analysis:

1. To get a measure of the mobility Y_{jk}, a force can be applied at location k (measuring it with a force transducer) and the vibration velocity measured at position j.
2. The right-hand side of equation (5.114) tells the reader several things:

   - That Y_{jk} is made from the sum of contributions from N different modes.
   - When the damping is low, and the excitation frequency ω tends towards ω_r (equation 5.113) then the product of the two displacements $j\phi_r$ and $k\phi_r \to \infty$ and hence $Y_{jk} \to \infty$.

So, a peak in the mobility plot (or frequency response function (FRF)) indicates the presence of one (or more) mode of vibration, each one with its own natural frequency, modal damping and mode shape. For simple structures with light damping and well-distributed modes (well-spaced peaks in the FRF) the modal properties of the mdof can easily be extracted, thus:

1. The modal frequencies can be read directly off the FRF (the frequencies corresponding to the peaks in the FRF).
2. The modal damping may be obtained from

$$\eta_r = \frac{\Delta f}{f_r} \tag{5.115}$$

η_r = loss factor of rth mode

Δ_f = bandwidth between the two half power points, each given by $\dfrac{Y_{peak}}{\sqrt{2}}$

f_r = modal frequency.

3. If the mobility of an sdof were plotted as a Nyquist diagram then the circle formed around the resonance would cross the imaginary axis twice – once at 0 Hz and once at ω_n. Therefore at resonance, the real part of $Y(\omega)$ goes to zero and the magnitude of $Y(\omega)$ is solely dependent on the imaginary part. So for a simple structure, excited at frequency $\omega = \omega_n$, the spatial distribution of $Im\left[Y_j\right]$ yields the mode shape. It is customary to make measurements Y_{jk} so that the relative phase of the $Im\left[Y_j\right]$ is preserved and so the 'true' mode shape may be found.

For more complex structures with higher damping, the modes get 'smeared' and several different modes may have significant impact on the motion at a particular frequency. In such a case, it is necessary to extract the modes from the FRF_{jk} (this could be mobility or often inertance (acceleration)):

$$accelerance = \frac{\ddot{x}_j(\omega)}{F_k(\omega)} \tag{5.116}$$

A common method of mode extraction is to use an sdof-fit method better known as a circle-fit method. The sdof produces a perfect circle on the Nyquist plot around the resonant peak in the FRF. In practice, the FRF measured around a resonance in the complex mdof will produce a limited number of points that sit roughly on a circle on the Nyquist plot. A perfect circle may be fitted to that data to obtain an sdof approximation to the true mdof behaviour at that resonance as shown in Figure 5.13. The position of the circle may be offset due to the influence of other nearly coincident modes.

If all the peaks in the FRF are considered in this way a new FRF may be constructed in accordance with equation (5.114). This new FRF can then be used to determine estimates of the modal properties in the way already discussed for simple structures.

An example of a mobility FRF generated from circle-fit data is shown in Figure 5.14.

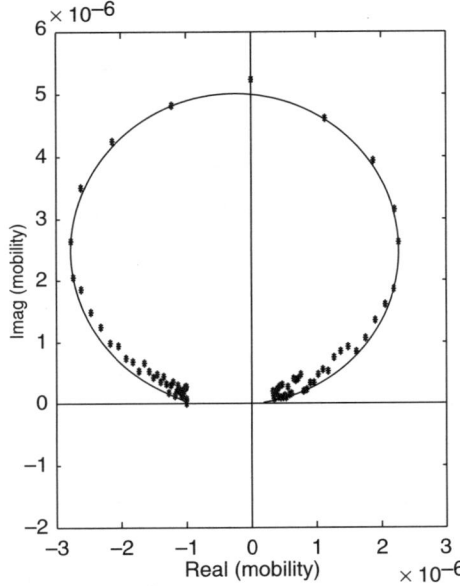

Figure 5.13 Circle fit to resonant data: after (Ewins, 1984).

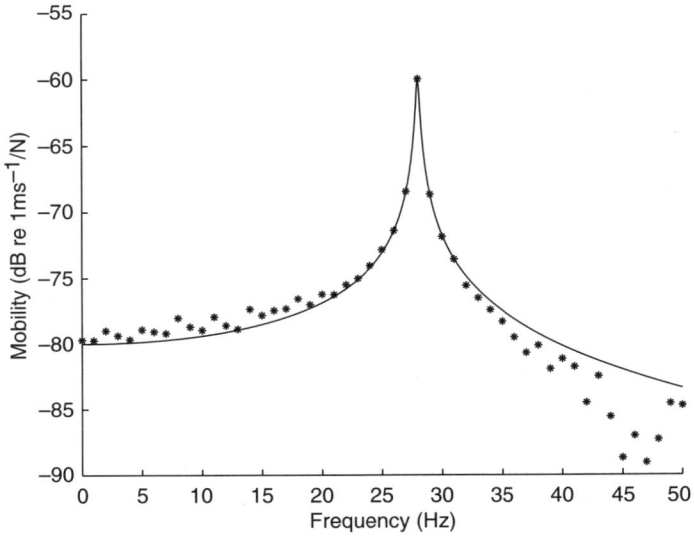

Figure 5.14 Regeneration of the mobility curve from circle-fit data: after (Ewins, 1984).

5.5.1 Experimental set-up for modal analysis

Rather than like transfer path analysis, model analysis generates a large quantity of measured data, all of which has to be analysed and manipulated. Commercial systems are available to make the whole process as efficient as possible, such as

- MODAL+
- STAR MODAL

A typical programme for the modal analysis of a body in white automotive body shell (just the basic bodyshell without trim or even paint – the term 'body in white' is explained by the common use of white primer on steel bodyshells) might be:

 (i) Mount or hang the bodyshell (on rubber cards or rubber mounts) so that its likely modal behaviour is as least constrained as possible.
 (ii) Determine the highest frequency (and hence shortest wavelength) of interest and use this to determine the spatial distribution of measuring points (two measuring points must be spaced at a distance much less than the shortest wavelength of interest).
(iii) Mark a grid of measuring locations (could be hundreds) on the bodyshell. Decide on an efficient method for labelling these.
(iv) Attach a shaker (or several shakers) fitted with a force transducer to a suitable point(s) (often the point at which vibration would normally enter the bodyshell, such as a suspension mounting). An instrumented hammer might be used for smaller modal surveys.
 (v) Using tri-axial accelerometers, measure all the FRFs required to characterise the motion of the structure. Long sample times with lots of averaging of FFTs are generally used. One accelerometer is kept at a reference location for all tests. An alternative is to use a pair of single-point laser doppler vibrometers to make rapid non-contact measurements of vibration velocity (see Section 5.2). One vibrometer is kept at the reference location.
(vi) Circle fit to every resonant peak in every FRF to produce a family of regenerated FRFs.
(vii) Use the regenerated FRFs to estimate the modal damping, modal frequency and mode shape for each mode extracted. The model shapes can be displayed graphically as in Figure 5.15.

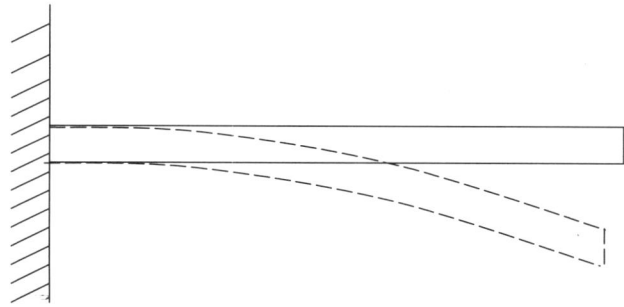

Figure 5.15 Deformed mode shape.

Note: The shaker should be attached with a wire rod (or stinger) that is stiff in the direction of intended force application but does not constrain the motion of the body under analysis in other directions.

References

Blevins, R.D., Formulas for natural frequencies and mode shape, Van Nostrand Reinhold Company, 1979

Bruel & Kjaer Product Data Sheet, Charge amplifier – Type 2635 Bruel & Kjaer, 1999

Bruel & Kjaer Product Data Sheet, Calibration exciter – Type 4294 Bruel, & Kjaer, 1999

Bruel & Kjaer, Product data sheet – torsional vibration meter Type 2523, Bruel & Kjaer, 2003

Bruel & Kjaer/Ormetron, Product data sheet – Type 8330 scanning laser doppler vibrometer, Bruel & Kjaer/Ormetron, 2003

Burgess, D.S., Photonics examines brake noise, Photonics Spectra, July 2003, p. 20, 2003

Callow, G.D., Havergill, D., Russell, M.F., Surface vibration measurements by scanning laser: case study of application to fuel injection pump, IMechE Paper No. C487/044/94, Printed in: 'Vehicle NVH and refinement'. International conference, Birmingham, UK, 3 – 5 May 1994, Mechanical Engineering Publications Ltd, 1994

Cox, P.E., Fliesser, W., Exploring the limits of spatial resolution of noise sources using sound intensity visualisation and laser doppler velocimetry, IMechE Paper No. C577/020/2000, Printed in: 'Vehicle noise and vibration 2000'. European conference, London, UK, 10–12 May 2000, Professional Engineering Publishing Ltd, Bury St Edmunds and London, UK, 2000.

Cremer, L., Heckl, M., Structure-Borne Sound: structural vibrations and sound radiation at audio frequencies-Second edition, Springer-Verlag, Berlin, 1988 (also 1973)

Ewins, D.J., Modal testing: theory and practice, Research Studies Press Ltd, 1984

Felske, A., Hoppe, G., Matthai, H., Oscillation in squealing brakes – analysis of vibration modes by holographic interferometry, SAE Paper No. 780333, 1978

King, P.D., Vehicle and powertrain refinement using laser-based interferometer techniques, IMEchE Paper No C420/014, Printed in: 'Quiet revolutions – powertrain and vehicle noise refinement'. International conference, Heathrow, UK, 9–11 October 1990, Mechanical Engineering Publications Ltd, 1990

Kinsler, L.E., Frey, A.R., Coppens, A.B., Sanders, J.V., Fundamentals of acoustics – Third edition, John Wiley & Sons Inc., 1982

Krupka, R., Waltz, T., Ettemeyer, A., Evans, R., Vibration measurement using ESPI, IMechE Paper No. C605/033/2002, 2002

Ormetron, Product data sheet – VS100 single point laser doppler vibrometer, Ormetron, 2003

Papinniemi, A., Lai, J.C.S., Zhao, J., Loader, L., Brake squeal: a literature review, Applied Acoustics 63, 391–400, 2002

Petniunas, A., Otto, N.C., Amman, S., Simpson, R., Door system design for improved closure sound quality, SAE Paper No. 1999-01-1681, 1999

Talbot, C., Fieldhouse, J.D., Graphical representation of disc brakes generating noise using data from laser holography, IMechE Paper No. C605/021/2002, 2002

Thompson, W.T., Theory of vibration with applications – Fourth Edition Chapman & Hall, 1993

Weltner, K., Grosjean, J., Schuster, P., Weber, W.J., Mathematics for engineers and scientists, Stanley Thomas (Publishers) Limited, 1986

6

Sources of vibration and their control

6.1 Introduction

Whereas Chapter 5 deals mostly with the vibration behaviour of undamped systems, this chapter considers:

- the effects of damping as a means of vibration control;
- the use of isolation and absorption techniques for vibration control;
- sources and control of engine drivetrain vibration;
- sources and control of whole vehicle vibrations.

6.2 Damping of vibrations

The term 'damping' is used to cover the phenomenon of a force inherent in a system which acts to restrict the oscillatory behaviour of the system. One might consider it to be analogous to the effect that sound absorbing material has on the reverberant field in a room (see Section 2.5). There are several forms of damping, the first being structural damping where the energy is dissipated internally within the material of the vibrating structure itself. In this case the energy dissipated per cycle is proportional to the square of the vibration displacement amplitude and is independent of frequency.

A second form of damping is known as Coulomb damping and is the result of the sliding of two dry surfaces. The damping force is equal to the product of the normal force and the coefficient of friction μ and is assumed to be independent of velocity once the motion is initiated.

The final form of damping is that produced by the disturbance of a viscous fluid, which is known as viscous damping. The algebra used in describing the motion of a system including viscous damping is more straightforward than for other forms of damping and so it is often convenient to consider general damping in terms of an equivalent viscous damping coefficient.

Viscous damping force is proportional to the velocity of vibration and is expressed by this equation:

$$F_d = c\dot{x} \tag{6.1}$$

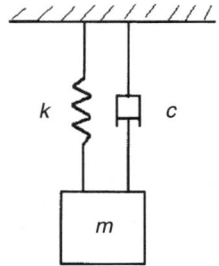

Figure 6.1 A simple sdof system with viscous damping.

where c is the viscous damping coefficient. Viscous damping is usually shown in diagrams as a dashpot in parallel with the stiffness as shown in Figure 6.1 which represents a typical sdof (single degree of freedom) system with damping.

The equation of motion for this system is

$$m\ddot{x} + c\dot{x} + kx = F(t) \tag{6.2}$$

which for the case of free vibration reduces to

$$m\ddot{x} + c\dot{x} + kx = 0 \tag{6.3}$$

which is a homogenous, second-order differential equation.

From observation one may easily deduce that the displacement amplitude of a viscously damped sdof system undergoing free vibration decays exponentially with time. This phenomenon can be expressed thus:

$$x = e^{st} \tag{6.4}$$

where s is a constant. Therefore

$$\dot{x} = se^{st} \tag{6.5}$$

$$\ddot{x} = s^2 e^{st} \tag{6.6}$$

and substitution into the equation of motion yields

$$ms^2 e^{st} + cse^{st} + ke^{st} = 0$$

$$(ms^2 + cs + k)e^{st} = 0$$

$$\left(s^2 + \frac{c}{m}s + \frac{k}{m}\right)e^{st} = 0 \tag{6.7}$$

Equation (6.7) is satisfied for all values of t when

$$s^2 + \frac{c}{m}s + \frac{k}{m} = 0 \tag{6.8}$$

an equation which is known as the characteristic equation.

It should be noted that the characteristic equation is a simple quadratic equation the roots of which may be deduced using the well-known method

$$ax^2 + bx + c = 0$$

$$\text{roots} = \frac{-b \pm \sqrt{b^2 - 4ac}}{2a}$$

so

$$s_{1,2} = \frac{\frac{-c}{m} \pm \sqrt{\frac{c^2}{m^2} - 4\frac{k}{m}}}{2}$$

$$s_{1,2} = \frac{-c}{2m} \pm \sqrt{\left(\frac{c}{2m}\right)^2 - \frac{k}{m}} \tag{6.9}$$

The general solution to the equation of motion takes the form

$$x = Ae^{s_1 t} + Be^{s_2 t} \tag{6.10}$$

where the constants A and B may be found from the initial conditions $x(0)$ and $\dot{x}(0)$ and

$$x = Ae^{\left(-\frac{c}{2m} + \sqrt{\left(\frac{c}{2m}\right)^2 - \frac{k}{m}}\right)t} + Be^{\left(-\frac{c}{2m} - \sqrt{\left(\frac{c}{2m}\right)^2 - \frac{k}{m}}\right)t} \tag{6.11}$$

or

$$x = e^{-\left(\frac{c}{2m}\right)t}\left(Ae^{\left(+\sqrt{\left(\frac{c}{2m}\right)^2 - \frac{k}{m}}\right)t} + Be^{\left(-\sqrt{\left(\frac{c}{2m}\right)^2 - \frac{k}{m}}\right)t}\right) \tag{6.12}$$

The term

$$e^{-\left(\frac{c}{2m}\right)t}$$

is an exponentially decaying function of time often referred to as the exponential envelope.
   The oscillatory behaviour is determined by the c, k and m in the inner parenthesis of equation (6.12). When

$$\left(\frac{c}{2m}\right)^2 > \frac{k}{m},$$

the exponent is a real number, oscillations are not formed and the system is said to be over damped.
   When $t = 0$, equation (6.10) becomes

$$x(0) = A + B \tag{6.13}$$

The time differential of equation (6.10) is

$$\dot{x} = As_1 e^{s_1 t} + Bs_2 e^{s_2 t} \tag{6.14}$$

When $t = 0$, equation (6.14) becomes

$$\dot{x}(0) = As_1 + Bs_2 \tag{6.15}$$

Re-arranging equation (6.13)

$$A = x(0) - B \tag{6.16}$$

Substituting equation (6.16) into equation (6.15)

$$\dot{x}(0) = (x(0) - B)s_1 + Bs_2$$
$$\dot{x}(0) = x(0)s_1 - Bs_1 + Bs_2$$
$$\dot{x}(0) = x(0)s_1 + B(s_2 - s_1)$$
$$B = \frac{\dot{x}(0) - x(0)s_1}{s_2 - s_1} \tag{6.17}$$

Now imagine an idealised laboratory experiment where a 5-kg mass is sitting at its equilibrium position on a spring of stiffness $200\,\mathrm{N\,m^{-1}}$ and a damper with viscous damping coefficient $80\,\mathrm{N\,s\,m^{-1}}$. At time $t = 0$, an instantaneous force is applied so that the mass instantaneously starts to move with a velocity of $5\,\mathrm{m\,s^{-1}}$. Applying equations (6.17) and (6.16) to equation (6.12) and solving for the first 5 seconds of time yields Figure 6.2.

Figure 6.2 shows no oscillatory motion. After initial excursion of the spring, the motion decays in an exponential manner dictated by the ratio of damping to mass that governs the exponential envelope.

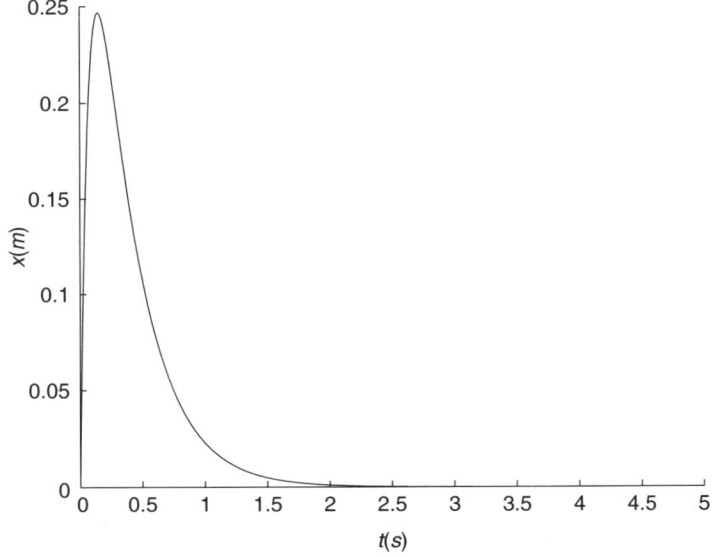

Figure 6.2 Over damped motion of a sdof.

When

$$\left(\frac{c}{2m}\right)^2 < \frac{k}{m}$$

the roots of the characteristic equation become the square root of a negative number, which requires the use of complex numbers.

$$s_{1,2} = \frac{-c}{2m} \pm \left(\sqrt{\left(\frac{c}{2m}\right)^2 - \frac{k}{m}}\right)$$

when negative, it can be written as

$$s_{1,2} = \frac{-c}{2m} \pm \left(i\sqrt{\frac{k}{m} - \left(\frac{c}{2m}\right)^2}\right) \tag{6.18}$$

Now applying equation (6.18) to equation (6.10), equation (6.12) for the under damped case becomes

$$x = e^{-\left(\frac{c}{2m}\right)t}\left(Ae^{+i\left(\sqrt{\frac{k}{m} - \left(\frac{c}{2m}\right)^2}\right)t} + Be^{-i\left(\sqrt{\frac{k}{m} - \left(\frac{c}{2m}\right)^2}\right)t}\right) \tag{6.19}$$

Now applying Euler's formula (Weltner et al., 1986)

$$e^{\pm i\sigma} = \cos\sigma \pm i\sin\sigma \tag{6.20}$$

to equation (6.19)

$$x = e^{-\left(\frac{c}{2m}\right)t}\left(A\left[\cos\left(\sqrt{\frac{k}{m} - \left(\frac{c}{2m}\right)^2}\right)t + i\sin\left(\sqrt{\frac{k}{m} - \left(\frac{c}{2m}\right)^2}\right)t\right]\right.$$
$$\left. + B\left[\cos\left(\sqrt{\frac{k}{m} - \left(\frac{c}{2m}\right)^2}\right)t - i\sin\left(\sqrt{\frac{k}{m} - \left(\frac{c}{2m}\right)^2}\right)t\right]\right) \tag{6.21}$$

Now equation (6.16) and (6.17) can be substituted into equation (6.21) and this can be used to predict the motion of the earlier idealised laboratory experiment, with all input data the same except for this time the coefficient of viscous damping is $8\,\mathrm{N\,s\,m^{-1}}$. The results of such a calculation are shown in Figure 6.3.

Figure 6.3 shows oscillatory motion at a frequency close to the natural frequency of the mass on the spring. After initial excursion of the spring, the motion oscillates and the magnitude of the oscillation decays in an exponential manner dictated by the ratio of damping to mass that governs the exponential envelope. The maximum excursion of the spring is larger than for the over damped case.

When

$$\left(\frac{c}{2m}\right)^2 = \frac{k}{m}$$

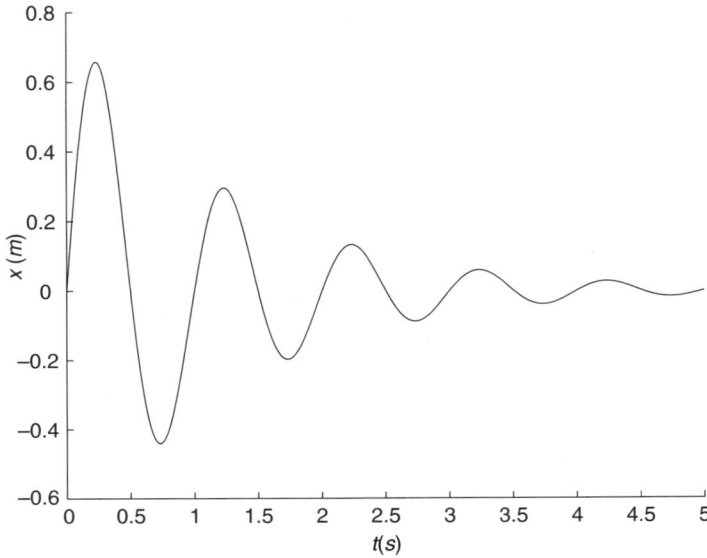

Figure 6.3 Motion of under-damped sdof.

the roots of the characteristic equation are identical:

$$s_{1,2} = \frac{-c}{2m} \tag{6.22}$$

and equation (6.10) becomes

$$x = e^{-\frac{c}{2m}t}(A+B) \tag{6.23}$$

This limiting condition is known as critical damping which is denoted using a damping coefficient c_c.

$$\left(\frac{c_c}{2m}\right)^2 = \frac{k}{m} \tag{6.24}$$

$$c_c = 2m\sqrt{\frac{k}{m}} = 2m\omega_n = 2\sqrt{km} \tag{6.25}$$

Now equations (6.16) and (6.17) can be substituted into equation (6.23) and this can be used to predict the motion of the earlier idealised laboratory experiment, with all input data the same except for this time the coefficient of viscous damping is $63\,\mathrm{N\,s\,m^{-1}}$. The results of such a calculation are shown in Figure 6.4.

Figure 6.4 shows a motion similar to that of the over-damped case, but the maximum excursion of the spring is larger in the critical damping case. Any damping condition can be described in terms of the critical damping using the damping ratio

$$\xi = \frac{c}{c_c} \tag{6.26}$$

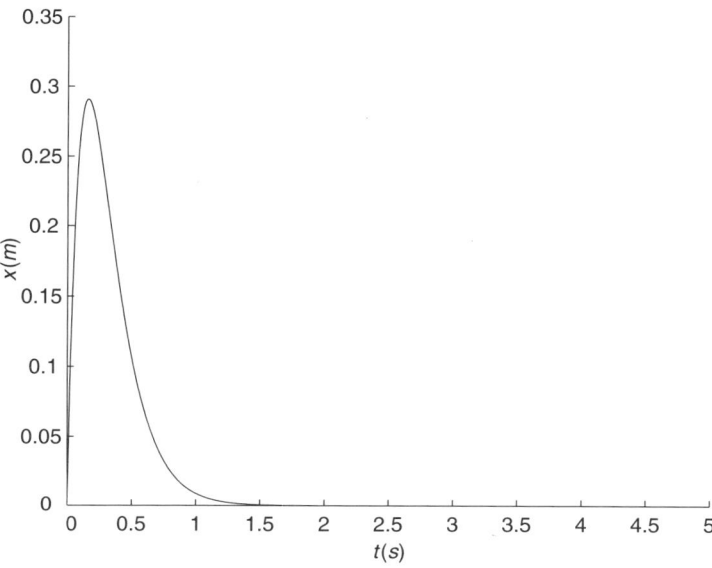

Figure 6.4 Critically damped motion of a sdof.

so $c_c = \dfrac{c}{\xi}$ and from equation (6.25)

$$\frac{c}{\xi} = 2m\omega_n \quad \text{and} \quad \frac{c}{2m} = \xi\omega_n \tag{6.27}$$

The roots to the characteristic equation become, after substitution of equation (6.27)

$$s_{1,2} = -\xi\omega_n \pm \sqrt{\xi^2\omega_n - \frac{k}{m}}$$

$$s_{1,2} = \left(-\xi \pm \sqrt{\xi^2 - 1}\right)\omega_n \tag{6.28}$$

Equation (6.28) can be rewritten as

$$s_{1,2} = \left(-\xi \pm i\sqrt{1 - \xi^2}\right)\omega_n \tag{6.29}$$

Now from inspection of equations (6.21) and (6.28) one can deduce that the frequency of damped oscillation is equal to

$$\omega_d = \omega_n\sqrt{1 - \xi^2} \tag{6.30}$$

The original equation of motion which was

$$m\ddot{x} + c\dot{x} + kx = F(t)$$

Assuming harmonic motion

$$x = Ae^{i\omega t} \tag{6.31}$$

$$\dot{x} = i\omega Ae^{i\omega t} \tag{6.32}$$

$$\ddot{x} = -\omega^2 Ae^{i\omega t} \tag{6.33}$$

$$F = F_0 e^{i\omega t} \tag{6.34}$$

equation (6.2) becomes on substitution

$$-\omega^2 m Ae^{i\omega t} + i\omega c Ae^{i\omega t} + k Ae^{i\omega t} = F_0 e^{i\omega t} \tag{6.35}$$

so

$$A = \frac{F_0}{(k - \omega^2 m) + i\omega c} \tag{6.36}$$

This can be separated into real and imaginary parts by multiplying numerator and denominator by the complex conjugate of the denominator, thus:

$$A = \frac{F_0 \left[(k - \omega^2 m) - (i\omega c) \right]}{(k - \omega^2 m)^2 + (\omega c)^2} \tag{6.37}$$

The forced response of a typical sdof with damping is shown in Figure 6.5. The effect of damping on:

- the peak response at resonance;
- the sharpness of the resonant peak;
- the lag in phase between the forcing and the response of the mass

is clearly shown in Figure 6.5. Also, Figure 6.5 clearly shows the negligible effect that damping has on the response at frequencies away from resonance.

Dividing numerator and denominator of equation (6.37) by k^2 gives

$$A = \frac{\dfrac{F_0}{k}\left[\left(1 - \dfrac{\omega^2}{\omega_n^2}\right) - \dfrac{i\omega c}{k}\right]}{\left(1 - \dfrac{\omega^2}{\omega_n^2}\right)^2 + \left(\dfrac{\omega c}{k}\right)^2} \tag{6.38}$$

Now, F_0/k is the static deflection due to the force F_0 and from

$$\frac{c}{k} = \frac{2\xi}{\omega_n} \tag{6.39}$$

the dynamic magnification factor is obtained:

$$\frac{A}{A_{st}} = \frac{\left(1 - \dfrac{\omega^2}{\omega_n^2}\right) - i2\xi\dfrac{\omega}{\omega_n}}{\left(1 - \dfrac{\omega^2}{\omega_n^2}\right)^2 + \left(2\xi\dfrac{\omega}{\omega_n}\right)^2} \tag{6.40}$$

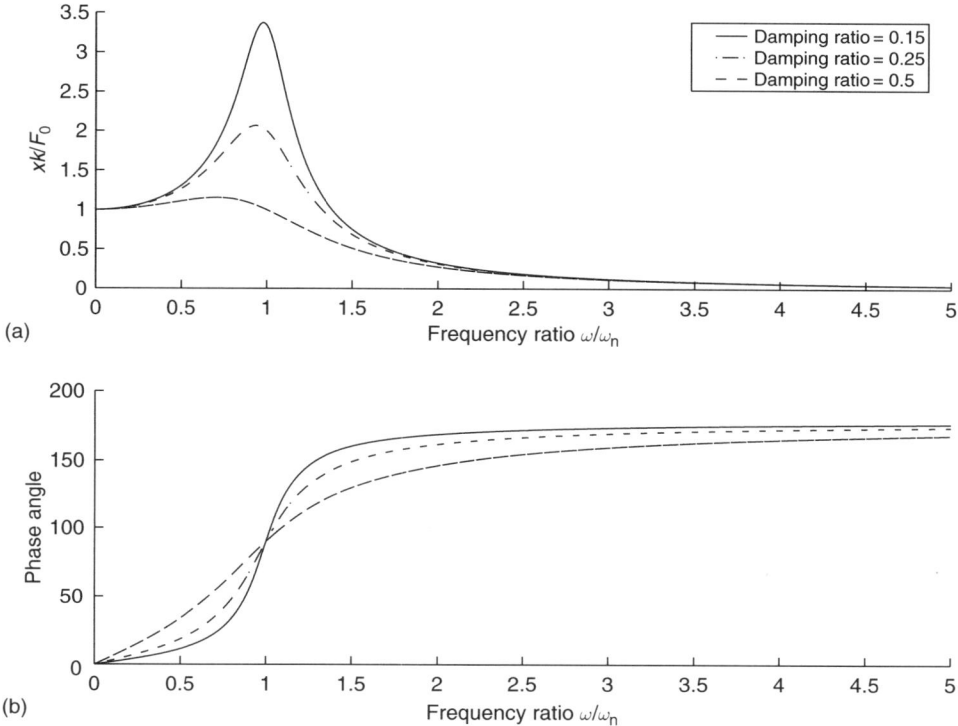

(a)

(b)

Figure 6.5 The response of the forced sdof, produced using equations (6.36) and (6.37) respectively.

The dynamic magnification factor is useful as it allows easy comparison of the response of a system to a dynamic (fluctuating) force compared with a static (constant) force. If a system is designed considering only static forces and loads, then the larger displacements that would occur for dynamic forces and loads may prove unacceptably high. The amplitude of the magnification factor is dependent on frequency (relative to natural frequency or frequencies of the system) and the damping of the system.

The forced response of a system is therefore frequency and damping-dependent. Re-capping on an earlier discussion (see Section 5.4.4), the following relationships were derived for an sdof with viscous damping:

$$\frac{xk}{F_0} = \frac{1}{\sqrt{\left[1 - \left(\frac{\omega}{\omega_n}\right)^2\right]^2 + \left[2\xi\left(\frac{\omega}{\omega_n}\right)\right]^2}} \tag{6.41}$$

$$\tan\varphi = \frac{2\xi\left(\frac{\omega}{\omega_n}\right)}{1 - \left(\frac{\omega}{\omega_n}\right)^2} \tag{6.42}$$

When $\omega/\omega_n \ll 1$ the inertia and damping forces are small which leads to a small φ. The spring force is nearly equal to the applied force. When $\omega/\omega_n = 1$ the phase angle is $90°$ and the inertia force balances the spring force. The displacement amplitude at this resonant condition is given by

$$X_{res} = \frac{F_0}{2\xi k} \tag{6.43}$$

At large values of $\omega/\omega_n \gg 1$ the phase lag approaches $180°$ and the applied force is expended almost entirely in overcoming the large inertia force. Equation (6.43) provides a simple means of finding the value of the damping ratio by experiment, if the stiffness and the forcing are known.

The sharpness of the resonance in plots such as the one in Figure 6.5 provides a more general way of determining the value of the damping ratio ξ by experiment (see Thompson (1993) or Nashif et al. (1985) or Harris and Crede (1961)). An analysis starts with equation (6.43) – when $\omega/\omega_n = 1$, the displacement amplitude at this resonant condition is given by equation (6.43).

There are two frequencies ω_1 and ω_2 on either side of ω_n where the amplitude of the displacement is equal to $X_{res}/\sqrt{2}$. These points are referred to as the half-power points (or -3 dB points) because

$$10\log_{10}\left(\frac{1}{\sqrt{2}}\right)^2 = -3.01 \text{ dB}$$

The frequency bandwidth between ω_2 and ω_1 is often referred to as the 3-dB bandwidth. Squaring equation (6.41) yields

$$\left(\frac{xK}{F_0}\right)^2 = \frac{1}{\left[1-\left(\frac{\omega}{\omega_n}\right)^2\right]^2 + \left[2\xi\left(\frac{\omega}{\omega_n}\right)\right]^2} \tag{6.44}$$

Letting

$$x = \frac{X_{res}}{\sqrt{2}} \tag{6.45}$$

and substituting this into equation (6.43)

$$x = \frac{F_0}{2\xi k}\frac{1}{\sqrt{2}} \tag{6.46}$$

Substituting equation (6.46) into equation (6.44) gives

$$\frac{1}{2}\left(\frac{1}{2\xi}\right)^2 = \frac{1}{\left[1-\left(\frac{\omega}{\omega_n}\right)^2\right]^2 + \left[2\xi\left(\frac{\omega}{\omega_n}\right)\right]^2} \tag{6.47}$$

Expanding both sides of equation (6.47) yields

$$\frac{1}{2}\left(\frac{1}{2\xi}\right)^2 = \frac{1}{1-2\left(\dfrac{\omega}{\omega_n}\right)^2 + \left(\dfrac{\omega}{\omega_n}\right)^4 + 4\xi^2\left(\dfrac{\omega}{\omega_n}\right)^2}$$

$$\frac{1}{8\xi^2} = \frac{1}{\left(\dfrac{\omega}{\omega_n}\right)^4 + (4\xi^2 - 2)\left(\dfrac{\omega}{\omega_n}\right)^2 + 1}$$

$$\left(\frac{\omega}{\omega_n}\right)^4 + (4\xi^2 - 2)\left(\frac{\omega}{\omega_n}\right)^2 + 1 = 8\xi^2$$

$$\left(\frac{\omega}{\omega_n}\right)^4 - 2(1 - 2\xi^2)\left(\frac{\omega}{\omega_n}\right)^2 + (1 - 8\xi^2) = 0 \tag{6.48}$$

Equation (6.48) is a quadratic in $\left(\dfrac{\omega}{\omega_n}\right)^2$ with roots given by

$$ax^2 + bx + c = 0$$

$$\text{roots} = \frac{-b \pm \sqrt{b^2 - 4ac}}{2a}$$

so

$$a = 1$$
$$b = -2(1 - 2\xi^2)$$
$$c = 1 - 8\xi^2$$

$$\text{roots} = \frac{2(1 - 2\xi^2) \pm \sqrt{4(1 - 4\xi^2 + 4\xi^4) - 4(1 - 8\xi^2)}}{2}$$

$$\text{roots} = \frac{2(1 - 2\xi^2) \pm 2\sqrt{4\xi^4 + 4\xi^2}}{2}$$

$$\text{roots} = (1 - 2\xi^2) \pm \sqrt{4\xi^4 + 4\xi^2}$$

$$\text{roots} = (1 - 2\xi^2) \pm 2\xi\sqrt{\xi^2 + 1} \tag{6.49}$$

When $\xi \ll 1$ the roots of the quadratic are approximately

$$\text{roots} \cong 1 \pm 2\xi$$

and hence

$$\left(\frac{\omega}{\omega_n}\right)^2 \cong 1 \pm 2\xi \tag{6.50}$$

Equation (6.50) itself has two roots

$$\omega_1^2 = \omega_n^2 - 2\omega_n^2\xi \tag{6.51}$$

$$\omega_2^2 = \omega_n^2 + 2\omega_n^2\xi \tag{6.52}$$

Subtracting equation (6.51) from equation (6.52) gives

$$\omega_2^2 - \omega_1^2 = 4\omega_n^2\xi$$

$$\frac{\omega_n}{\omega_2 - \omega_1} \cong \frac{1}{2\xi} = Q \tag{6.53}$$

The Q-factor or quality factor is a single index description of the sharpness of the resonant peak in the response spectrum and can be used to obtain the damping ratio.

Another way to measure the damping ratio by experiment is to measure the rate of decay of free oscillations (see Thompson (1993), for example). Remember the general damped case, where the response of the system was given by

$$x = Ae^{s_1 t} + Be^{s_2 t}$$

where

$$s_{1,2} = \left(-\xi \pm i\sqrt{1-\xi^2}\right)\omega_n$$

Equation (6.10), with substitution of equation (6.29), can be thus rewritten:

$$x = Xe^{-\xi\omega_n t} \sin\left(\sqrt{1-\xi^2}\,\omega_n t + \phi\right) \tag{6.54}$$

There is no algebraic trick here, one simply expects an exponentially decaying sine wave with a particular phase ϕ relative to the time origin.

A term called the 'logarithmic decrement' δ is now introduced, being the natural logarithm of the ratio of any two successive peak amplitudes, one at time t_1 and one at time $t_1 + \tau_d$.

$$\delta = \ln\frac{x_1}{x_2} = \ln\frac{e^{-\xi\omega_n t_1} \sin\left(\sqrt{1-\xi^2}\,\omega_n t_1 + \phi\right)}{e^{-\xi\omega_n(t_1+\tau_d)} \sin\left(\sqrt{1-\xi^2}\,\omega_n(t_1+\tau_d) + \phi\right)} \tag{6.55}$$

When x_1 and x_2 are separated by an exact period of oscillation, the two sine terms are identical and equation (6.55) reduces to

$$\delta = \ln\frac{x_1}{x_2} = \ln\frac{e^{-\xi\omega_n t_1}}{e^{-\xi\omega_n(t_1+\tau_d)}} = \ln e^{\xi\omega_n \tau_d} = \xi\omega_n \tau_d \tag{6.56}$$

Substituting equation (6.30)

$$\omega_d = \omega_n\sqrt{1-\xi^2}$$

$$\tau_d = \frac{2\pi}{\omega_n\sqrt{1-\xi^2}} \tag{6.57}$$

and so

$$\delta = \frac{2\pi\xi}{\sqrt{1-\xi^2}} \tag{6.58}$$

The damping ratio is then calculated from the measured logarithmic decrement. Several periods of oscillation may be used to improve the measurement, and equation (6.57) becomes equation (6.59) and is subsequently substituted into equation (6.56) as before.

$$\tau_d = \frac{2\pi n}{\sqrt{1-\xi^2}} \quad n = 1, 2, 3 \ldots \tag{6.59}$$

6.2.1 Material damping

When materials undergo cyclic stressing, energy is dissipated within the material itself. For most engineering metals, the energy dissipated per cycle is:

- independent of the frequency of excitation;
- proportional to the square of the vibration displacement amplitude.

Damping fitting this description is known as structural damping (Thompson, 1993). Other material damping, known as hysteretic damping, has the damping force in phase with the velocity (as in the viscous damping considered so far) but the magnitude of the damping force is proportional to the displacement amplitude (Fahy and Walker, 1998).

The adoption of the viscous form of damping is advantageous as it results in a characteristic equation (equation (6.8)) that can be readily solved to yield the displacement time history as already done. Other forms of damping do not produce that convenient characteristic equation. In the case of viscous damping, the amplitude at resonance of an sdof was found to be

$$X_{res} = \frac{F_0}{2\xi k} \tag{6.60}$$

or

$$X_{res} = \frac{F_0}{c\omega_n} \tag{6.61}$$

For other types of damping, no such simple expression exists (Thompson, 1993). It is possible, however, to define the resonant amplitude by using an equivalent damping coefficient c_{eq} in place of the viscous damping coefficient. However, care must be taken with this substitution as the response curves of systems with viscous and other forms of damping differ and so they may only be made equal at one frequency (Fahy and Walker, 1998).

The energy lost per cycle due to a damping force F_d is given by (Thompson, 1993)

$$W_d = \oint F_d dx \tag{6.62}$$

with viscous damping in the sdof

$$F_d = c\dot{x} \tag{6.63}$$

with displacement given by

$$x = X \sin(\omega t - \phi) \tag{6.64}$$

the velocity is given by

$$\dot{x} = \omega X \cos(\omega t - \phi) \tag{6.65}$$

So, from equation (6.62) the energy dissipated by viscous damping per cycle is given by

$$W_d = \oint c\dot{x}dx = \oint c\dot{x}\frac{dx}{dt}dt = \oint c\dot{x}^2 dt$$

$$W_d = \int_0^{\frac{2\pi}{\omega}} c\omega^2 X^2 \cos^2(\omega t - \phi)\, dt$$

$$W_d = c\omega^2 X^2 \int_0^{\frac{2\pi}{\omega}} \cos^2(\omega t - \phi)\, dt$$

$$W_d = c\omega^2 X^2 \int_0^{\frac{2\pi}{\omega}} \frac{1}{2}(\cos 2\omega\tau + 1)\, d\tau \quad \text{(standard integral)}$$

$$W_d = c\omega^2 X^2 \frac{1}{2}\left[\frac{\sin(2\omega\tau)}{2\omega} + \tau\right]_0^{2\pi/\omega} \quad \text{(standard integral)}$$

$$W_d = \pi c\omega X^2 \tag{6.66}$$

Equation (6.66) can be written in general terms for any form of damping as

$$W_d = \alpha X^2 \tag{6.67}$$

where α is a constant.
By substituting c_{eq} for c in equation (6.66)

$$c_{eq} = \frac{\alpha}{\pi\omega} \tag{6.68}$$

Remembering

$$-\omega^2 m A e^{i\omega t} + i\omega c A e^{i\omega t} + k A e^{i\omega t} = F_0 e^{i\omega t} \tag{6.35}$$

and substituting c_{eq} for c gives

$$-\omega^2 m A e^{i\omega t} + i\frac{\alpha}{\pi} A e^{i\omega t} + k A e^{i\omega t} = F_0 e^{i\omega t} \tag{6.69}$$

and letting

$$\eta = \frac{\alpha}{\pi k} = \text{loss factor} \tag{6.70}$$

so that the complex stiffness k' is given by

$$k' = k(1+i\eta) \tag{6.71}$$

and

$$A = X = \frac{F_0}{(k-\omega^2 m)+i\eta k} \tag{6.72}$$

Resonance occurs when the spring and inertial forces cancel each other and so the amplitude at resonance is given by

$$|X| = \frac{F_0}{\eta k} \tag{6.73}$$

Comparing equation (6.73) with equation (6.43) gives

$$\eta = 2\xi \tag{6.74}$$

Dividing numerator and denominator of equation (6.72) by k

$$X = \frac{F_0/k}{\left(1-\dfrac{\omega^2}{\omega_n^2}\right)+i\eta} \tag{6.75}$$

or

$$H(\omega) = \frac{X}{F_0/k} = \frac{1}{\left(1-\dfrac{\omega^2}{\omega_n^2}\right)+i\eta} \tag{6.76}$$

This can be separated into real and imaginary parts by multiplying numerator and denominator by the complex conjugate of the denominator, thus:

$$H(\omega) = \frac{X}{F_0/k} = \frac{\left(1-\dfrac{\omega^2}{\omega_n^2}\right)-i\eta}{\left(1-\dfrac{\omega^2}{\omega_n^2}\right)^2+\eta^2} \tag{6.77}$$

The real and imaginary parts of equation (6.77) can be presented together on a Nyquist plot as shown in Figure 6.6.

This is a circle, with centre at point $-1/2\eta$ that crosses the real axis once at zero frequency and once at the resonant frequency. At resonance, equation (6.77) reduces to

$$H(\omega) = \frac{-i}{\eta}$$

Therefore, the loss factor can be simply read off the Nyquist diagram. In the case of Figure 6.6, $\eta = 0.2$.

This resonant method (along with the earlier one using the half-power bandwidth) may be used to estimate the loss factor of elastomeric components at a particular frequency achieved by varying the additional mass placed on top of the component.

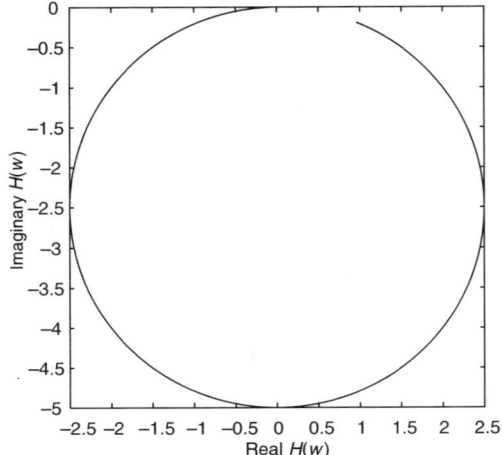

Figure 6.6 Nyquist plot of equation (6.77).

6.2.2 Hysteresis loops

Remembering that the energy lost per cycle due to a damping force F_d is given by

$$W_d = \oint F_d dx \tag{6.62}$$

and writing the velocity as (Thompson, 1993)

$$\mathring{x} = \omega X \cos(\omega t - \phi) = \pm \omega X \sqrt{1 - \sin^2(\omega t - \phi)} = \pm \omega \sqrt{X^2 - x^2} \tag{6.78}$$

the damping force becomes

$$F_d = c_{eq}\mathring{x} = \pm c_{eq}\omega\sqrt{X^2 - x^2} \tag{6.79}$$

Rearranging equation (6.79) thus

$$F_d^2 = c_{eq}^2 \omega^2 \left(X^2 - x^2\right)$$

$$\frac{F_d^2}{c_{eq}^2 \omega^2} = X^2 - x^2$$

$$\left(\frac{F_d}{c_{eq}\omega X}\right)^2 = 1 - \frac{x^2}{X^2}$$

$$\left(\frac{F_d}{c_{eq}\omega X}\right)^2 + \left(\frac{x}{X}\right)^2 = 1 \tag{6.80}$$

Equation (6.80) is that of an ellipse, and from equation (6.62) it can be seen that the area enclosed by that ellipse gives the energy dissipated per cycle and from this c_{eq} can be found. X is the maximum excursion of the sample.

The loss factor η is defined as

$$\eta = \frac{W_d}{2\pi U} \tag{6.81}$$

where U is the peak strain energy.

$$U = \frac{1}{2}\sigma\varepsilon \quad \text{per unit volume (Benham and Crawford, 1987)} \tag{6.82}$$

with σ denoting stress and ε denoting strain.

Often the total cyclic force applied to a component is plotted against the cyclic excursion as sketched in Figure 6.7. This is called a hysteresis loop and the area yields the loss factor. For linear damping, an ellipse is formed, but for non-linear damping the loop is not an ellipse.

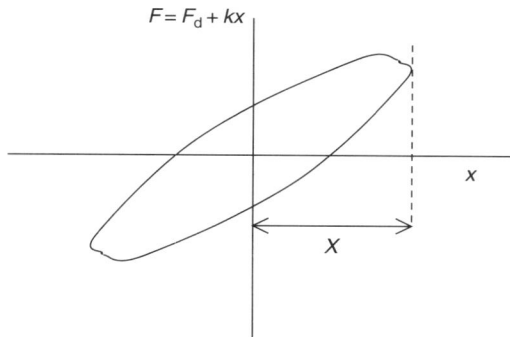

Figure 6.7 Sketched hysteresis loop.

6.2.3 Complex modulus

Most elastomers are not perfectly elastic and Young's Modulus does not adequately describe the relationship between stress and strain. When a sinusoidal stress is applied to a common elastomer, the resulting strain lags the stress by a phase angle θ (Gade et al., 1995).

Let σ be a sinusoidal stress, thus:

$$\sigma = \hat{\sigma}e^{i\omega t} \tag{6.83}$$

The resulting strain will be

$$\varepsilon = \hat{\varepsilon}e^{i(\omega t - \theta)} \tag{6.84}$$

The real part of the complex modulus of elasticity E' will be given by the in-phase ratio of stress and strain

$$E' = \left(\frac{\hat{\sigma}}{\hat{\varepsilon}}\right)\cos\theta \tag{6.85}$$

The imaginary part represents the out-of-phase relationship between stress and strain is given by

$$E'' = \left(\frac{\hat{\sigma}}{\hat{\varepsilon}}\right) \sin\theta \tag{6.86}$$

The complex modulus of elasticity E^* is given by

$$E^* = E' + iE'' \tag{6.87}$$

And the loss factor by

$$\eta = \frac{E''}{E'} = \tan\theta \tag{6.88}$$

So

$$E^* = E'(1 + i\eta) \tag{6.89}$$

As stress is force divided by area, and strain is deflection divided by original length, equations (6.85) and (6.86) can be re-written as

$$E' = \frac{|F|/A}{|d|/L}\cos\theta = \text{real}\left(\frac{F}{d}\right) \times \frac{L}{A} \tag{6.90}$$

$$E'' = \frac{|F|/A}{|d|/L}\sin\theta = \text{imag}\left(\frac{F}{d}\right) \times \frac{L}{A} \tag{6.91}$$

where F/d is the dynamic stiffness. For random excitation, it may be measured as the frequency response function between the force and the displacement (acceleration integrated twice with respect to time). This method is known as the non-resonant method for estimating complex modulus and loss factor. At low frequencies it works well (B&K, 1996) but at higher frequencies the inertial effects of the specimen and of the measuring apparatus affect the result. B&K (1996) offers an improved non-resonant method that takes account of these effects.

Another method for measuring the high frequency dynamic stiffness is given by Thompson et al. (1998). This method places the resilient component between two large blocks, and the vibrations of each block are measured. This method allows the separation of lateral and rotational components as well as the imposition of a pre-load on the component.

6.2.4 Friction damping

Friction damping (or Coulomb damping) results from the sliding of two dry surfaces. It is taken to be independent of velocity once motion is initiated. It commonly arises in riveted, jointed and clamped systems.

Table 6.1 Loss factors of engineering materials (Beranek, 1988)

Material	Range of loss factor η at 1000 Hz
Aluminium*	10^{-4} to 10^{-2}
Steel*	10^{-4} to 10^{-2}
Concrete	0.005–0.02
Glass	0.001–0.01
Lead	0.015

* Sensitive to construction techniques and edge conditions.

6.2.5 Surface damping treatments

Engineering materials commonly have rather low loss factors as shown in Table 6.1 (Beranek, 1988).

There are surface treatments available (trade names include Antiphon and Mandolex amongst others) that can be sprayed or painted onto engineering materials to increase the total loss factor in the range 10^{-3} to near unity depending on the thickness of the added material (Nashif et al., 1985; Cremer and Heckl, 1988). The damping performance of these materials will often be relatively poor at low temperatures. The damping of the material is generally greatest at low frequencies (Nashif et al., 1985) although when used as a layered treatment it is generally taken to be frequency-independent (Cremer and Heckl, 1988).

Surface damping treatments are effective in controlling response over a wide range of frequencies for structures with overlapping modes. This is due to the structure being always at or near resonance for any frequency above a certain lower limiting frequency.

A frequency-selective damping treatment is obtained when a layer of viscoelastic material is trapped between the engineering structure and a stiff but light covering plate. This is known as constrained layer damping and it can be engineered to be more effective than single layer treatments, but only over a more narrow range of frequencies.

6.3 Vibration isolation and absorption

Elsewhere in this chapter and also in Chapter 5 the vibration of sdof systems and mdof (multiple degree of freedom) systems have been considered with and without damping. The case where the support to the system has a force applied to it rather than the mass will now be analysed to see how the mass may be isolated from that force (Figure 6.8).

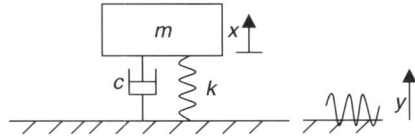

Figure 6.8 An sdof being excited at the base support.

In the case shown in Figure 6.8, both the support and the mass are moving rather than just the mass and so it is the relative displacement between the mass and the support that is important. The equation of motion for free vibration becomes:

$$m\ddot{x} = -k(x - y) - c(\dot{x} - \dot{y}) \qquad (6.92)$$

Declaring that

$$z = x - y, \qquad \dot{z} = \dot{x} - \dot{y}, \qquad \ddot{z} = \ddot{x} - \ddot{y}$$

then

$$m(\ddot{z} + \ddot{y}) = -k(z) - c\dot{z}$$
$$m\ddot{z} + kz + c\dot{z} = -m\ddot{y} \qquad (6.93)$$

and if it is for harmonic excitation

$$y = Y \sin \omega t \qquad (6.94)$$

then

$$\dot{y} = \omega Y \cos \omega t \qquad (6.95)$$
$$\ddot{y} = -\omega^2 Y \sin \omega t \qquad (6.96)$$

and so

$$\omega^2 mY \sin \omega t = m\ddot{z} + kz + c\dot{z} \qquad (6.97)$$

Now if in this case it is assumed that

$$z = Z \sin(\omega t) \qquad (6.98)$$

then

$$\dot{z} = \omega Z \cos(\omega t) = i\omega Z \sin(\omega t) \qquad (6.99)$$
$$\ddot{z} = -\omega^2 Z \sin(\omega t) \qquad (6.100)$$

so

$$\omega^2 mY \sin(\omega t) = -m\omega^2 Z \sin(\omega t) + kZ \sin(\omega t) + i\omega cZ \sin(\omega t) \qquad (6.101)$$

and therefore

$$\omega^2 mY = \sqrt{(Zk - Zm\omega^2)^2 + (Zc\omega)^2}$$
$$\omega^2 mY = Z\sqrt{(k - m\omega^2)^2 + (c\omega)^2}$$
$$Z = \frac{\omega^2 mY}{\sqrt{(k - m\omega^2)^2 + (c\omega)^2}} \qquad (6.102)$$

and the phase angle between the motion of the base and that of the mass is

$$\phi = \text{atan} \frac{cw}{k - m\omega^2} \tag{6.103}$$

This describes the relative motion of the mass and the base. To yield the absolute motion of the mass, the following substitution must be made:

$$x = z + y \tag{6.104}$$

where

$$y = Ye^{i\omega t} \tag{6.105}$$

$$z = Ze^{i(\omega t - \phi)} = (Ze^{-i\phi})e^{i\omega t} \tag{6.106}$$

$$x = Xe^{i(\omega t - \psi)} = (Xe^{-i\psi})e^{i\omega t} \tag{6.107}$$

Now remembering equation (6.93)

$$m\ddot{z} + kz + c\dot{z} = -m\ddot{y}$$

where

$$\dot{z} = i\omega(Ze^{-i\phi})e^{i\omega t} \tag{6.108}$$

$$\ddot{z} = -\omega^2(Ze^{-i\phi})e^{i\omega t} \tag{6.109}$$

$$\dot{y} = i\omega Ye^{i\omega t} \tag{6.110}$$

$$\ddot{y} = -\omega^2 Ye^{i\omega t} \tag{6.111}$$

then

$$-m\omega^2(Ze^{-i\phi})e^{i\omega t} + i\omega c(Ze^{-i\phi})e^{i\omega t} + k(Ze^{-i\phi})e^{i\omega t} = m\omega^2 Ye^{i\omega t}$$

$$-m\omega^2(Ze^{-i\phi}) + i\omega c(Ze^{-i\phi}) + k(Ze^{-i\phi}) = m\omega^2 Y \tag{6.112}$$

and

$$Ze^{-i\phi}[-m\omega^2 + i\omega c + k] = m\omega^2 Y$$

$$Ze^{-i\phi} = \frac{m\omega^2 Y}{-m\omega^2 + i\omega c + k} \tag{6.113}$$

Now $x = y + z$, so

$$x = (Ze^{-i\phi} + Y)e^{i\omega t} \tag{6.114}$$

so that

$$x = \left(\frac{m\omega^2 Y}{k - m\omega^2 + i\omega c} + Y\right)e^{i\omega t}$$

$$x = \left(\frac{kY + i\omega cY}{k - m\omega^2 + i\omega c}\right)e^{i\omega t}$$

$$x = \left(\frac{k + i\omega c}{k - m\omega^2 + i\omega c}\right)Ye^{i\omega t} \tag{6.115}$$

Now $x = \left(Xe^{-i\psi}\right)e^{i\omega t}$

so

$$\frac{Xe^{-i\psi}}{Y} = \frac{k+i\omega c}{k-m\omega^2 + i\omega c} \tag{6.116}$$

The modulus of this function can be written from inspection as

$$\left|\frac{X}{Y}\right| = \sqrt{\frac{k^2 + (\omega c)^2}{(k-m\omega^2)^2 + (\omega c)^2}} \tag{6.117}$$

Dividing numerator and denominator by k^2

$$\left|\frac{X}{Y}\right| = \sqrt{\frac{1+\left(\frac{\omega c}{k}\right)^2}{\left(1-\frac{m\omega^2}{k}\right)^2 + \left(\frac{\omega c}{k}\right)^2}} \tag{6.118}$$

Now, recalling that (from Section 5.4.4)

$$\frac{m}{k} = \left(\frac{1}{\omega_n}\right)^2 \tag{6.119}$$

and

$$\frac{c}{k} = \frac{2\xi}{\omega_n} \tag{6.120}$$

$$\left|\frac{X}{Y}\right| = \sqrt{\frac{1+\left(\frac{2\xi\omega}{\omega_n}\right)^2}{\left(1-\frac{\omega^2}{\omega_n^2}\right)^2 + \left(\frac{2\xi\omega}{\omega_n}\right)^2}} \tag{6.121}$$

The phase is obtained from equation (6.116), thus:

$$\frac{Xe^{-i\psi}}{Y} = \frac{k+i\omega c}{k-m\omega^2 + i\omega c} = \frac{(k+i\omega c)\left(k-m\omega^2 + i\omega c\right)}{(k-m\omega^2 + i\omega c)(k-m\omega^2 - i\omega c)}$$

$$\frac{Xe^{-i\psi}}{Y} = \frac{k^2 - km\omega^2 - i\omega ck + i\omega ck - im\omega^3 c + (\omega c)^2}{(k-m\omega^2)^2 + (\omega c)^2}$$

$$\frac{Xe^{-i\psi}}{Y} = \frac{k\left(k-m\omega^2\right) + (\omega c)^2 - im\omega^3 c}{(k-m\omega^2)^2 + (\omega c)^2}$$

$$\tan(-\psi) = \frac{-mc\omega^3}{k\left(k-m\omega^2\right) + (\omega c)^2}$$

$$\tan\psi = \frac{mc\omega^3}{k\left(k-m\omega^2\right) + (\omega c)^2} \tag{6.122}$$

Dividing both numerator and denominator by k

$$\tan \psi = \frac{\dfrac{mc\omega^3}{k}}{(k - m\omega^2) + \dfrac{1}{k}(\omega c)^2} \tag{6.123}$$

and substituting equation (6.120)

$$\tan \psi = \frac{\dfrac{2\xi m\omega^3}{\omega_n}}{(k - m\omega^2) + \dfrac{1}{k}(\omega c)^2} \tag{6.124}$$

Dividing denominator and numerator by k again

$$\tan \psi = \frac{\dfrac{2\xi m\omega^3}{k\omega_n}}{\left(1 - \dfrac{m\omega^2}{k}\right) + \left(\dfrac{\omega c}{k}\right)^2} \tag{6.125}$$

and substituting equations (6.119) and (6.120)

$$\tan \psi = \frac{\dfrac{2\xi\omega^3}{\omega_n^3}}{\left(1 - \dfrac{\omega^2}{\omega_n^2}\right) + \left(\dfrac{2\xi\omega}{\omega_n}\right)^2} \tag{6.126}$$

Equations (6.121) and (6.126) have been used to generate Figure 6.9.
The transmissibility TR is the ratio of the transmitted force to the applied force and

$$TR = \left|\frac{F_\tau}{F_0}\right| = \left|\frac{X}{Y}\right| \tag{6.127}$$

TR will be less than 1 (isolation) providing that $\dfrac{\omega}{\omega_n} > \sqrt{2}$

When damping is negligible, equation (6.121) reduces to

$$TR = \frac{1}{1 - \left(\dfrac{\omega}{\omega_n}\right)^2} = \frac{1}{1 - \omega^2\dfrac{\Delta}{g}} \tag{6.128}$$

as from Section 5.4.2, $\omega_n = \sqrt{\dfrac{g}{\Delta}}$

where Δ is the static deflection of the mass m on the spring stiffness k.

To reduce the transmissibility at a given frequency, the stiffness may be reduced. This will give rise to an increased deflection of the mass. To avoid this, the mass m may be increased by placing the mass on top of a much larger mass (often known as a seismic

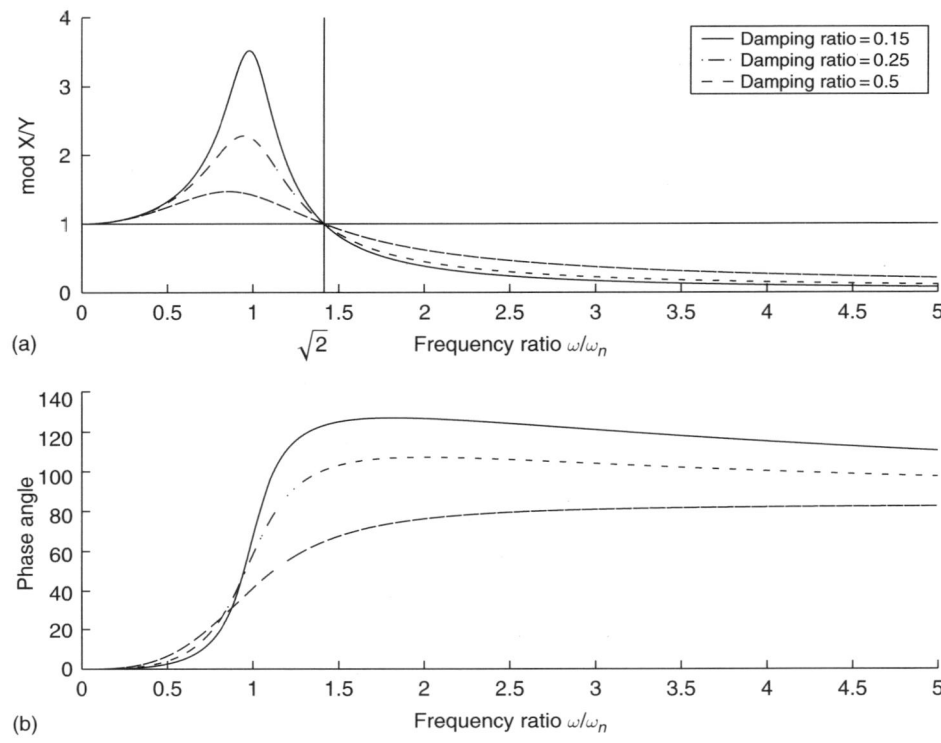

Figure 6.9 The response of the sdof when forced at its support. Generated from equations (6.121) and (6.126).

block), and then the stiffness k can be increased whilst ratio ω/ω_n remains constant. The effect of increasing the stiffness is that for a given force transmitted from the isolated mass to the base, the deflection of the mass will reduce with increasing stiffness.

6.3.1 The vibration absorber

By tuning the two degree of freedom system shown in Figure 6.10 to the frequency of the exciting force so that

$$\omega = \frac{k_2}{m_2} \tag{6.129}$$

the system acts as a vibration absorber and in the ideal case reduces the motion of the main mass m_1 to zero.

From inspection of equation (5.110) in Section 5.4.5 the equation of motion for this system can be written down in matrix form as

$$\left[-\omega^2\left[M\right]+\left[k\right]\right]\{a\} = \left\{\begin{array}{c} F_0 \\ 0 \end{array}\right\} \tag{6.130}$$

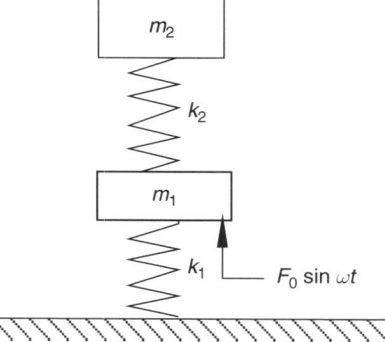

Figure 6.10 The vibration absorber.

where

$$M = \begin{bmatrix} m_1 & 0 \\ 0 & m_2 \end{bmatrix}, \quad k = \begin{bmatrix} k_1 + k_2 & -k_2 \\ -k_2 & k_2 \end{bmatrix} \text{ and } x_1 = a_1 e^{i\omega t}$$

so equation (6.130) can be expanded as

$$\begin{bmatrix} k_1 + k_2 - \omega^2 m_1 & -k_2 \\ -k_2 & k_2 - \omega^2 m_2 \end{bmatrix} \begin{Bmatrix} a_1 \\ a_2 \end{Bmatrix} = \begin{Bmatrix} F_0 \\ 0 \end{Bmatrix} \tag{6.131}$$

The inversion of equation (6.131) to yield the displacement amplitudes a_1 and a_2 can be achieved, by using determinant of the left-hand matrix (see Weltner et al. [1986] for instance), thus:

$$\text{Det } K = (k_1 + k_2 - \omega^2 m_1)(k_2 - \omega^2 m_2) - k_2^2$$

$$\begin{Bmatrix} a_1 \\ a_2 \end{Bmatrix} = \frac{1}{\text{Det } K} \begin{bmatrix} k_2 - \omega^2 m_2 & k_2 \\ k_2 & k_1 + k_2 - \omega^2 m_1 \end{bmatrix} \begin{Bmatrix} F_0 \\ 0 \end{Bmatrix} \tag{6.132}$$

This gives

$$a_1 = \frac{(k_2 - \omega^2 m_2) F_0}{(k_1 + k_2 - \omega^2 m_1)(k_2 - \omega^2 m_2) - k_2^2} \tag{6.133}$$

$$a_2 = \frac{k_2 F_0}{(k_1 + k_2 - \omega^2 m_1)(k_2 - \omega^2 m_2) - k_2^2} \tag{6.134}$$

Dividing the numerator and the denominator of equation (6.133) by $k_1 k_2$ gives

$$a_1 = \frac{\left(\dfrac{1}{k_1} - \omega^2 \dfrac{m_2}{k_1 k_2} \right) F_0}{\left(\dfrac{1}{k_2} + \dfrac{1}{k_1} - \omega^2 \dfrac{m_1}{k_1 k_2} \right) \left(\dfrac{1}{k_1} - \omega^2 \dfrac{m_2}{k_1 k_2} \right) - \dfrac{k_2}{k_1}}$$

Letting

$$\omega_{11} = \frac{k_1}{m_1} \quad \text{and} \quad \omega_{22} = \frac{k_2}{m_2}$$

$$a_1 = \frac{\dfrac{F_0}{k_1}\left(1 - \dfrac{\omega^2}{\omega_2^2}\right)}{\left(\dfrac{1}{k_2} + \dfrac{1}{k_1} - \dfrac{\omega^2}{k_2\omega_{11}^2}\right)\left(\dfrac{1}{k_1} - \dfrac{\omega^2}{k_1\omega_{22}^2}\right) - \dfrac{k_2}{k_1}}$$

$$a_1 = \frac{\dfrac{F_0}{k_1}\left(1 - \dfrac{\omega^2}{\omega_2^2}\right)}{\left(1 + \dfrac{k_2}{k_1} - \dfrac{\omega^2}{\omega_{11}^2}\right)\left(1 - \dfrac{\omega^2}{\omega_{22}^2}\right) - \dfrac{k_2}{k_1}} \tag{6.135}$$

When $\omega = \omega_{22}$, amplitude $a_1 = 0$, but the absorber mass has a displacement equal to

$$a_2 = -\frac{F_0}{k_2} \tag{6.136}$$

k_2 and m_2 depend on the maximum allowable a_2.

It should be noted that although the vibration absorber is effective at $\omega = \omega_{22}$ there are two natural frequencies of the system on either side of ω_{22} and these have the effect of increasing a_1 at those frequencies. These increases can be controlled to some extent by the addition of damping, but this will decrease the effectiveness of the absorber at ω_{22}.

6.3.2 Isolation of three-dimensional masses

Consider the mass suspended on two springs shown in Figure 6.11.

The motion of the mass is likely to be 'coupled' – both vertical displacement and a rotation. Solutions to the coupled two degree of freedom model may be sought (see Section 5.4.3).

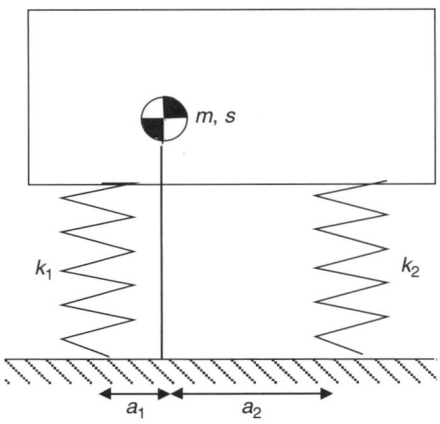

Figure 6.11 The isolated mass.

The uncoupled vertical natural frequency is given by

$$\omega_1 = \sqrt{\frac{k_1 + k_2}{m}} \tag{6.137}$$

and the corresponding uncoupled natural frequency of rotation is given by:

$$\omega_2 = \sqrt{\frac{k_1 a_1{}^2 + k_2 a_2{}^2}{J}} \tag{6.138}$$

where J is the polar moment of area through the centre of mass and perpendicular to the page.

In reality, the two natural frequencies differ from the uncoupled cases as the motion involves both vertical displacement and a rocking motion. One of the actual natural frequencies will fall below both f_1 and f_2 and the other will fall above both f_1 and f_2. The effect of coupling therefore increases the spread in natural frequencies.

The presence of coupled motion makes the choice of spring stiffness difficult if one wants to ensure that all natural frequencies fall below the excitation frequency. To avoid this complication, the spring rates may be chosen so that they undergo the same static deflection under the loads that they experience (different height springs keep the machine level). Then ω_1 and ω_2 are the actual natural frequencies, and vertical forces at the springs or at the centre of mass produce no rotational motion.

6.3.3 Real non-ideal isolators

Real isolators are not massless as assumed in the foregoing classical analysis. Transmissibility will increase (isolation decrease) at frequencies corresponding to resonances of the isolator. Springs are not likely to be linear for large deflections (they become stiffer with increasing deflection).

The dynamic stiffness of rubber is greater than the static stiffness. Rubber is stiffer in compression than in shear (the use of this property is illustrated in the resilient mountings sketched in Figure 6.12). Voids in rubber blocks allow for a two-stage stiffness. Hydraulic mounts give additional damping (see Section 6.4.6). Typical isolator stiffness values are shown in Table 6.2.

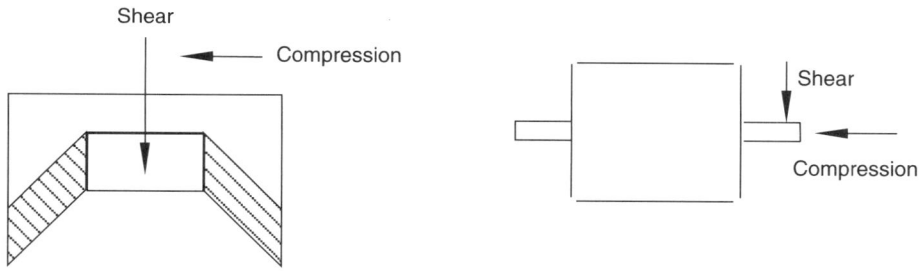

Figure 6.12 Some rubber mounts exploiting the difference between shear and compression stiffness.

Table 6.2 Typical isolator stiffness values

	Deflection (mm)	f_o (Hz)
Cork or Felt	0.1–0.5	50–25
Rubber	0.1–10	50–5
Metal Springs	5–50	5–1

Spring stiffness can be calculated for wire-wound springs, thus:

$$k = \frac{Gd^4}{64nR^3} \quad G = 80\,\text{GN m}^{-2} \text{ for steel (shear modulus)}$$

Where R is the radius of the coil and d is the thickness of the wire.
To calculate total spring stiffness for springs in series:

$$\frac{1}{k} = \frac{1}{k_1} + \frac{1}{k_2} \tag{6.139}$$

To calculate total spring stiffness for springs in parallel:

$$k = k_1 + k_2 \tag{6.140}$$

6.3.4 Vibration isolation at higher frequencies

The classical methods for determining the effectiveness of vibration isolation discussed so far assume:

- a perfectly rigid mass;
- a perfectly rigid (albeit moving) support to the resilient mounts;
- a stiffness for the isolator that does not vary with frequency.

These simplifying assumptions hold at low frequencies but become increasingly unrealistic with increasing frequency.

The effectiveness of an isolator at high frequencies can be determined using the differential mobility across the isolator (Fahy and Walker, 1998).

$$\text{Differential mobility} = \frac{\tilde{v}_2 - \tilde{v}_1}{\tilde{F}} \tag{6.141}$$

where $\tilde{v}$ is the complex vibration velocity and $\tilde{F}$ is the complex force as shown in Figure 6.13.
In the general case,

$$m\ddot{x} + c\dot{x} + kx = F_0 e^{i\omega t} \tag{6.142}$$

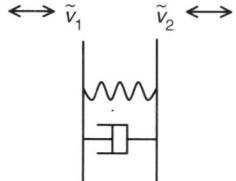

Figure 6.13 The differential mobility of an isolator.

and as

$$x = Xe^{i\omega t} \tag{6.143}$$

$$\dot{x} = i\omega Xe^{i\omega t} \tag{6.144}$$

$$\ddot{x} = -\omega^2 Xe^{i\omega t} \tag{6.145}$$

then

$$-\omega^2 mx + i\omega cx + kx = F_0 \tag{6.146}$$

Now for a massless isolator as in Figure 6.13, $\tilde{F}_1 = \tilde{F}_2$ and

$$i\omega cx + kx = \tilde{F}_1$$

or

$$c\dot{x} + \frac{k}{i\omega}\dot{x} = \tilde{F}_1 \tag{6.147}$$

where $\dot{x} = \tilde{v}_2 - \tilde{v}_1$

Therefore, the mobility of the isolator is

$$\tilde{Y}_I = \frac{\tilde{v}_2 - \tilde{v}_1}{c + \dfrac{k}{i\omega}(\tilde{v}_2 - \tilde{v}_1)} = \frac{1}{c + \dfrac{k}{i\omega}} \tag{6.148}$$

Now consider a general source of vibration S which is being excited by some applied forces. When the source is in a vacuum, its velocity response to the forces is called the free normal velocity $\tilde{v}_f$ of the surface S.

If S is now attached to an isolator with differential mobility $\tilde{Y}_I$ then a reflected vibrational wave $\tilde{v}_r$ is produced within S. The total normal velocity on the surface S is now $\tilde{v}_f - \tilde{v}_r$. The reflected velocity must be the product of the force and the source mobility $\tilde{Y}_s$

$$\tilde{v}_r = F\tilde{Y}_s \tag{6.149}$$

Therefore,

$$\tilde{v}_f - \tilde{v}_r = \tilde{v}_f - F\tilde{Y}_s \tag{6.150}$$

The velocity at the other end of the isolator $\tilde{v}_t$ is given by

$$\tilde{v}_t = F\tilde{Y}_r \qquad (6.151)$$

where $\tilde{Y}_r$ is the mobility of the body receiving the vibration from the source, transmitted through the isolator.

From the definition of the differential mobility of the isolator one can now write

$$\tilde{Y}_I = \frac{(\tilde{v}_f - \tilde{v}_r) - \tilde{v}_t}{F} \qquad (6.152)$$

Re-arranging equation (6.152)

$$\tilde{Y}_I F = (\tilde{v}_f - \tilde{v}_r) - \tilde{v}_t \qquad (6.153)$$

and substituting equation (6.150) into equation (6.153)

$$\tilde{Y}_I F = \tilde{v}_f - F\tilde{Y}_s - \tilde{v}_t \qquad (6.154)$$

then substituting equation (6.151) into equation (6.154)

$$\tilde{Y}_I F = \tilde{v}_f - F\tilde{Y}_s - F\tilde{Y}_r$$

so,

$$F = \frac{\tilde{v}_f}{\tilde{Y}_I + \tilde{Y}_r + \tilde{Y}_s} \qquad (6.155)$$

Substituting equation (6.151) into equation (6.155) gives

$$\frac{\tilde{v}_t}{\tilde{Y}_r} = \frac{\tilde{v}_f}{\tilde{Y}_I + \tilde{Y}_r + \tilde{Y}_s}$$

so,

$$\frac{\tilde{v}_t}{\tilde{v}_f} = \frac{\tilde{Y}_r}{\tilde{Y}_I + \tilde{Y}_r + \tilde{Y}_s} \qquad (6.156)$$

When the isolator is replaced by a perfectly rigid connection $\tilde{Y}_I = 0$ and then

$$\frac{\tilde{v}_t}{\tilde{v}_f} = \frac{\tilde{Y}_r}{\tilde{Y}_r + \tilde{Y}_s} \qquad (6.157)$$

Taking the effectiveness of the isolator as

$$E = \frac{\left(\dfrac{\tilde{v}_t}{\tilde{v}_f}\right)_{\tilde{Y}_I=0}}{\left(\dfrac{\tilde{v}_t}{\tilde{v}_f}\right)_{\tilde{Y}_I \neq 0}} \qquad (6.158)$$

$$E = \frac{\dfrac{\tilde{Y}_r}{\tilde{Y}_r + \tilde{Y}_s}}{\dfrac{\tilde{Y}_r}{\tilde{Y}_I + \tilde{Y}_r + \tilde{Y}_s}} = \frac{\tilde{Y}_r}{\tilde{Y}_r + \tilde{Y}_s}\left[\frac{\tilde{Y}_I + \tilde{Y}_r + \tilde{Y}_s}{\tilde{Y}_r}\right] = \frac{\tilde{Y}_I + \tilde{Y}_r + \tilde{Y}_s}{\tilde{Y}_r + \tilde{Y}_s} = \left|1 + \frac{\tilde{Y}_I}{\tilde{Y}_r + \tilde{Y}_s}\right| \qquad (6.159)$$

So for good isolation $|\tilde{Y}_I| \gg |\tilde{Y}_r + \tilde{Y}_s|$. Remembering that

$$\tilde{Y}_I = \frac{1}{c + \dfrac{k}{i\omega}}$$

this requires

- low stiffness for the isolator;
- low damping in the isolator.

Now from equation (6.146)

$$-\omega^2 m x + i\omega c x + kx = F_0$$

and

$$x = \frac{\dot{x}}{i\omega}$$

the impedance Z is thus obtained:

$$Z = \frac{F}{x} = \frac{1}{Y} = c + \frac{k}{i\omega} - \frac{m\omega}{i}$$

and so the mobility of a rigid mass m is given by

$$Y_M = \frac{1}{i\omega m} \qquad \text{in accordance with equation (6.148)}.$$

The impedance of an infinite, thick plate of thickness h, Young's Modulus E and density ρ is given approximately by (Fahy, 1985)

$$Z_\infty = 8\sqrt{B'm''} \tag{6.160}$$

where B' is the bending stiffness per unit width $B' = \dfrac{h^3}{12}E$ and m'' is the mass per unit area of the panel ($m'' = \rho h$).

6.4 Engine and drivetrain vibrations

At vehicle ride frequencies (commonly taken to be frequencies below the first torsional mode of the chassis, say below 30 Hz) the vehicle acts as a rigid six degree of freedom system (vertical, lateral, longitudinal displacements along with pitch, roll and yaw motions). In addition, the engine and transmission assembly behaves as a simple six degree of freedom system at frequencies below the first powertrain-bending mode (commonly below 150 Hz). Both engine and vehicle have their own sets of natural (eigen) frequencies. If rectilinear forces act on either system then a vibration response is caused. When a frequency component of force coincides in frequency with a system eigen frequency then a particularly large (resonant) response is initiated, controllable only by the level of damping present in the system at that frequency.

Rules of thumb have evolved for tuning the eigen frequencies for these two systems, such as:

- In order to isolate the vehicle from powertrain vibrations, the powertrain eigen frequencies should be outside the frequency range of the main excitation force.
- For acceptable secondary ride (commonly taken to be vehicle/powertrain vibration phenomena in the frequency range of 5–30 Hz) quality, engine eigen frequencies should be separated from wheel/hub/suspension eigen frequencies unless hydraulic engine mounts are used (with high levels of damping in the region of one or more eigen frequencies) when the engine can be used to act as a dynamic absorber to the resonant suspension motion.
- For acceptable durability, engine roll eigen frequencies should not coincide with the vehicle longitudinal (the so-called shuffle) eigen frequencies.

Lee et al. (1995) extend these rules to the point where engine/powertrain eigen frequencies should not coincide with any flexible eigen frequencies of the chassis. The powertrain and its driveline are rotational systems and therefore there are rotational eigen frequencies to consider in addition to the rectilinear eigen frequencies discussed so far. Balfour et al. (2000) suggest that of the many parameters that influence the dynamics of the rotating powertrain/driveline system, the following are most important:

- engine block inertia;
- the inertia of the flywheel and reciprocating mass;
- the inertia of axles and wheels;
- the stiffness of the three effective interconnecting shafts that link the three inertias above;
- overall gear ratio.

Hodgetts (1982) adds the torsional stiffness of the tyres to the above list and Ambrosi and Orofino (1992) add the stiffness and damping characteristics of the clutch. The torsional driveline system is excited by a fluctuating torque at the crankshaft, and as transferred to the flywheel. The magnitude of this fluctuation is directly affected by:

- the number of cylinders in the engine;
- the firing order;
- the firing intervals;
- combustion stability and cycle-to-cycle variability in mean effective pressures in each cylinder.

The cycle-to-cycle variability is affected in turn by the flywheel inertia (the flywheel is an energy-storage device that regulates the crankshaft rotational speed) and at low mean engine speeds this can help combustion quality by maintaining at all times the turbulent intensity of the air–fuel mixture above the minimum value required to avoid engine stall.

A fluctuating torque at the flywheel causes oscillations of the rotating driving wheels once the clutch is fully engaged. This in turn causes the so-called shuffling of the vehicle in the longitudinal direction with an accompanying pitching motion. This can be sustained by heavy vehicles slowly climbing hills in low gears (Hodgetts, 1982) or during on–off throttle cycles (the so-called tip-in–tip-out maneuvers) (Balfour et al., 2000).

When considering the effects of engine-generated forcing of the powertrain and chassis assemblage, the following issues must be addressed:

- the vertical and lateral forces on the engine structure caused by the gas and inertial forces in the cylinders;
- the moments along the length of the crankshaft caused by these gas and inertial forces;
- the effects of fluctuating torque at the flywheel;
- the effects of fluctuating flywheel speed.

All of the above issues will be addressed in the sections that follow.

6.4.1 Why balance an engine?

When a machine is placed on bearings and is free to rotate, unwanted inertia forces and moments may arise. In the extreme cases these may shake the machine with dangerously high amplitude. In the less extreme cases, vibration response to these forces and moments may cause premature failure due to fatigue and the additional loads placed on the bearings may reduce bearing life.

In addition to these durability issues, imbalance in the engine (in the crank mechanism and in the valve actuation mechanism including the valve timing system) is one contributor to the lower frequency (<500 Hz) forcing of the engine and this forcing can give rise to both tactile vibration and audible noise within the vehicle cabin.

In this book, only the imbalance of the crank mechanism will be examined analytically as this mechanism produces greater forcing than the valve gear. Interested readers should refer to Norton (1999) for a treatment of the valve train as a source of vibration and that should be read in conjunction with Section 6.3 of this book.

It should be noted that the crank mechanism may cause inertial forces and moments that result from the layout of the mechanism. In addition, further forces and moments may be caused by non-perfect tolerances in the manufacturing process. These non-perfect tolerances may be deliberate as it is often cheaper to produce a mechanism with naturally varying tolerance, and then pay the modest on cost of balancing each mechanism separately than to invest heavily in producing perfect mechanisms each and every time.

6.4.2 Static and dynamic balance

There are two types of imbalance that may arise with a crankshaft (or any other rotating element): static and dynamic imbalance. To understand static imbalance, imagine a crankshaft taken from an inline four-cylinder engine (Figure 6.14).

Imagine each end of the crankshaft placed on a long knife-edge. If the crankshaft is given a gentle push, it will roll along the knife-edges before coming to rest. If each time this is done, a mark is made on one end of the crankshaft to show where it stopped, a family of marks is quickly constructed. If the marks are bunched together then the

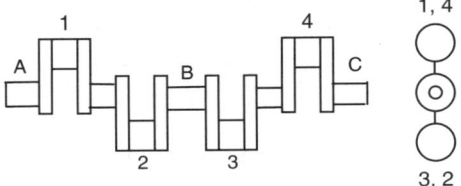

Figure 6.14 Schematic of an inline four-cylinder crankshaft.

crankshaft is in static imbalance. If the marks are randomly distributed, that crankshaft is in static balance.

Now imagine that a small hole is drilled in the outer web of throw #1, and an equal-sized hole is drilled in the same position in the outer web of #3 throw. This would not affect the static balance of the crankshaft. However, when the crankshaft rotates a moment will be set up along the crank axis caused by the heavier throws (#2 and #4) being on opposing sides of the crank centreline and different distances from bearings A and B. For dynamic equilibrium of the crankshaft, this moment must be resisted by additional force at bearings A and B. Crankshaft balance is achieved (at least in part) by using counterweights on either side of each throw as seen in Figure 6.15.

Figure 6.15 Photograph showing a close-up of crankshaft counterweights.

6.4.3 The forces acting on an individual crank throw

It is convenient to assume four different forces act on each throw of a crankshaft. These are:

1. the gas force;
2. the inertia force due to the mass of the piston assembly;
3. the inertia force due to the mass of the conrod assigned to the piston-pin end of the conrod;
4. the centrifugal force due to the rotating mass of the crank assembly (the crankshaft plus the proportion of the conrod mass assigned to the crank assembly).

When the behaviour of the crank system is thought to be linear, these forces add together by superposition with their relative phases being dictated by the geometry of the crank assembly. The principle of superposition may also be used to sum the forces and moments caused by individual crank throws on a multi-cylinder crankshaft.

In this section, the magnitude and phase of the forces (1–4) will be quantified for a single-throw crankshaft. In the next section these forces will be summed for a multi-throw crankshaft and any moments caused will be introduced. The analysis follows the pattern adopted by Uicker et al. (2003) and the interested reader can find alternative patterns in Taylor (1985) and in Zweiri et al. (1999).

6.4.3.1 Gas force

By inspection of Figure 6.16 the following relationship is found

$$x = a+b \tag{6.161}$$

$$= r \cos \omega t + \ell \cos \varphi \tag{6.162}$$

Also it can be written that

$$r \sin \omega t = c \tag{6.163}$$

$$\ell \sin \varphi = c \tag{6.164}$$

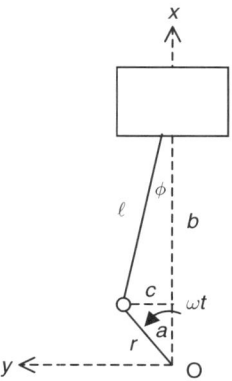

Figure 6.16 An individual crank throw.

and therefore

$$r \sin \omega t = \ell \sin \varphi \tag{6.165}$$

$$\sin \varphi = \frac{r}{\ell} \sin \omega t \tag{6.166}$$

Remembering that

$$(\cos \varphi)^2 = 1 - (\sin \varphi)^2 \tag{6.167}$$

$$\cos \varphi = \sqrt{1 - (\sin \varphi)^2} \tag{6.168}$$

From equation (6.162)

$$x = r \cos \omega t + \ell \left(1 - (\sin \varphi)^2\right)^{1/2} \tag{6.169}$$

Substituting equation (6.166) into equation (6.169)

$$x = r \cos \omega t + \ell \sqrt{1 - \left(\frac{r}{\ell} \sin \omega t\right)^2} \tag{6.170}$$

The lateral piston ring force F_{PR} is given by (see Figure 6.17)

$$F_{PR} = PA_P \tan \varphi \tag{6.171}$$

Remembering that

$$\tan \varphi = \frac{\sin \varphi}{\cos \varphi} \tag{6.172}$$

and from equations (6.166) and (6.168)

$$\tan \varphi = \frac{\frac{r}{\ell} \sin \omega t}{\sqrt{1 - \left(\frac{r}{\ell} \sin \omega t\right)^2}} \tag{6.173}$$

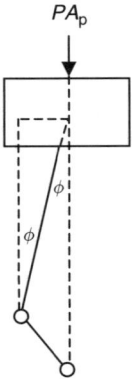

$$PA_P$$

Figure 6.17 The geometry of gas forces acting on a piston.

then

$$F_{PR} = P A_P \left[\frac{\frac{r}{\ell} \sin \omega t}{\sqrt{1 - \left(\frac{r}{\ell} \sin \omega t\right)^2}} \right] \tag{6.174}$$

The torque delivered to the crankshaft by the gas force is given by:

$$T_g = F_{PR} x \tag{6.175}$$

Substituting equations (6.174) and (6.170) into equation (6.175)

$$T_g = P A_P \left[\frac{W}{\sqrt{1 - W^2}} \right] \left[r \cos \omega t + \ell \sqrt{1 - W^2} \right] \tag{6.176}$$

where

$$W = \frac{r}{\ell} \sin \omega t \tag{6.177}$$

The force in the x direction on the small-end bearing F_{bs} is given by

$$F_{bs} = \frac{P A_P}{\cos \varphi} \tag{6.178}$$

From equations (6.176), (6.177) and (6.168)

$$F_{bs} = \frac{P A_P}{\sqrt{1 - W^2}} \tag{6.179}$$

In accordance with Newton's third law, the magnitude of the big-end bearing force is the same as that of the small end with opposing direction. It should be noted that whilst this is true for the case of gas-generated forces, it is not strictly true for the case of inertial forces, due to the influence of the simplifying assumption of apportioning some of the conrod mass to the piston and the remainder to the crankshaft.

Equations (6.174), (6.179) are forces and (6.176) a torque due to the gas loads alone. There are also inertial loads to be considered and these are discussed next. Sometimes the gas forces relieve the bearing inertial forces and sometimes they reinforce them.

6.4.3.2 Inertial forces

There are two classes of inertial force:

1. Those due to the total reciprocating force (caused by the mass of the piston and some of the conrod). These cannot be overcome using crankshaft counterweights but they can be minimised.
2. Those due to the effective rotating mass (including a portion of the conrod mass) assuming that the crankshaft itself is dynamically balanced and so generates no net force itself. These can be counteracted using counterbalance weights on the crankshaft.

The commonly used method of partitioning the conrod mass is as follows. The partition of mass is an approximation but one that is very frequently made as it allows for a considerable simplification in the calculation of the inertia forces. The reason for this is that the big end of the conrod moves on a circle and the little end moves on a straight line and both of these motions are easy to analyse. The centre of the mass of the conrod is of course somewhere between the big end and the small end and its motion is more complex. In addition, the true conrod has a particular moment of inertia, and this cannot be replicated accurately when the conrod mass is partitioned and so care should be taken when calculating the big- and small-end bearing forces caused by the inertia of the piston and its conrod.

The conrod shown in Figures 6.16 and 6.18 has mass and rotational inertia. There is a centre of gravity at which the mass can be considered to be centred. With the total mass of the conrod partitioned at the big and small ends respectively, Figure 6.19 is redrawn as Figure 6.20.

The following relationships can be written directly:

$$m_{cr} = m_{se} + m_{be} \tag{6.180}$$

$$m_{be}\ell_a = m_{se}\ell_b \tag{6.181}$$

Solving equations (6.180) and (6.181) simultaneously

$$m_{se} = m_{cr} - m_{be}$$

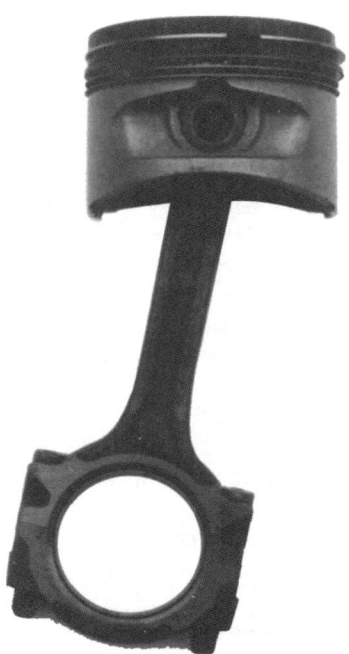

Figure 6.18 Photograph showing a typical piston and conrod assembly for a gasoline engine.

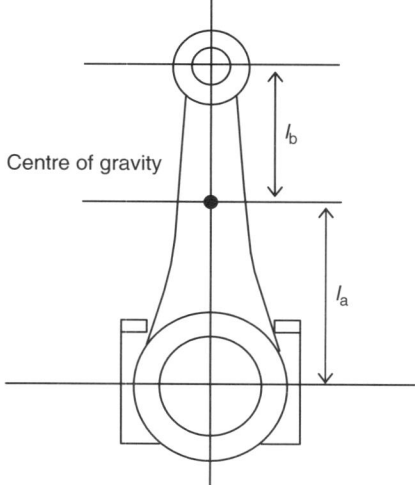

Figure 6.19 The geometry of a conrod.

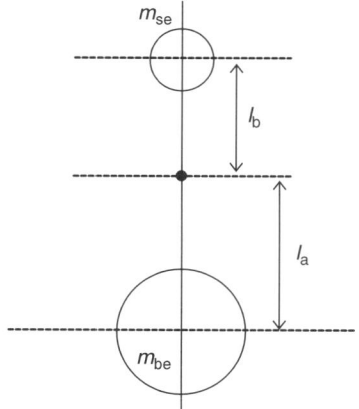

Figure 6.20 Distributing the effective mass of a conrod.

$$m_{be}\ell_a = m_{cr}\ell_b - m_{be}\ell_b$$

$$m_{be}[\ell_a + \ell_b] = m_{cr}\ell_b$$

$$m_{be} = m_{cr}\frac{\ell_b}{\ell_a + \ell_b} \tag{6.182}$$

and by inspection

$$m_{se} = m_{cr}\frac{\ell_a}{\ell_a + \ell_b} \tag{6.183}$$

Often it is assumed that two-thirds of the conrod mass is concentrated at the big end and one-third at the small end. The one disadvantage with the above analysis is that there is no attempt to ensure that the conrod rotational inertia is correctly described. Uicker et al. (2003) attempted to include this in their analysis but it was later excluded. Taylor (1985) includes a discussion on the effects of the true moment of inertia of the conrod but only when calculating the torque generated by the angular acceleration of the conrod.

The rotation of the big-end partitioned mass m_{be} is given by

$$x(t) = r\cos\omega t\,\bar{i} + r\sin\omega t\,\bar{j} \qquad (6.184)$$

where r is the crank throw, $\bar{i}$ is the unit vector in the x direction and $\bar{j}$ is the unit vector in the y direction.

Differentiating equation (6.184) with respect to time gives the velocity of the big end

$$\bar{v}(t) = -r\omega\sin\omega t\,\bar{i} + r\omega\cos\omega t\,\bar{j} \qquad (6.185)$$

Assuming that $\bar{v}(t)$ is kept constant in the analysis and differentiating equation (6.185) with respect to time gives the acceleration of the big end mass

$$\bar{a}(t) = -r\omega^2\cos\omega t\,\bar{i} - r\omega^2\sin\omega t\,\bar{j} \qquad (6.186)$$

Remembering that the linear displacement of the small-end mass is

$$x = r\cos\omega t + \ell\sqrt{1 - \left(\frac{r}{\ell}\sin\omega t\right)^2}$$

this can be simplified as follows; first, the term under the square root is expanded using the binomial theorem

$$\left(1 - \left(\frac{r}{\ell}\sin\omega t\right)^2\right)^{1/2} = \left(1 - \frac{r^2}{\ell^2}\sin^2\omega t\right)^{1/2}$$

$$\left(1 - \frac{r^2}{\ell^2}\sin^2\omega t\right)^{1/2} = 1^{1/2} + \frac{1}{2}(1)^{-1/2}\left[-\frac{r^2}{\ell^2}\sin^2\omega t\right]$$

$$+ \left[\frac{\frac{1}{2}\left(\frac{1}{2}-1\right)}{2!}\right](1)^{-1.5}\left(-\frac{r^2}{\ell^2}\sin^2\omega t\right)^2 + \cdots$$

Neglecting all but the first two terms

$$\left(1 - \left(\frac{m}{\ell}\sin\omega t\right)^2\right)^{1/2} = 1 - \frac{r^2}{2\ell^2}\sin^2\omega t \qquad (6.187)$$

and using the standard trigonometry relation

$$2\sin^2\omega t + \cos 2\omega t = 1$$

equation (6.187) becomes

$$\left(1 - \left(\frac{r}{\ell}\sin\omega t\right)^2\right)^{1/2} = 1 - \frac{r^2}{2\ell^2}\left[\frac{1-\cos 2\omega t}{2}\right]$$

$$= 1 - \frac{r^2}{4\ell^2} + \frac{r^2}{4\ell^2}\cos 2\omega t \qquad (6.188)$$

Equation (6.170) can now be written as

$$x = r \cos \omega t + \ell - \frac{r^2}{4\ell} + \frac{r^2}{4\ell} \cos 2\omega t$$

$$x = \ell - \frac{r^2}{4\ell} + r \left(\cos \omega t + \frac{r}{4\ell} \cos 2\omega t \right) \qquad (6.189)$$

Differentiating equation (6.189) with respect to time gives the velocity of the small-end mass

$$\dot{x} = -r\omega \sin \omega t - \frac{r^2}{2\ell} \omega \sin 2\omega t \qquad (6.190)$$

Differentiating equation (6.190) again gives the acceleration

$$\ddot{x} = -r\omega^2 \cos \omega t - \frac{r^2 \omega^2}{\ell} \cos 2\omega t \qquad (6.191)$$

assuming that the velocity does not change with time in the analysis.

The rotational acceleration of the big-end mass (equation [6.186]) as well as the linear acceleration of the small-end mass plus the mass of the piston (equation [6.191]) are now known. Therefore, the magnitude of the total inertia force in the x direction is

$$F_{xi} = -m_{be} \, r\omega^2 \cos \omega t - (m_{se} + m_p) \left[r\omega^2 \cos \omega t + \frac{r^2 \omega^2}{\ell} \cos 2\omega t \right]$$

$$F_{xi} = -(m_{be} + m_{se} + m_p)(r\omega^2 \cos \omega t) - (m_{se} + m_p) \left(\frac{r^2 \omega^2}{\ell} \cos 2\omega t \right) \qquad (6.192)$$

The total inertia force in the y direction is

$$F_y = -m_{be} \, r\omega^2 \sin \omega t \qquad (6.193)$$

The inertia torque imposed on the crankshaft is due to the reciprocating part only and is equal to

$$T_i = -(m_{se} + m_p) \ddot{x} \tan \varphi x \qquad (6.194)$$

Substituting equations (6.173), (6.189) and (6.191) into equation (6.194) yields

$$T_i = -(m_{se} + m_p)(r\omega^2) \left(\cos \omega t + \frac{r}{\ell} \cos 2\omega t \right) \left[\frac{W}{\sqrt{1 - W^2}} \right] \times \left[\ell - \frac{r^2}{4\ell} \right.$$

$$\left. + r \left(\cos \omega t + \frac{r}{4\ell} \cos 2\omega t \right) \right] \qquad (6.195)$$

where

$$W = \frac{r}{\ell} \sin \omega t$$

To summarise, the x direction forces acting for a single throw of a crankshaft are:

Gas force

$$F_{bs} = \frac{PA_P}{\sqrt{1-W^2}} \qquad (6.179)$$

Rotational inertia force

$$F_{rot,x} = -m_{be}\, r\omega^2 \cos \omega t \qquad (6.196)$$

Reciprocating inertia force

$$F_{rec,x} = -\left(m_{se} + m_p\right)\left[r\omega^2 \cos \omega t + \frac{r^2\omega^2}{\ell}\cos 2\omega t\right] \qquad (6.197)$$

The rotational inertial force also has a component in the horizontal direction given by

$$F_{rot,y} = -m_{be}\, r\omega^2 \sin \omega t \qquad (6.198)$$

When considering the balancing of multi-cylinder engines, equations (6.196), (6.197) and (6.198) can be written as

$$F_{total,\,rot,\,x} = \sum_{n=1}^{n} m_{be}\, r\omega^2 \cos\left(\theta_1 - \delta_n\right) \qquad (6.199)$$

where θ_1 is the crank angle at cylinder #1 and δ_n is the angle between the nth crank and crank #1.

$$F_{total,rot,\,y} = \sum_{n=1}^{n} m_{be}\, r\omega^2 \sin\left(\theta_1 - \delta_n\right) \qquad (6.200)$$

$$F_{total,rec,x} = \sum_{n=1}^{n} \left(m_{se} + m_p\right)\left(r\omega^2 \cos\left(\theta - \delta_n\right) + \frac{r^2\omega^2}{\ell}\cos 2\left(\theta_1 - \delta_n\right)\right) \qquad (6.201)$$

Equations (6.199) and (6.201) are multiplied by $\cos(\alpha/2)$ for the case of a $\alpha°$ vee-engine. Equation (6.200) is multiplied by $\sin(\alpha/2)$ for the cylinders on one bank and by $-\sin(\alpha/2)$ for the cylinders on the other bank.

6.4.3.3 *Moments along the crankshaft*

In addition to these inertial forces (equations (6.199), (6.200) and (6.201)) there are moments about both the x and y axes. These are taken about the centreline of the crankshaft in both vertical and horizontal directions. By way of example the total forces and moments can be found for the inline, four-cylinder crankshaft shown in Figure 6.21. From equation (6.199) for $\theta_1 = 0$

$$F_{total,\,rot,\,x} = m_{be}\, r\omega^2 \left[\cos 0° + \cos\left(0° - 540°\right) + \cos\left(0° - 180°\right) + \cos\left(0° - 360°\right)\right]$$

$$= m_{be}\, r\omega^2 \left[1 - 1 - 1 + 1\right] = 0$$

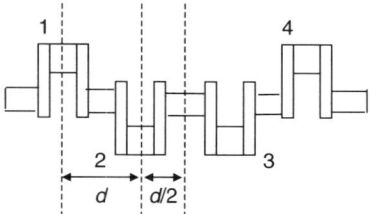

Figure 6.21 Crankshaft geometry details.

From equation (6.200)

$$F_{total, rot, y} = m_{be}\, r\omega^2\, [\sin(0°) + \sin(0° - 540°) + \sin(0° - 180°) + \sin(0° - 360°)]$$
$$= m_{be}\, r\omega^2\, [0 + 0 + 0 + 0] = 0$$

From equation (6.201)

$$F_{total, rec, x} = (m_{se} + m_p)\, [r\omega^2]\, [\cos(0°) + \cos(0° - 540°) + \cos(0° - 180°)$$
$$+ \cos(0° - 360°)] + (m_{se} + m_p)\left(\frac{r^2\omega^2}{\ell}\right)[\cos(0°) + \cos(0° - 1080°)$$
$$+ \cos(0° - 360°) + \cos(0° - 720°)]$$

$$F_{total, rec, x} = (m_{se} + m_p)\, (r\omega^2)\, [1 - 1 - 1 + 1] + (m_{se} + m_p)\left(\frac{r^2\omega^2}{\ell}\right)[1 + 1 + 1 + 1]$$
$$= 4\, (m_{se} + m_p)\left(\frac{r^2\omega^2}{\ell}\right)$$

As there is no net rotational inertia force in the x or y planes there will be no net couple either. However, there may be a couple due to the secondary reciprocating force. Referring to Figure 6.21:

$$C_{total, rec, vertical} = (m_{se} + m_p)\left(\frac{r^2\omega^2}{\ell}\right)\left[\frac{-3d}{2}\cos(0°) - \frac{d}{2}\cos(0° - 1080°)\right.$$
$$\left. + \frac{d}{2}\cos(0° - 360°) + \frac{3d}{2}\cos(0° - 720°)\right] = 0$$

Therefore, this example shows that for the inline, four-cylinder crankshaft, there are no couples and only the secondary reciprocating force is left out of balance. This can be balanced by adding two extra contrarotating shafts with a small eccentric mass to each one. The size of each eccentric mass $mb/2$ depends on the radius at which it is placed. The shafts run at a frequency 2ω

$$F_{balancing, x} = m_b\, r_b\, (2\omega)^2 = 4 m_b\, r_b\, \omega^2$$

$$F_{total, rec, x} = 4\, (m_{se} + m_p)\left(\frac{r^2\omega^2}{\ell}\right)$$

$$m_b\, r_b = \frac{(m_{se} + m_p)}{\ell}\, r^2 \qquad\qquad (6.202)$$

Table 6.3 Presence of net forces and moments for common crankshaft configurations with even cylinder spacing and even firing intervals on four-stroke engines

Configuration	Primary forces	Secondary forces	Primary moments	Secondary moments	Firing interval (degrees)
Inline two-cylinder		Vertical	Vertical		360
Inline three-cylinder			Vertical	Vertical	240
Inline four-cylinder		Vertical			180
Inline five-cylinder			Vertical	Vertical	144
Inline six-cylinder					120
90° V 4		Vertical	Horizontal		180
60° V 6					
1-5-3-6-2-4					120
60° V 6			Vertical and	Vertical and	120
1-4-2-5-3-6			horizontal	horizontal	
90° V 8		Horizontal		Vertical	90
single-plane crankshaft					
90° V 8			Vertical and		90
two-plane crankshaft			horizontal		
72° V 10			Vertical and	Vertical and	72
			horizontal	horizontal	

The above analysis method can be repeated for any engine configuration. A summary of results is shown in Table 6.3. The primary forces and moments (those that fluctuate with radial frequency ω) have been separated from the secondary forces and moments (those that fluctuate with radial frequency 2ω) as is customary (Adler, 1986).

6.4.4 Controlling engine torque and engine speed: the flywheel

The torque generated by the gas force was given earlier as

$$T_g = PA_P \left[\frac{W}{\sqrt{1-W^2}} \right] \left[r\cos\omega t + \ell\sqrt{1-W^2} \right] \tag{6.176}$$

where

$$W = \frac{r}{\ell}\sin\omega t \tag{6.177}$$

The inertial torque was also given earlier as:

$$T_i = -\left(m_{se} + m_p \right) \left(r\omega^2 \right) (\cos\omega t$$
$$+ \frac{r}{\ell}\cos 2\omega t) \left[\frac{W}{\sqrt{1-W^2}} \right] \times \left[\ell - \frac{r^2}{4\ell} + r\left(\cos\omega t + \frac{r}{4\ell}\cos 2\omega t \right) \right] \tag{6.195}$$

The total torque delivered by the engine is the sum of gas torque and inertia torque. As both of these include terms in ωt they both are time-varying and as a result the total torque is also time-varying. The total torque delivered by a simple crankshaft in a four-cylinder engine can easily vary by plus or minus 300% every revolution of the crankshaft. Such large torque fluctuations will cause significant fluctuations in engine speed throughout the engine cycle and may also provide strong excitation for potential torsional resonances in the driveline.

The use of a simple mass flywheel is the well-known solution to the control of both fluctuating output torque and fluctuating engine speed.

The ideal engine would deliver a constant torque and hence would operate at a constant speed. The value of that constant torque would be the time average value for the engine cycle, that is the time average of the sum of:

- positive torque delivered by the combustion process; and
- the negative torque required to scavenge the engine, compress the fresh air–fuel mixture prior to combustion and overcome the inherent friction and other parasitic action in the engine.

The flywheel is an energy storage device. It is charged up with energy during the periods when the instantaneous torque is greater than the time average torque (the energy stored being equal to the area under the torque/crankangle diagram which is above the average torque line) and it releases energy to power the engine through periods when the instantaneous torque is less than the time average torque.

The kinetic energy in a rotating flywheel with a moment of inertia I is given by:

$$E = \frac{1}{2} I \omega^2 \tag{6.203}$$

During a period where the flywheel is being charged with energy, the additional energy causes an acceleration of the flywheel α (rad s^{-2}) and writing Newton's second law for such a case (Norton, 1999):

$$T - T_{avg} = I\alpha \tag{6.204}$$

Now as:

$$\alpha = \frac{d\omega}{dt} = \frac{d\omega}{dt} \left(\frac{d\theta}{d\theta} \right) = \omega \frac{d\omega}{d\theta} \tag{6.205}$$

equation (6.204) becomes on substitution

$$\left(T - T_{avg} \right) d\theta = I\omega \, d\omega \tag{6.206}$$

Both sides of equation (6.206) can be integrated between two extreme limits:

1. the crankangle at which ω is a minimum (after a period of sustained energy release by the flywheel);
2. the crankangle at which ω is a maximum (after a period of sustained energy absorption by the flywheel).

Thus:

$$\int_{\theta_1}^{\theta_2} (T - T_{avg}) \, d\theta = \int_{\omega_{min}}^{\omega_{max}} I\omega \, d\omega = \frac{1}{2} I \left(\omega_{max}^2 - \omega_{min}^2 \right) \tag{6.207}$$

where θ_1 is the crankangle at which the speed is a minimum and θ_2 is the crankangle at which the speed is a maximum. Equation (6.197) gives the change in kinetic energy ΔE of the flywheel between the extremes of minimum and maximum rotational speed. Equation (6.207) can be thus factored:

$$\Delta E = \frac{1}{2} I \left(\omega_{max}^2 - \omega_{min}^2 \right) = \frac{1}{2} I \left(\omega_{max} + \omega_{min} \right) \left(\omega_{max} - \omega_{min} \right) \tag{6.208}$$

A fairly reliable estimate of the average speed of an engine is found from

$$\omega_{avg} = \frac{(\omega_{max} + \omega_{min})}{2} \tag{6.209}$$

Substituting equation (6.209) into equation (6.208)

$$\Delta E = I\omega_{avg} \left(\omega_{max} - \omega_{min} \right) \tag{6.210}$$

Declaring a coefficient of fluctuation (k) as

$$k = \frac{(\omega_{max} - \omega_{min})}{\omega_{avg}} \tag{6.211}$$

Equation (6.211) can be substituted into (6.210) and re-arranged to give

$$I = \frac{\Delta E}{k\omega_{avg}^2} \tag{6.212}$$

This simple expression, along with a numerical integration of equation (6.207) allows the engine designer to select a flywheel inertia that will produce a fluctuation in speed equal to

$$k\omega_{avg} = \omega_{max} - \omega_{min} \tag{6.213}$$

Typical values for k are in the range of 0.01–0.1. A modern lightweight flywheel from a one litre I-4 gasoline engine is shown in Figure 6.22. It should be noted that the flywheel effect can be achieved by adding balanced inertia to either end of the crankshaft or to both ends simultaneously. The lightweight flywheel shown in Figure 6.22 is possible because the engine has a large timing belt pulley fitted to the free end of the crankshaft and this incorporates a torsional damper as well and also because a massive and balanced clutch assembly (also shown in Figure 6.22) adds to the flywheel inertia.

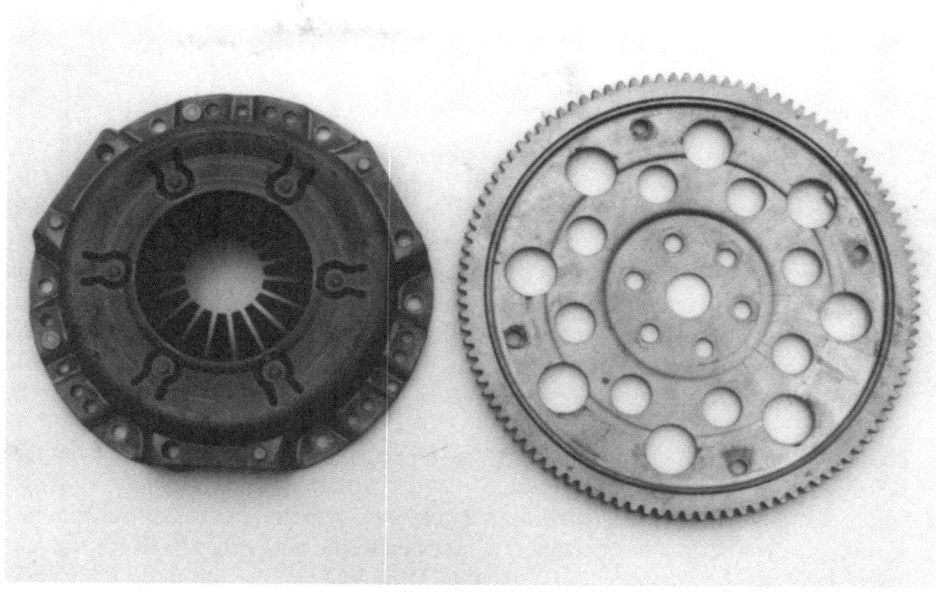

Figure 6.22 A modern lightweight flywheel and clutch assembly for a small I-4 gasoline engine.

Inspection of equation (6.212) reveals that for a given value of k, the inertia required to reduce the speed fluctuations to the correct level is greater at low engine speeds than at higher engine speeds. Therefore, with a practical engine design, the flywheel should be sized in accordance with some sensible compromise between a large inertia to smooth out speed and torque fluctuations at low speeds that is excessively large at higher speeds and an alternative where a smaller flywheel is inadequate at low speeds.

The coefficient of fluctuation can be used to quickly compute the effect that a flywheel would have on the torque fluctuations of a given engine. Instantaneous values of flywheel-smoothed torque can be calculated from:

$$T_{\text{smooothed}}(\theta) = T_{\text{avg}} + k\left(T(\theta) - T_{\text{avg}}\right) \qquad (6.214)$$

The ratio between instantaneous values of flywheel-smoothed torque and the average torque can be thus used to find instantaneous values of engine speed:

$$\frac{T_{\text{smoothed}}(\theta)}{T_{\text{avg}}} = \frac{\omega(\theta)}{\omega_{\text{avg}}}$$

$$\omega(\theta) = \omega_{\text{avg}}\left(\frac{T_{\text{smoothed}}(\theta)}{T_{\text{avg}}}\right) \qquad (6.215)$$

The sometime complex relationships shown in equations (6.161)–(6.215) are best illustrated using a case study.

Case study: The 2-litre four-cylinder gasoline engine vs the two-litre six-cylinder gasoline engine

The two engines selected for this case study have been chosen carefully for the following reasons:

- Firstly, in Europe, the inline four-cylinder (I-4) gasoline engine has been the workhorse in family cars for decades, whereas traditionally the six-cylinder (I-6 or V-6) gasoline engine has been synonymous with power and smoothness and hence its adoption has been relatively infrequent. This presents an interesting contrast for a case study from the European perspective, a contrast between the everyday and the special. The 2-litre swept volume has been chosen as a compromise, being at the large end of the spectrum for the I-4 (0.5 litres per cylinder) and at the small end for the I-6 or V-6 (0.33 litres per cylinder).

- Secondly, the six-cylinder engine (the V-6 in particular) has become familiar to those accustomed to full-size family cars in the US Federal market although the swept volume of 2 litres is rather small for this class of vehicle (2.5 or 3 litres would be more typical). The V-6 has become the more fuel-efficient replacement for the much loved 4–5-litre V-8 that would have traditionally powered the full-sized US car. The US consumer is also well accustomed to modestly sized four-cylinder engines (such as a 2.2-litre I-4) as power for mid-sized and compact cars. This background presents an interesting contrast for a case study from the US perspective, a contrast between the driving experience in the full-sized car and the driving experience in the compact. Although the 2-litre swept volume is a little small to be fully typical, it has been retained in order to allow for the European perspective.

- The advent of the so-called world-car (take the Ford Mondeo launched in 1991 as a good early example) meant that the same vehicle platform could be shared between US Federal and European markets. Initially, manufacturers prepared a range of engines where the small ones were destined solely for Europe and the big ones solely for the US and in the middle of the range were a number of engines to be utilised in both markets. The 2.5-litre V-6 gasoline engine is a representative example of such a shared engine as is the 2.2-litre I-4. The case study presented here could be viewed in terms of two viable (if a little small) engine options for the same world-car.

- Until the recent rise in popularity of the 2.5 litre V-6, European six-cylinder engines were mostly inline engines to be used in sports-cars and smaller executive cars. It can be argued that it took the arrival of the Sports Utility vehicles and People Carriers of the 1990s for European customers to grow accustomed to not only a six-cylinder engine in their family car but a Vee engine to boot. This acceptance also coincided with the rise of the world-car which offered manufacturers the opportunity of economies of scale in engine manufacture. Therefore, the case study presented here could be viewed as the European tradition represented by the I-4 opposed by a new-found acceptance for the V-6.

The specifications of the two engines adopted for this case study are shown in Table 6.4. The same swept volume and the same stroke has been retained for both engines in order to make simple comparisons between the vibration performance of each one.

Table 6.4 Comparison of engine data modelled

Inline four-cylinder gasoline engine	*Vee six-cylinder gasoline engine*
2.0-litre swept-volume	2.0-litre swept-volume
Firing order: 1 - 3 - 4 - 2	firing order: 1 - 5 - 3 - 6 - 2 - 4
86-mm bore	three-throw (120-120-120) crankshaft
86-mm stroke	70-mm bore
143.5-mm conrod	86-mm stroke
aluminium alloy piston (mass 372 g)	143.5-mm conrod
steel conrod (mass 647 g)	aluminium alloy piston (mass 303 g)
	steel conrod (mass 647 g)

Equations (6.179), (6.196) and (6.197) have been used in accordance with the methodology described in this chapter to calculate the shaking force produced by the two engines at 6000 rev min^{-1}. The results are shown in Figure 6.23. The results agree with those summarised in Table 6.2 where the I-4 is predicted to produce only secondary shaking force (at a frequency equal to twice the rotational frequency of the crankshaft or second-order (2E) component as it is known) and the V-6 is predicted to be free of all shaking force. The magnitude of the shaking force in the I-4 should be considered as large, being 12 000 N produced in an engine with a mass of only say 100 kg. Such a force-to-inertia ratio would produce considerable engine motion if it were not for the fact that the frequency of excitation is relatively high (200 Hz in this case) and the effects of the engine mount stiffness providing a restraining counter-force. Notwithstanding this, there will be considerable vibration at the engine side of the mounts on the I-4 to cause structure-borne engine noise in the cabin and this will be completely absent with the (idealised) V-6.

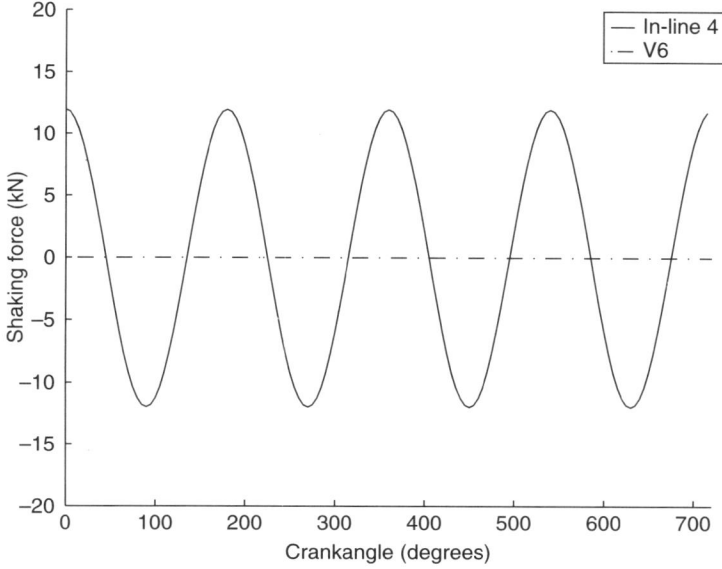

Figure 6.23 I-4 vs V-6 shaking forces (6000 rev min^{-1}).

It is timely to point out that although the engine vibration analysis presented in this chapter provides useful insight into the problem, it is by no means a complete treatment which would merit several chapters. In particular, the analysis neglects the following:

- The effects of rotational imbalance. The results shown in Figure 6.23 in effect are based on the assumption of perfect balancing of the crankshaft.
- The effects of the rotational inertias presented in the system by the timing gear, the camshaft, the valve gear, ancillaries such as alternators and pumps.
- The effects of friction forces.

Notwithstanding these simplifications, the relative refinement of the V-6 compared with the equivalent I-4 is obvious.

The V-6 is commonly associated with 'smoothness'. The attribute of 'smoothness' is most likely linked to the relative weakness of torsional vibrations in the driveline in the V-6 when compared with the equivalent I-4. The relative weakness is due to the fluctuations in torque inherent in any engine being smaller in the V-6 than in the I-4. In order to illustrate this, the gas and inertia torques have been calculated for both engines. In order to calculate the gas torques, engine performance simulation has been used to generate realistic gas-pressure time histories in each cylinder. The engine performance simulation has been undertaken using AVL's *Boost* program, a time-marching simulation software that allows for the calculation of the fluid and thermodynamics of all types of IC engine. Figure 6.24 shows a schematic diagram of the Boost model used for the I-4 engine. The intake system is laid out on the left and top of the diagram and this includes an air-cleaner, a Helmholtz resonator, a plenum chamber and a centre-feed intake plenum. The four cylinders are shown in the centre of the diagram. The exhaust system with its two silencer volumes and four-into-two-into-one exhaust manifold is shown at the bottom and on the right of the diagram. The Boost model predicted peak cylinder pressures of around 50 bar at $6000\,\mathrm{rev\,min}^{-1}$.

Equations (6.176) and (6.195) have been used in accordance with the methodology described in this chapter to calculate the total torque (the sum of gas torque and inertia torque) produced by the two engines at $6000\,\mathrm{rev\,min}^{-1}$. No flywheel-smoothing effect is included in this analysis. The results are shown in Figure 6.25. A few observations are offered. Firstly, the time average torque from the two engines is almost the same. Secondly, it is clear that the torque fluctuates less in the V-6 than in the I-4, confirming the perception of smoothness in the V-6 (although with a firing order different to that carefully chosen here, it is possible to get more fluctuations in a V-6 than an I-4 at certain engine speeds). Thirdly, for the I-4, the torque fluctuates predominantly at 2E (200 Hz at $6000\,\mathrm{rev\,min}^{-1}$) with traces of other frequency components whilst the torque from the V-6 fluctuates quite purely at 3E or 300 Hz.

Equation (6.214) has been used to calculate the total torque produced by the two engines at $6000\,\mathrm{rev\,min}^{-1}$ and smoothed using a flywheel with coefficient of fluctuation of 0.05. The result is shown in Figure 6.26.

The case study ends here.

The calculated flywheel-smoothed torques shown in Figure 6.26 lead to a discussion of the torsional response of the powertrain (driveshafts, gearbox, crankshaft) to fluctuating torque.

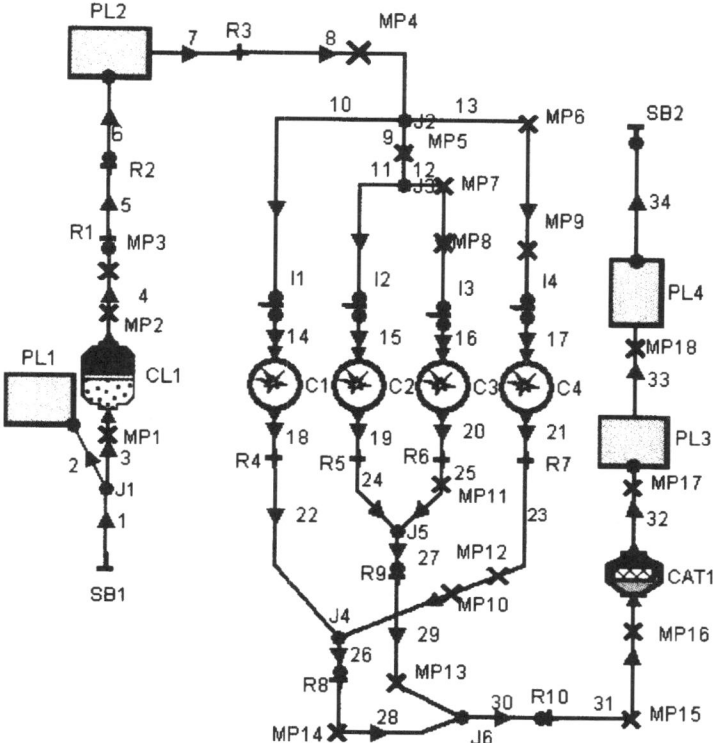

Figure 6.24 Boost model for the I-4 engine.

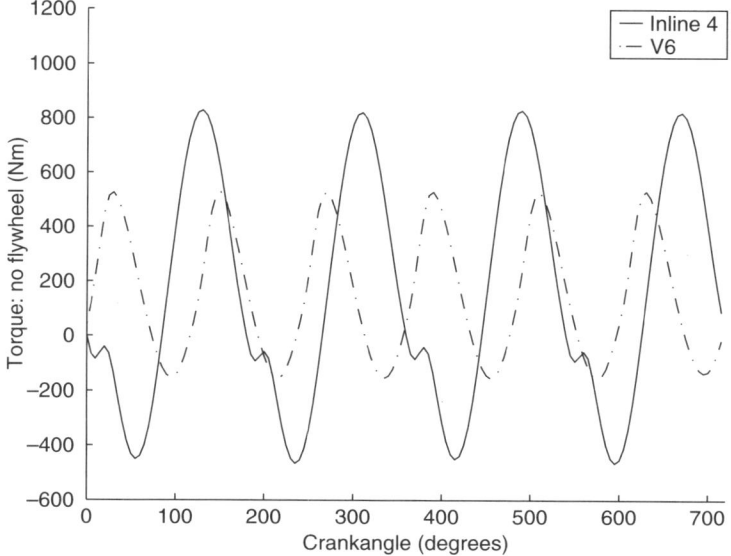

Figure 6.25 I-4 vs V-6 fluctuating torque: no flywheel (6000 rev min^{-1}).

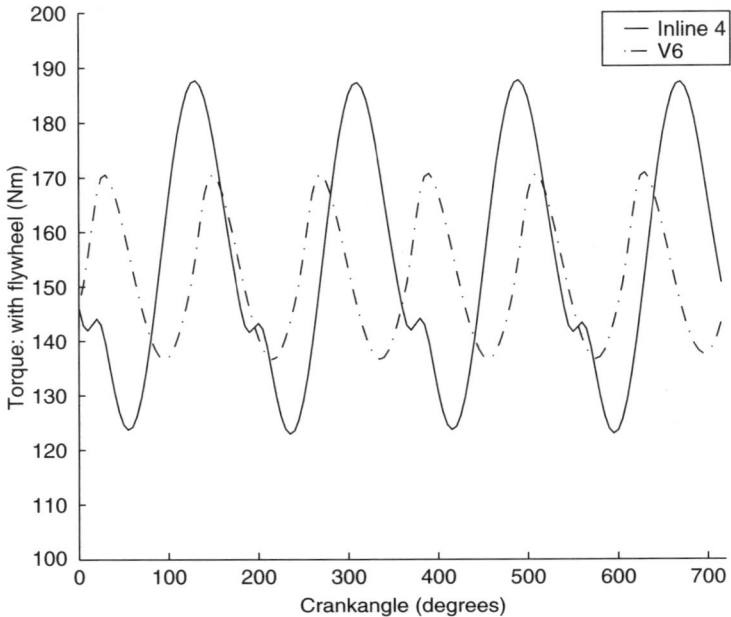

Figure 6.26 I-4 vs V-6 fluctuating torque: with $k = 0.05$ flywheel (6000 rev min^{-1}).

6.4.5 Modes of drivetrain and crankshaft vibration

The two engines investigated in the earlier case study both produced fluctuating torque levels as indeed would any other practical engine. Such fluctuating torque can excite torsional vibrations in the crankshaft, in the engine as an assembly and in the drivetrain as a whole. The drivetrain as a whole is taken to include (in strict sequence):

- the crankshaft (including the effects of any vibration dampers, the timing gear, the camshaft(s) and ancillaries;
- the flywheel (and clutch or torque converter);
- the input shaft to the gearbox (including the gears attached and any auxiliary gears or shafts);
- the output shaft of the gearbox (including the gears attached and any propeller shaft);
- the differential;
- the driveshafts;
- the driven hubs, brakes, road wheels and tyres.

Drivetrain oscillations (torsional vibrations) reduce the driveability of the vehicle as discussed earlier in this chapter. Perhaps more seriously, crankshaft oscillations can produce cyclic stresses in the shaft great enough to cause the crankshaft to break. If an engine that suffers an intense crankshaft torsional resonance at a particular speed is kept running under load at that speed, it may break. The failed shaft is usually cleaved quite cleanly at a point near to the flywheel-end main bearing. In the very worst cases, such a failure can occur within even the first 100 miles of part load city driving.

In drivetrain analysis, it is customary to reduce each of the rotating stages in the drivetrain to an equivalent rotating disk, and to have these disks connected together by simple shafts with stiffness and damping properties as close as possible to the shafts found in the physical system (see for example Norton (1999)). The ratio of shaft stiffness to the equivalent inertias of neighbouring disks determines the resonant frequencies of the drivetrain.

The drivetrain includes two main stages of gearing: in the gearbox and then again at the differential. This gearing has a pronounced effect on the effective inertias of the rotating components and hence on the frequencies at which the drivetrain exhibits resonance. The effect of speed on inertia must be understood and is best accounted for by using the speed of rotation of one driven tyre as a reference speed.

Regardless of the gear selected in the gearbox, the driven tyre on a typical vehicle will have a lower rotational speed than the crankshaft due to the inherent gearing in the differential. Therefore, tracing the torque path back up to the engine from the tyre:

- the reference driven road wheel will rotate at the reference speed along with its hub and driveshaft;
- the other drive wheel may rotate at a slightly different speed due to the action of the differential;
- the propeller shaft and gearbox output shaft will rotate at an intermediate speed;
- the gearbox input shaft, the clutch, the flywheel and the crankshaft will rotate at the highest speed on the drivetrain;
- the timing gear, the camshaft(s) and the ancillaries will rotate at their own geared speeds.

The mass moment of inertia of a rotating element is given by:

$$I = mr^2 \tag{6.216}$$

The effective, combined, mass moment of inertia of the pair of driven road wheels in a two-wheel-drive vehicle (neglecting for simplicity the speed difference at the driving wheels caused by the differential), including the mass loading due to the vehicle, plus the associated hub, tyre and driveshaft assembly (denoted here as I_{RW}) is given by (Wong, 2001):

$$I_{RW} = m_{veh} r_{roll}^2 + \frac{(I_W + I_{DS})}{m_{veh} r_{roll}^2} \tag{6.217}$$

where I_W is the mass moment of inertia of the wheel, tyre and hub assembly, I_{DS} is the mass moment of inertia of the driveshaft assembly, m_{veh} is the mass of the vehicle and r_{roll} is the rolling radius of the driven tyre(s).

When calculating the effective rotational inertias of the disks chosen to represent the final drive (the system comprising differential, propeller shaft and gearbox output shaft), the final drive gear ratio ξ_0 must be taken into account:

$$\xi_0 = \frac{\omega_{f\_drive}}{\omega_{RW}} \tag{6.218}$$

where ω_{f_drive} is the rotational frequency of the final drive and ω_{RW} is the radial frequency of the driven road wheels.

For drivetrain analysis, it is convenient to represent the differential, propeller shaft and gearbox output shaft as a simple pair of gears. The input gear has a mass (m_{in}) equal to the sum of output shaft mass, propeller shaft mass and differential cage mass, and the output gear has a mass given by:

$$m_{out} = \frac{I_{RW}}{r_{roll}^2} \tag{6.219}$$

The effective radius of the input gear is denoted by r_{in} and the effective radius of the output gear by r_{out} so that:

$$\xi_0 = \frac{\omega_{f\_drive}}{\omega_{RW}} = \frac{\omega_{in}}{\omega_{out}} = \frac{r_{out}}{r_{in}} \tag{6.220}$$

Now

$$I_{in} = m_{in} r_{in}^2 \tag{6.221}$$

$$I_{out} = m_{out} r_{out}^2 \tag{6.222}$$

Dividing equation (6.221) by equation (6.222) and substituting equation (6.220)

$$\frac{I_{in}}{I_{out}} = \frac{m_{in}}{m_{out}} \times \xi_0^2 \tag{6.223}$$

By equivalence, and on substitution of equation (6.219)

$$I_{in} = I_{RW} \times \frac{m_{f\_drive}}{\left(\dfrac{I_{RW}}{r_{roll}^2}\right)} \times \xi_0^2$$

$$I_{in} = I_{f\_drive} = r_{roll}^2 \times m_{f\_drive} \times \xi_0^2 \tag{6.224}$$

The same approach can be used to find the equivalent inertia of the crankshaft, flywheel, clutch, gearbox input assembly denoted here as I_{eng}:

$$I_{eng} = \left(r_{roll}^2 \times m_{f\_drive} \times \xi_0^2\right) \times \frac{m_{eng}}{m_{f\_drive}} \times \xi_n^2 \tag{6.225}$$

where m_{eng} is the combined mass of the crankshaft, flywheel, clutch and gearbox input assembly, and the gear selected is given by

$$\xi_n = \frac{\omega_{eng}}{\omega_{f\_drive}} = \frac{r_{f\_drive}}{r_{eng}} \tag{6.226}$$

Because of the inclusion of the vehicle mass in equation (6.217) I_{RW} is large. Because of the large combined mass of the crankshaft and flywheel in particular, and because of the term ξ_n^2, $I_{eng} \gg I_{f\_drive}$. As a result, the lowest resonant frequency in torsion of the whole drivetrain can be readily found by treating the drivetrain as a two degree of freedom system as shown in Figure 6.27 where K_{DS} is the parallel rotational stiffness of the driveshafts (typically a much lower stiffness than any of the other shafts in the drivetrain).

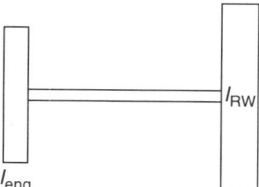

Figure 6.27 A simple representation of the drivetrain suitable for estimating the lowest resonant frequency in torsion.

By inspection of Figure 6.27 and as $I_{RW} \gg I_{eng}$ the first mode in torsion will have a node at the engine (where the fluctuating torque is being applied to the system) and an anti-node at the road wheels. The anti-node will manifest itself as a rotational oscillation of the road wheels and this will of course cause a fore–aft shuffling motion of the vehicle accompanied with a small degree of pitching motion. This so-called 'shuffle mode' will typically occur at 1–3 Hz in the lowest (first) gear and at a frequency between 10 and 30 Hz in top gear.

Higher-order rotational modes of the drivetrain involve more elements (like the gearbox shafts, crankshaft, etc.) and are calculated by employing more degrees of freedom in the modelling, solved using a numerical method such as the Holtzer method (Thompson, 1993). However, experience shows that the order of the modes may be placed in sequence of ascending natural frequency according to the relative stiffness of the various connecting shafts:

- In nearly all cases, the lowest torsional mode will be the shuffle mode described earlier. The impact of this mode on passenger comfort is affected by the stiffness of the vehicle suspension (thus influencing the pitching response) and by the clutch properties (influencing the fluctuating torque input) and the driveshaft stiffness. The flexure will be mainly in the driveshafts except when non-typical clutches are adopted.
- The next torsional mode (typically at a frequency around 50 Hz) will often be influenced by the rotational stiffness of the tyre, this being the next most compliant element in the drivetrain after the driveshafts.
- Higher-order modes at frequencies in the range of 100–300 Hz tend to involve the gearbox. They may manifest themselves as gear rattles.
- The highest-order torsional modes mostly involve the crankshaft as this is typically the stiffest shaft in the drivetrain. Norton (1999) shows the first torsional mode of a typical I-4 crankshaft to be at 350 Hz. Taylor (1985) suggests the following empirical relationship for the frequency of the first torsion mode:

$$f = \frac{140}{L} \tag{6.227}$$

where f is the frequency (Hz) and L is the crankshaft length in metres. The length L can be estimated from :

$L = b\,(1.1N_c)$ for an inline engine,
$L = b\,(0.55N_c + 0.19)$ for a vee engine
where b is the bore (m) and N_c denotes the number of cylinders.

Testing this out for an I-4 engine with 86-mm bore, the first torsion mode is predicted at 370 Hz, agreeing well with Norton (1999). For an I-6 with the same bore it would be 247 Hz and it would be 466 Hz for a V-6. Although first torsion mode in the I-4 or the V-6 would probably not cause too many problems, in the I-6 it would be excited by the dominant third-order (3E) component at 4940 rev min^{-1} and would be of grave concern if the engine were operated for any period at that speed.

Because the crankshaft is relatively stiff in torsion compared with other elements in the drivetrain, the natural frequencies of the crankshaft in torsion are typically at least one order of magnitude higher than the natural frequencies found in other drivetrain elements. As a result, it is reasonable to analyse the vibration characteristics of the crankshaft in isolation, without including other elements in the drivetrain.

Traditionally, lumped mass models were used to calculate the torsional, axial and bending vibration in crankshafts. Examples of this include Kabele (1984), Hodgetts (1986) and Genta (1998). For an I-4 crankshaft, these methods would typically show the first bending mode of the crankshaft at around 150 Hz and the first torsion mode at around 350 Hz. In more recent times, these lumped mass models have been replaced by Finite Element Analyses (Ambrosi and Orofino (1992) for example).

Some simple arithmetic is useful for illustration here. If the first bending mode of the I-4 crankshaft occurs at 150 Hz and the main source of excitation for this is the 2E torque fluctuation expected for the I-4 engine then forced torsional resonance will occur at 4500 rev min^{-1}. Such a condition is known as a 'critical speed'. A combination of rotation and bending will cause the ends of the crankshaft to whirl in their bearings. This can manifest itself as a flywheel wobble, end bearing wear and potentially crankshaft failure. Crankshaft bending modes usually result in intense engine structure-radiated noise at the critical speeds.

The natural frequencies of the crankshaft vary with the reciprocal of shaft length. If the I-4 crankshaft discussed earlier was extended by 50% to make it into an I-6 shaft then the first bending mode would be expected at 100 Hz. With 3E excitation being dominant in the I-6 this means a critical speed for first bending mode of the crankshaft of only 2000 rev min^{-1}. With further bending modes expected at 4000 and 6000 rev min^{-1} and first torsion mode at 4900 rev min^{-1} the operational envelope of a typical I-6 is seen to be packed with critical speeds. The worst case would be the high-speed diesel I-6 where gas forces are high, inertia forces are high (due to heavier pistons) and hence the torque will fluctuate strongly through the cycle.

By way of contrast with the I-6, the I-4 and V-6 engines are likely to have only one critical speed in the normal operating speed range for the engine (taken to be below 6000 rev min^{-1}). Certain engine configurations can be identified as problem cases in terms of crankshaft vibration. In order, listing the worst problem cases first:

1. Very high-speed I-6 gasoline (arguably)
2. I-6 Tdi diesel
3. High-speed I-4 Tdi diesel
4. V-8 gasoline (arguably)
5. V-6 Tdi diesel
6. I-4 gasoline
7. V-6 gasoline

Figure 6.28 Photograph of a five bearing I-4 crankshaft.

There are various solutions to these problem cases:

- Using five main bearings on the I-4 crankshaft rather than the three that historically have been used as standard. A five-bearing crankshaft is shown in Figure 6.28.
- Use a so-called dual mass flywheel (DMFW) for the worst cases: 1–4.
- Use a so-called 'crankshaft damper' (although this is more likely to be a vibration absorber than a damper) on any of the engines in (1–7) as appropriate.

These 'crankshaft dampers' (sometimes known as torsional vibration dampers or TV dampers) have been used on automotive engines ever since the earliest six-cylinder cars developed before the First World War. Such 'dampers' are still fitted to most V-8, V-6 and I-6 engines to this day. There are many variants but all developed so far fall into one of three categories:

1. friction or viscous dampers;
2. tuned absorbers (with or without friction or viscous damping added);
3. pendulous absorbers.

Members of the first two categories tend to look like 150-mm diameter pulleys that are 25–50-mm thick and they tend to be fitted to the free end (the timing end) of the crankshaft. They can easily be mistaken for regular timing pulleys, even more so these days when timing pulleys are frequently designed with TV dampers/absorbers built within them.

Members of the first two categories are mounted external to the engine block/crankcase. However, some engine manufacturers are well known for using internal absorbers (pendulous absorbers usually). Continental and Lycoming follow aircraft-derived practice and incorporate pendulous absorbers into the crankshaft counterweights. Different examples from the three categories of TV damper/absorber will be discussed in turn.

6.4.5.1 *Friction and viscous dampers*

These devices are true dampers as they remove energy from torsional vibration through dry friction or visco-thermal action. As a result, they tend to get very hot and their effectiveness may reduce with prolonged use. This might be either due to wear of friction surfaces over time or loss of fluid viscosity at elevated temperatures.

The oldest type of friction damper consists of two flywheels pressed together with high friction surfaces where they contact. Such a damper featured on the I-6 Rolls Royce engine of 1904. One flywheel is fixed to the free end of the crankshaft and the other is free to rotate relative to the first. The action of dry friction makes the behaviour non-linear (the coefficient of static friction is always higher than the coefficient of dynamic or slipping friction). These devices naturally wear and need periodic maintenance.

An improved variant is known as the Lanchester type damper and this is still in use today. It looks like a motorcycle multi-plate clutch assembly with several pairs of friction disks lightly squeezed together and immersed in oil for cooling.

A third variant, known historically as the 'Houdaille damper' but more recently as a 'fluid damper' works on the principle of energy dissipation by viscous drag. A metal can is fitted to the free end of the crankshaft. Inside there is a close-fitting metal inertia ring. This floats in a viscous silicon fluid and is free to rotate. As the outer can vibrates along with the crankshaft in torsion the inertia ring rotates in sympathy but with a phase lag, and a relative velocity between ring and casing is developed. Shear forces in the connecting viscous fluid dissipate energy in the form of heat.

The advantage with all three of these variants is that they function at all engine speeds and therefore will have beneficial effect on any form of torsional vibration in the crankshaft. The Lanchester type damper is the most powerful device but not the most durable. The viscous fluid damper is probably the most popular variant due to its small size and the fact that it is maintenance free.

6.4.5.2 *Tuned absorbers for torsional vibration*

These are rotational versions of the rectilinear vibration absorber whose behaviour was described earlier in this chapter using equations (6.135) and (6.136). They are tuned to one particular frequency in torsion (one critical engine speed) and through the introduction of at least one additional degree of rotational freedom to the drivetrain, their adoption will cause at least two additional torsional resonances at frequencies on either side of the tuned frequency. At the tuned frequency of course, vibration levels are attenuated.

The simplest type of tuned absorber has an outer inertia ring bonded to a central hub by a thick elastomeric ring. The central hub is attached rigidly to the free end of the crankshaft. The elastomer packaging can be designed so that the inertia ring has degrees of freedom in the axial and radial directions as well as in torsion. The radial vibration can be tuned to absorb vibration at the first bending mode frequency of the crankshaft.

Without the incorporation of some damping, such devices are so highly tuned to a rather narrow frequency/speed band to be of little practical use on automotive engines (although they might be ideal for fixed-speed power generation engine). However, they are commonly found on automotive engines because they are cheap, and because several effective damping strategies have evolved. The most common strategy is to adopt an elastomer with high material damping properties. Care must be taken with these to avoid overheating as this reduces the damping and can cause the bond between the elastomer and the inertia ring to break thus rendering the unit useless and putting the engine at risk of failure.

Figure 6.29 shows a photograph of the crankshaft pulley of an I-6 TDI engine fitted with both an elastomeric damper and a viscous fluid damper. The adoption of these two devices (along with careful combustion management using common-rail fuel injection) allowed the use of an unusually lightweight flywheel for an I-6 diesel engine.

Baker et al. (1994), report the fitting of elastomeric ring crankshaft dampers to a small (less than 1.5 litre) I-4 gasoline engine and to a medium-sized (1.5–2.0 litre) I-4 gasoline engine. Neither engine had problems with torsional vibration so they had been developed without dampers. The application of the damper in both cases reduced broadband engine noise by 1.5–3.5 dB across the speed range when running under full load. The mechanism by which this noise reduction was achieved is not clear.

More complex alternatives incorporate metal springs that link the inertia ring to the hub in ingenious ways. The action of these springs under torsional vibration of the inertia ring causes oil to be pumped from one internal chamber to another via narrow connecting passages thus providing damping through visco-thermal action. The most complex variants have an inertia ring comprising dozens of separate small masses co-joined together by

(a)   (b)

Figure 6.29 (a) Crank timing pulley incorporating an elastomeric torsional damper and a viscous fluid torsional damper as fitted to a medium-sized I-6 turbo-diesel engine. (b) Lightweight flywheel fitted to the other end of the crankshaft: made possible by the high inertia pulley with its two torsional dampers.

fluid-pumping lever springs. These are extremely effective devices but rather large and costly and they need a continuous supply of high-pressure oil for their operation.

6.4.5.3 Pendulous absorbers

The period of oscillation T for small amplitudes of swing in a pendulum is given by the Penguin dictionary of physics (Illingworth, 1990) as:

$$T = 2\pi \sqrt{\frac{L}{g}} \qquad (6.228)$$

where L is the effective length of the pendulum 'string' and g is the acceleration due to gravity.

Equation (6.228) can be re-written as:

$$f = \frac{1}{2\pi} \sqrt{\frac{g}{L}} \qquad (6.229)$$

$$\omega = \sqrt{\frac{g}{L}} \qquad (6.230)$$

where f is the frequency of oscillation (Hz) and ω is the radial frequency (rad s^{-1}).

Now, the acceleration on a body a (m s^{-2}) caused by circular rotation about a point in space is given by the well-known relationship:

$$a = \Omega^2 r \qquad (6.231)$$

where Ω is the radial frequency of circular rotation about a point on a radius of curvature r as shown in Figure 6.30.

Substitution of equation (6.231) into equation (6.230) yields the final result:

$$\omega = \Omega \sqrt{\frac{r}{L}} \qquad (6.232)$$

It is clear from inspection of equation (6.232) that the frequency ω at which the pendulum is tuned and hence will absorb torsional vibration is a function of the designed ratio of r/L and rises linearly with rotational frequency Ω. This is a very convenient

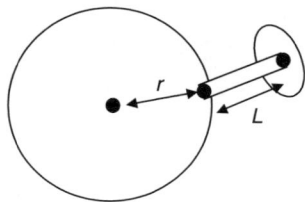

Figure 6.30 Critical dimensions of the simplest pendulous absorber.

feature for use on variable-speed rotating systems such as automotive drivetrains. The ratio ω/Ω is an order (the ratio of two rotational frequencies). If this ratio is set to 2.0, then the pendulum will swing at 2E. This could be used to absorb some of the marked 2E torsional vibration expected from an I-4 crankshaft at high speeds. The pendulum could be set to 3E for an I-6 engine and 4E for a V-8 and so on.

There is a practical difficulty to overcome however. If the device is the standard 150-mm diameter as for other types of T.V. absorber, then $r = 75$ mm. If $\omega/\Omega = 2.0$ then L must equal 18.75 mm. This is inconveniently short but several elegant incarnations of the pendulous absorbers were developed for radial aircraft engines in the period 1934–1939 that overcome this difficulty. The Hispano-Suiza, Pratt and Whitney and Curtiss-Wright companies all employed vibration specialists to develop pendulous absorbers at around the same time. These could be fitted to the counterweights of the crankshaft, internal to the engine. Pratt and Whitney produced (arguably) the best example of these (see Figure 6.31 for a sketch) and all competitors are variants of the same so-called 'Chilton damper' or 'bifular damper'. The principle is always the same, the pendulum length L is the difference between the radius of the pins that restrain the pendulum and the radius of the holes in which the pins sit. In this way L can be made very small. A further variant is the 'balanced pendulous absorber' which has four bifular elements mounted on a balanced plate and can be used external to the engine by fixing it to the free end of the crankshaft thus avoiding the design of a new crankshaft.

In more recent times (since the mid-1980s, a complete alternative to the TV damper has been developed and this is known as the 'dual mass flywheel'. The device may replace a conventional TV damper or it may be used in addition. It consists of two flywheels: a primary flywheel and a secondary flywheel. The primary flywheel is attached to the crankshaft in the normal way for flywheels and has the toothed ring on its perimeter for the starter motor. The secondary flywheel is pressed against the primary and is free (in a limited way) to rotate relative to that primary. The torsion load path from primary to secondary flywheel is through a set of wire-wound steel springs that act as a torsional

Figure 6.31 Sketch of a pendulous absorber incorporated into a crankshaft counterweight.

vibration isolating element. The two flywheels are also in light contact with each other over a wide high-friction surface.

The DMFW is a torsional vibration isolator with added damping and therefore is a variant of the rectilinear case described by equations (6.121) and (6.126). The operating principle is to make the springs compliant enough so that the DMFW assembly is tuned to a critical speed of around 200–400 rev min^{-1} and therefore will have a useful isolating effect at the idle crankshaft speed and above. During engine startup, the 200–400 rev min^{-1} critical speed will be briefly passed through but otherwise the DMFW is only being forced at frequencies well above its natural frequency in torsion.

The DMFW effectively performs two functions:

1. Through its action as an isolator it isolates the transmission from torque fluctuations in the crankshaft. The isolation is not perfect but the amplitude of the highest instantaneous torque level (as might be caused at a critical speed or by rough use of the clutch) is reduced. Without the DMFW the transmission might be damaged and gear rattle would most likely increase.
2. Through its action as a torsional damper, the amplitudes of torsional vibration in the transmission are further reduced, reducing gear rattle.

There are well-documented problems with some of the DMFWs developed to date:

- The friction surfaces wear out. The DMFW should be replaced once the clutch has been replaced for the second time and the DMFW is an expensive item to replace. The wear rate increases significantly if drivers are rough with the clutch.
- Overheating the secondary flywheel will cause it to lose important bearing lubrication and also may crack the disk. Overheating is most commonly caused by drivers who slip the clutch for long periods.

Notwithstanding these difficulties, the DMFW has gained in popularity for use in diesel-powered pickup trucks and high performance V-8 and flat-6 engined sports cars although there are a small but growing number of (generally higher performance) family cars fitted with them. Umeyama et al. (1990) reported that Toyota adopted DMFWs on all of their diesel engines fitted in manual transmission cars in order to minimise gear-rattle. The resonant speed of the DMFW is shown to be around 200 rev min^{-1}. They reported on elastically mounted friction damping elements used in DMFWs and on elastically mounted viscous dampers. The elastically mounted viscous damper was the most effective device for the DMFW but its performance deteriorated with prolonged actuation due to the effects of heat and so the elastically mounted friction damper was adopted for production use.

6.4.6 Engine mounts

The engine and transmission assembly can be considered as a six degree of freedom system exhibiting:

- axial, lateral, vertical displacements;
- yaw, pitch and roll rotations.

It is forced into combinations of these motions by the forces, moments and torques caused by the cranktrain and the cyclic variations in gas pressures discussed in detail earlier in this chapter. A simple (if not truly accurate) guide is that:

- vertical shaking forces encourage vertical displacements of the powertrain;
- horizontal shaking forces encourage lateral motion;
- moments in the vertical plane encourage pitching;
- moments in the horizontal plane encourage yawing;
- torques encourage roll.

The above guide is not truly accurate because the motions will be almost never pure but coupled (vertical motion coupled with pitching for example) and excited by a combination of forces, moments and torques.

There are some governing principles, however. The purpose of the engine mounts is threefold:

1. To constrain the motion of the powertrain whilst the car is in motion in order to minimise the risk of damage caused by high cyclic stresses on the powertrain or on the chassis.
2. To constrain the motion of the powertrain whilst the car is in motion in order to preserve driveability.
3. To reduce structure-borne noise and vibration in the vehicle.

Consider the third purpose. The behaviour of an sdof vibration isolator is described by equations (6.121) and (6.126). Inspection of Figure 6.9 suggests that an engine mounting should have a natural frequency well below the lowest frequency of excitation and there should be little damping in the isolator. If the idle speed of an engine is $850\,\text{rev}\,\text{min}^{-1}$ then the 1E (first order) frequency is $14.2\,\text{Hz}$. Setting the natural frequency to

$$\frac{14.2}{\sqrt{2}} = 10\,\text{Hz}$$

from equation (6.128)

$$\omega_n = \sqrt{\frac{g}{\Delta}}$$

$$\Delta = \frac{g}{\omega_n^2} = 0.0025\,\text{m or } 2.5\,\text{mm.}$$

Generally four engine mounts are used to suspend a powertrain. If the powertrain has a mass of $250\,\text{kg}$, the stiffness of each mount would be of the order of $250\,\text{N}\,\text{mm}^{-1}$.

Some simplified calculations are helpful at this point. Assume that the powertrain is such that it has identical rotational inertia in all three planes, $I_{xx} = I_{yy} = I_{zz} = 15.8\,\text{N}\,\text{ms}^2$, and that each of the four mounts had a stiffness of $250\,\text{N}\,\text{mm}^{-1}$. Taking pure rotation about the xx plane, from equation (6.138):

$$\omega = \sqrt{\frac{k_1 a_1^2 + k_2 a_2^2}{I_{xx}}}$$

and as $k_1 = k_2 = 500\,\text{N}\,\text{mm}^{-1}$, $a = 250\,\text{mm}$.

If each of these engine mounts is placed in a different corner of the powertrain, and each is 250 mm from the centre of gravity then to a first approximation all six natural frequencies of the system would coincide at 10 Hz which is a crankshaft critical speed of 600 rev min^{-1}(1E). The advantage of such an approach is that there is only one critical speed; however, at 600 rev min^{-1} the powertrain would be very lively.

The speed of 600 rev min^{-1} is below the idle speed and above the cranking speed. It will be passed through only briefly as the engine is first started or as it is switched off. Shaking forces and moments are low at 600 rev min^{-1} due to part load operation restricting the gas forces and the low speed not producing much reciprocating inertia force. Fluctuating torque is the main excitation at this speed. This will promote roll at 10 Hz although this can be damped if one mount is replaced by one with lower stiffness some distance further away from the centre of gravity and that mount has high levels of damping at 10 Hz. The so-called 'hydromount' is perfectly suited to this application (Bernuchon, 1984).

A sketch of a typical hydromount is shown in Figure 6.32 and a photograph of one example mount is shown in Figure 6.33. A large rubber element is placed at the base of the unit. This carries the load of the engine in shear and thus the overall stiffness of the unit is comparably low. A metal can is bonded above the rubber and contains two chambers separated by a diaphragm and filled with a mixture of water and antifreeze. The two chambers are connected by a circular passage. The metal can is attached to the engine whilst the rubber element below is attached to the chassis. Relative displacement between the engine and chassis cause water to be pumped back and forth between the two chambers and damping occurs through visco-thermal action. High levels of damping are obtained at the natural frequency of the powertrain mass on the mount stiffness but at other frequencies, the damping level tends to that present in the form of material damping in the rubber element.

Often it is difficult to find suitable positions for engine mounts. From equation (6.159) it is clear that the mobility of both engine and chassis at the mounting position must

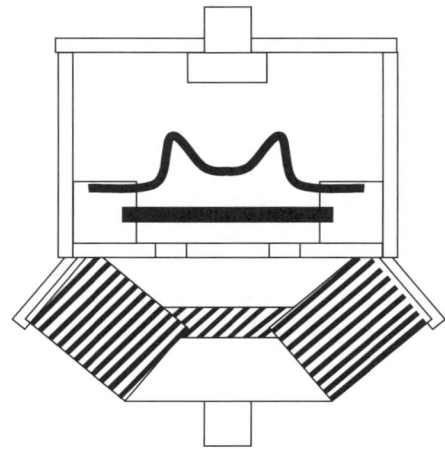

Figure 6.32 Sketch of the internal workings of a hydromount.

Figure 6.33 Photograph of a hydromount as fitted to a 4.0-litre V-6 gasoline engine in a full-sized luxury sedan.

be low for the mount to be a successful vibration isolator. Only a few locations around the powertrain fulfill these criteria (the use of mounting brackets should be avoided as they are almost always of high mobility) and these are adopted wherever possible. Variable distances from the centre of gravity result and the mount stiffness are tuned to compensate for this. Compromise is inevitable and more than one critical speed is likely to result.

The current fashion with transverse mounted front-wheel drive engines is to use two mounts very high up on the engine block and to hang the engine from these like a pendulum. A third mount is positioned low down to restrain roll and a fourth one is used to support the transmission and to restrict pitching. At least one of these lower mounts is likely to be a hydromount. Figure 6.34 shows this type of mounting in use.

The rational way to design engine mounting systems is to use:

- engine cycle analysis to determine the forcing due to forces, moments and torque;
- solid body dynamic modelling to solve the equations of motion for the six degree of freedom system;
- finite-element analysis to calculate the mobility of the engine and chassis at the mounting points as these mobilities provide realistic boundary conditions for the solid body modelling.

(a)

(b)

Figure 6.34 (a) Pendulum-type top mounts on a small I-4 gasoline engine. (b) Lower engine mount on the same engine.

6.5 Vehicle and chassis vibrations: ride quality

Section 6.4 discusses the subject of engine-induced vibrations, their cause and propagation and the effect that they have on driver comfort in terms of durability and engine noise. In this section the same approach will be taken although this time the subject is vehicle and chassis vibration.

The main consequence of vehicle and chassis vibration is a reduction in passenger comfort. According to the definitions proposed in Chapter 1:

Vibration   shall be used to describe tactile vibration, with particular attention paid to
frequencies in the range of 30–200 Hz.

And the three classes of vibration that most affect passenger comfort are:

Shakes	shall be taken to be relatively large amplitude vibration of vehicle components or elements of the vehicle structure with particular attention paid to frequencies in the range of 6–30 Hz.
Primary ride	taken to be the rigid body motion of the passenger compartment relative to the road. Typical frequency range is 0–6 Hz.
Secondary ride	taken to be the relatively large amplitude motion of sub-elements of the vehicle such as individual wheels, axles or elements of the powertrain. Typical frequency range is 6–30 Hz.

According to ISO 2631 Part 1 (1985) (see Section 5.3.4) the effects of these three classes of vibration include:

- reduction in passenger comfort;
- fatigue leading to decreased proficiency with tasks;
- a threat to health and safety.

The onset and severity of these depend on the amplitude of vibration, the frequency content thereof, the position of the passenger (seated, standing, lying down) and the length of exposure time. Other effects, specific to the experience of driving a car can be identified with a few moments thought:

- impaired proficiency with fine motor skills (operating fine switches, tuning the radio, etc.) not related to fatigue but simply caused by vibration of the hand;
- back-pain, neck-pain, blurred vision (for higher speed off-road driving in particular);
- motion sickness;
- general dissatisfaction with the driving experience (which might influence future car-purchase decisions).

ISO 2631 Part 1 (all revisions) discusses the sensitivity of humans to vibration at different frequencies. This is a function of body position and is different in each of the three spatial axes but generally the human subject is most sensitive to vibration in the frequency range of 4–8 Hz. This is irrespective of whether the subject is travelling in a vehicle or not.

In contrast to this, NASA (Leatherwood and Barker, 1984) propose that the human travelling in a vehicle is most sensitive in the frequency range of 1.5–6 Hz. In addition, NASA provide different sensitivity spectra for pitch and roll motions as well as the three axes of rectilinear motion adopted in ISO 2631. It appears that the vehicle occupant is most susceptible to longitudinal combined with pitching motion in the frequency range of 1.5–6 Hz. This frequency range covers many of the drivetrain vibration phenomena discussed in Section 6.4 (shunt, shuffle, etc.).

Other chassis vibration phenomena are discussed elsewhere in this book:

- Section 5.4.3 shows with a simplified analytical model that the bounce mode of the car on the suspension is around 1.0 Hz and the pitch mode is around 1.5 Hz. According to the NASA sensitivity information, it would seem prudent to suppress vehicle pitching wherever possible in order to maintain acceptable ride quality.

- Section 5.4.5 shows with another simplified analytical model that the bounce mode of the driver on the seat is around 4 Hz and that wheel hop occurs at around 12 Hz. It is clear that the driver bounce will have greater impact on ride quality than wheel hop, although wheel hop is important due to its negative effect on tyre forces and hence vehicle handling.
- Section 6.4 shows that the shuffle mode of the chassis and powertrain combination will be in the range of 1–3 Hz for the lowest gear in the transmission and in the range of 10–30 Hz in the highest gear. The so-called 'drive-away' phenomena, as the driver engages the lowest gear and starts away from rest, is clearly a great contributor to ride quality.
- Section 6.4.6 suggests that the six fundamental modes for the powertrain on its mounts will be clustered around 10 Hz. As a result, according to the NASA sensitivity information, the powertrain modes will have less effect on ride quality than the modes involving the chassis itself.

Sections 1.6.3.3 and 4.1.4 present a commonly used method for the subjective rating of vehicle refinement attributes including ride quality, shakes and tactile vibration. For modern passenger cars, these three should be detectable in the car but only lightly.

An alternative to this is to make objective measurement of chassis vibration levels and then to rate these against some objective criteria. Such measurements can be made at either of two levels of complexity:

- The simplest level where a tri-axial accelerometer is fixed to the driver's seat-rail and a single-axis accelerometer is fixed to the rim of the steering wheel.
- The more complex levels where piezo-resistive pads are placed on the seat squab and on the lumbar support cushion and then the driver sits on these whilst driving.

The advantages of the second method are:

- The vibration levels are measured at the interface between the driver's body and the supporting seat and hence are those experienced by the driver, whereas in the first method the vibration levels are not those experienced by the driver.
- The piezo-resistive pads measure ultra-low frequencies (including DC) whereas accelerometers do not.

Whatever the choice of instrumentation, the vibration levels will most frequently be gathered with the vehicle running in a straight line but at different fixed speeds and over different qualities of road surface. This approach provides plenty of near stationary vibration data to yield reliable spectral estimates at very low frequencies (refer to the discussion in Section 2.4.1.2, equation (2.78)) but will not correlate well with the subjective impression of ride quality obtained by driving around a course that involves cornering, acceleration and braking events.

Measured vibration levels are rated according to objective criteria. The most commonly used criteria are offered in:

- ISO 2631 Part 1 (1985);
- BS 6841 (1987);
- NASA discomfort level index (Leatherwood and Barker, 1984).

Experience of using all three has led to the personal conclusion that a family class vehicle for either the European or US Federal market) will be ready for sale when the appropriately frequency-weighted seat-rail vibration levels measured at 80 km hr^{-1} on a straight road with 5–10-year old tarmac and a few spot repairs (in other words a typical inter-urban road) are either:

- Somewhere close to the four-hour reduced comfort boundary in the vertical direction as defined in ISO 2631 Part 1 (1985);
- Have an rms level less than 0.63 m s^{-2} (classed as better than 'a little uncomfortable' according to BS 6841 (1987));
- Have a NASA discomfort rating below 4.0 (Bosworth et al., 1995).

Both ISO 2631 and BS 6841 are relatively easy to implement. They offer simple vibration thresholds based on frequency-weighted acceleration levels in the three principal axes of translation (vertical, longitudinal and lateral) and digital forms of the various frequency-weighting networks are given in detail.

By comparison, the NASA method is complex, involving many stages of computation and additional measurements of pitch and roll motions as well as interior sound pressure levels. One advantage of the method is that all of these data are combined to yield a single number as output (a ride quality 'metric') making comparisons between competitor vehicles easy. A low number denotes a low level of ride discomfort (or put another way, a high level of ride quality).

Although no longer current, and not offering a single number output, ISO 2631 Part 1 (1985) is still commonly used for ride quality assessment as the form of data presentation offers a more complete diagnosis of the ride quality than the single-number output used in the NASA and BS 6841 methods. Figure 6.35 shows the frequency-weighted vertical seat-rail vibration levels measured on a small executive-class car fitted with a 3.0-litre V-6 gasoline engine and automatic transmission. The vehicle is travelling in a straight line at 80 km hr^{-1} over a moderate quality asphalt road. As expected, the frequency-weighted accelerations are close to the four-hour reduced comfort boundary. This form of data presentation allows the engineer to gain an impression of the nature of the vehicle ride experience in this car. The influence of vibration in the range <10 Hz is clear.

As well as being routinely measured, vehicle chassis vibration levels are frequently calculated. This allows the vehicle designer to assess the impact on ride quality of say vehicle mass, or inertia characteristics or suspension design or suspension compliance and damping.

De Carbon (1950), Robson (1979), Jolly (1983) and Sharp and Hassan (1986) used multi-body vehicle models of the type shown in Section 5.4.5 to predict ride vibration levels. They used road roughness spectra as input. Cebon (1999) has become a gold-standard reference for such data. Improvements to the understanding of the behaviour of the human exposed to whole body vibration in cars is offered by Demic (1984) and Fairley and Graffin (1984).

Dunn et al. (1979) and Takahashi et al. (1987) extended their ride vibration calculations to include the prediction of interior road noise levels as well as ride levels.

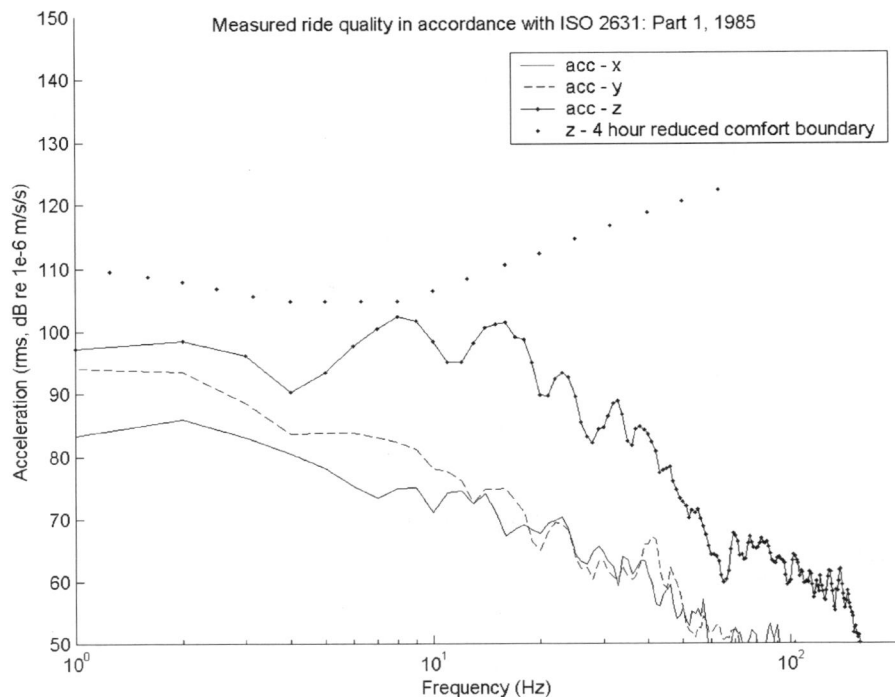

Figure 6.35 Vertical seat-rail vibration levels for a small executive-class vehicle at 80 km hr^{-1} in accordance with ISO 2631 Part 1 (1985).

References

Adler, U., (ed.), Automotive Handbook – Second edition, Robert Bosch GmbH, 1986

Ambrosi, G., Orofino, L., Driveline vibration simulation in a four-wheel drive vehicle, SAE Paper No. 925088, 1992

Baker, J.M., Bazeley, G., Harding, R., Needham, P., Refinement benefits of engine ancillary dampers, IMechE Paper C487/041/94, Printed in: 'Vehicle NVH and refinement'. International conference, Birmingham, UK, 3–5 May 1994, Mechanical Engineering Publications Ltd, 1994

Bruel & Kjaer Application Note, Improved method for complex modulus estimation, Bruel & Kjaer, 1996

Balfour, G., Dupraz, P., Ramsbottom, M., Scotson, P., Control of driveline oscillations for improved refinement – from concept to production, European Conference on Vehicle Noise and Vibration, 10–12 May 2000, London, IMechE Conference Transactions, 2000

Benham, P.P., Crawford, R.J., Mechanics of engineering materials, Longman Scientific & Technical, 1987

Beranek, L.L. (ed.) Noise and vibration control, Institute of Noise Control Engineering, 1988

Bernuchon, M., A new generation of engine mounts, SAE Paper No. 840259, 1984

Bosworth, R., Trinick, J., Smith, T., Horswill, S., Rover's system approach to achieving first class ride comfort for the new Rover 400, IMechE Paper No. C498/25/111/95, 1995

Ch. Boursier de Carbon, A theory and an effective design of the damped suspension of ground vehicles, III FISITA Congress, 1950

Cebon, D., Handbook of vehicle-road interaction, Swets & Zeitlinger, 1999

Cremer, L., Heckl, M., Structure-borne sound – structural vibrations and sound radiation at audio frequencies – Second edition, Springer-Verlag, 1988

Demic, M., Assessment of random vertical vibration on human body fatigue using a physiological approach, IMechE Paper No. C153/84, 1984

Dunn, J.W., Olantunbosun, O.A., Mills, B., Realistic prediction and control of vehicle noise resulting from road inputs ASME, Design Engineering Technical Conference, St Louis, USA, 10–12 September 1979

Fahy, F.J., Sound and structural vibration – radiation, transmission and response, Academic Press, 1985

Fairley, T., Griffin, M.J., Modelling a seat-person system in the vertical and fore-and-aft axes, IMechE Paper No. C149/84, 1984

Fahy, F., Walker, J., (ed.) Fundamentals of noise and vibration, E&FN Spon, 1998

Gade, S., Zaveri, K., Konstantin-Hansen, H., Herlufsen, H., Complex modulus and damping measurements using resonant and non-resonant methods, Paper No. 951333, SAE Noise & Vibration Conference & Exposition, May 1995

Genta, G., Vibrations of structures and machines; practical aspects – Third Edition, Springer Verlag, Berlin, 1998

Harris, C.M., Crede, C.E., Shock and vibration handbook – Vol. 1, McGraw-Hill Book Company, 1961

Hodgetts, D., Notes to accompany the course on piston engines, MSc in Automotive Product Engineering – Cranfield University, 1982

Hodgetts, D., The dynamic response of crankshafts and camshafts, SAE Paper No. 865025, 1986

Illingworth, V., (ed.), Penguin dictionary of physics, Penguin Books, London, 1990

Jolly, A., Study of ride comfort using a nonlinear mathematical model of a vehicle suspension, *International Journal of Vehicle Design*, 4(3), pp. 233–244, 1983

Kabele, D.F., A new approach in the simulation of crankshaft torsional vibration, IMechE Paper No. C140/84, 1984

Leatherwood, J.D., Barker, L.M., A user-orientated and computerized model for estimating vehicle ride quality, NASA Technical Paper 2299, 1984

Lee, J.M., Yim, H.J., Kim, J.H., Flexible chassis effects on dynamic response of engine mount systems, SAE Paper No. 951094, 1995

Nashif, A.D., Jones, D.I.G., Henderson, J.P., Vibration damping, John Wiley & Sons, 1985

Norton, R.L., Design of machinery - an introduction to the synthesis and analysis of mechanisms of machines – Second edition, McGraw-Hill International Editions, 1999

Robson, J.D., Road surface description and vehicle response, *International Journal of Vehicle Design*, 1(1), pp. 25–35, 1979

Sharp, R.S., Hassan, S.A., The fundamentals of passive automotive suspension systems Soc. of Environmental Engineers, Conference on Dynamics in Automotive Engineering, Cranfield, pp. 104–115, 1986

Takahashi, F., Watanabe, G., Nakada, M., Sakata, H., Ikeda, M., Tyre/suspension system modelling for investigation of road noise characteristics, *International Journal of Vehicle Design*, 8(4/5/6), pp. 588–597, 1987

Taylor, C.F., The internal combustion engine in theory and practice – (Volume 2: combustion, fuels, materials, design) – Revised edition, MIT Press, Massachusetts, 1985

Thompson, D.J., Van Vliet, W.J., Verheij, J.W., Developments of the indirect method for measuring the high frequency dynamic stiffness of resilient elements, *Journal of Sound and Vibration*, 213(1), 169–188, 1998

Thompson, W.T., Theory of vibration with applications – Fourth edition, Chapman & Hall, 1993

Uicker, J.J., Pennock, G.R., Shigley, J.E., Theory of machines and mechanisms, Oxford University Press, Oxford, 2003

Umeyama, M., Kobayashi, K., Hounoki, S., Otake, T., The basic consideration of the two-mass flywheel with the torsional damper IMechE Paper No. C420/020, Printed in: 'Quiet revolutions – powertrain ad vehicle noise refinement'. International conference, Heathrow, UK, 9–11 October 1990, Mechanical Engineering Publications Ltd, 1990

Weltner, K., Grosjean, J., Shuster, P., Weber, W.J., Mathematics for Engineers and Scientists, Stanley Thornes (Publishers) Limited, 1986

Wong, J.Y., Theory of ground vehicles – Third edition, Wiley John & sons, New York, 2001

Zweiri, Y.H., Whidborne, J.F., Senevirtne, L.D., Dynamic simulation of a single cylinder diesel engine including dynamometer modelling and friction, Proceedings of the Institution of Mechanical Engineers, 213, Part D, 391–402, 1999

Index